Recherches de
Théologie et Philosophie médiévales

Bibliotheca 4.1.1

Averrois Cordubensis commentum magnum super libro De celo et mundo Aristotelis

ex recognitione Francis James Carmody †
in lucem edidit Rüdiger Arnzen

Editioni praefatus est
Gerhard Endress

Tomus I: Praefatio, Liber I

PEETERS
2003

D. 2002/0602/13
ISBN 90-429-1087-9

PREFACE TO THE SERIES AVERROIS COMMENTARIA MAGNA IN ARISTOTELEM

It is well-known that the Latin translations of Averroes' writings, especially of his commentaries on the works of Aristotle, gave an immense impetus to medieval and Renaissance Western philosophy by introducing a rationalistic philosophical system methodologically independent of Muslim or Christian theology. Since H. A. Wolfson's "Plan for the Publication of a *Corpus Commentariorum Averrois in Aristotelem*" (1931) tremendous efforts have been made to produce reliable critical editions of these Latin translations, among which the project AVERROES LATINUS, based for the past two decades at the Thomas Institute of the University of Cologne, deserves particular mention. Nonetheless, those who do research in the field have had to rely up to the present day in many cases upon thoroughly insufficient Renaissance editions, the most widespread of which is the 1562-74 edition by the Giunta brothers of Venice, a revised reprint of their 1550-52 edition, reprinted as *Aristotelis Opera cum Averrois Commentariis*, Frankfurt am Main: Minerva, 1962. These editions not only fail to meet the typographical and editorial standards of modern scientific publications, but they are also restricted in at least one other respect crucially important for the study of medieval philosophy: They often do not give the text as read during the thirteenth and fourteenth centuries, offering instead a revised, sometimes augmented or otherwise restructured version based on earlier editions as well as on different Arabic-Latin (or even Hebrew-Latin and Graeco-Latin) traditions.

The present edition is the first of three editions, the publication of which has been undertaken by the Thomas Institute of the University of Cologne in cooperation with the De Wulf-Mansion Center of the Catholic University of Leuven and with the support of Peeters Publishers. Apart from the Commentary on Aristotle's *De Celo*, contained in this volume, the series comprises two further very influential works of the *Corpus Commentariorum Averrois*: the Latin versions of Averroes' Long Commentaries *(Commentaria magna)* on Aristotle's *Physica* and *Metaphysica.* Among the five Long Commentaries composed by Averroes, each of these three addresses central issues of medieval philosophy such as Nature and God (being as moved or unmoved being and the First Unmoved Mover, *natura naturata* and *natura naturans*), cosmic order and the eternity of the world (being as eternally moved and celestial being), and (prime) matter, substance and form (being as being). Averroes himself gave evidence for the significance of the doctrines and problems treated in the Aristotelian texts not only by the sheer extents of his commentaries, but also by the fact that the commented texts are the only ones to which he devoted three different kinds of philosophical interpretation each, namely: an Epitome, a Middle Commentary and a Long Commentary.

In order to make these important texts available to researchers of medieval philosophy in modern editions as quick as possible, the lengthy editorial process of a critical *editio maior* has been replaced by a careful choice of a limited number of testimonies sufficient for the preparation of a reliable semi-critical edition, with the exception of the present volume, which, due to the careful philological work accomplished by Francis James Carmody (†), almost meets the criteria of a critical edition. After completion of all three editions, these *Commentaria magna* will be included in an indexed and searchable CD-ROM, which is intended to comprise, as a further surplus, the Latin texts of the following editions and prints:

1 *Commentum medium super libro Peri hermeneias Aristotelis*: translatio Wilhelmo de Luna attributa. Ed. Roland Hissette. Leuven 1996.
2 *Averrois Cordubensis Commentarium medium in Aristotelis De generatione et corruptione libros.* Rec. Franciscus Howard Fobes adiuvante Samuele Kurland. Cambridge, Mass., 1956.
3 *Averrois Cordubensis Commentarium magnum in Aristotelis De anima libros.* Rec. F. Stuart Crawford. Cambridge, Mass., 1953.
4 *Averrois Cordubensis Compendia librorum Aristotelis qui Parva naturalia vocantur.* Rec. Aemilia Ledyard Shields adiuvante Henrico Blumberg. Cambridge, Mass., 1949.

5 *Averrois Cordubensis sermo De substantia orbis*. Venetiis apud Iunctas 1562. (Vol. IX, fol. 3ra-14vb.)

The editors of this series would like to express their deepest thanks to Peeters Publishers for their grateful support and intense cooperation.

Jan A. Aertsen, Cologne

Carlos Steel, Leuven

Rüdiger Arnzen, Cologne

CONTENTS

TOMUS I

PREFACE

Ibn Rušd (Abū-l-Walīd Muḥammad ibn-Aḥmad ibn-Muḥammad, A.H. 520–595 / A.D. 1126–98, Averroes in the Hispano-Latin reading of the West), defended philosophy by returning to the text of Aristotle. In his secular effort to explain and to revive the true doctrine of Aristotle, cosmology took a place of special importance. Aristotle's book 'On the Heavens' (*De caelo*), the *Kitāb al-Samā' wa-l-'ālam* 'Book on the Heaven and the Universe' of the Arabic tradition, is one of the few to which he devoted an early *Epitome* (*Ǧawāmi'*), a *Paraphrasis*, or 'Middle' commentary (*Talḫīṣ*), and a *Commentarium magnum* (*Tafsīr*, or *Šarḥ*). Ibn Rušd encountered here the essential rationalism of Hellenistic philosophy where reason is actual and visible in the reality of the cosmic order. The concepts and principles of this philosophy, ancient and 'modern', had become a field of unrelentant controversy: the dogma of the eternity of the world in physics, the universalist, rationalist theology of metaphysics, and the claim to absolute knowledge based on apodeictic logic. But the Aristotelian philosopher had a hard time to hold his own. He had to master particular difficulties as a commentator in that Aristotle's book is not without contradictions in itself and with regard to his later works. His aspiration after a coherent interpretation of this work accompanied all of his life.

The doctrinal development has been analyzed in G. Endress: 'Averroes *De caelo*: Ibn Rushd's cosmology in his commentaries on Aristotle's On the Heaven', *Arabic Sciences and Philosophy*, 5 (Cambridge 1995): 1–41.

All three of Ibn Rušd's treatments of the *De caelo* have come down to us:

a) The *Epitome* is the second part of Ibn Rushd's compendium (*Ǧawāmi'*) of Aristotle's four works on natural philosophy, intended to give "an analysis of Aristotle's demonstrative statements", and completed in 554/1159. The economy of the genre results in a strictly systematic layout; the subject of book III is relegated to *De Generatione et corruptione* (*Kitāb al-Kawn wa-l-fasād*). The printing of Hyderabad (*Rasā'il Ibn Rušd*, Dā'irat al-Ma'ārif al-'Uṯmāniyya, 1948) is based on two inferior MSS of the late Ṣafavid tradition, and should be replaced by a critical edition giving due consideration to the MSS of Madrid (Biblioteca Nacional, Gayangos 5000) and Cairo (Dār al-Kutub al-Miṣriyya, ḥikma wa-falsafa 5) and taking into regard other copies stemming from the school of Isfahan (sixteen MSS of which are known at present) as well as the Hebrew version of Mōšē b. Tibbōn.

b) The *Middle Commentary* or *Paraphrasis* (Arabic, *Talḫīṣ*), dated 566/1171 according to the Hebrew version, viz. between the Middle commentaries on the *Physica* of 565 and on *De Generatione* of 567, and before *De substantia orbis* (574/1178-9); a meticulous systematization of Aristotle's topics, statements and proofs. Ǧamāl-al-Dīn al-'Alawī's edition (Casablanca 1984), based on the two available MSS of the original Arabic, provides a reliable text and a valuable apparatus similium from the *Epitome* and the *Commentarium magnum.*

c) The *Commentarium magnum* (in the MS: *Tafsīr*, also called *Šarḥ*, a literal commentary on the full text of the Aristotelian work quoted in sections (*textus*, followed by *commentaria*), is one of Averroes' later works, probably written around 584/1188, i. e. after the *šurūḥ* on the *Analytica posteriora* and the *Physica*, but before those on the *De anima* and the *Metaphysica*. Although the main endeavour is towards literal exegesis, here again, and even more markedly than in the early interpretations, the consistency of Aristotle's doctrine is stressed by numerous cross-references (notably to book VIII of the *Physica*).

> In the course of his work, the commentator mentions three dated events, one of them a personal observation, in Almohad Morocco: (a) In 514 H./ 1120–21 A.D., date of "domus regis de Maroche quam erexit Almudi de Haceno [v.l. de Aceuç]" (i. e., the year of the advent of Ibn Tūmart in Marrakesh, and his erection of the Almohad régime), a giraffe (*ceurafe*, < Arabic *zarāfa*) was observed there (II, c. 111, commenting on II.14, 298a13 τεκμαιρόμενοι καὶ τοῖς ἐλέφασιν)[1]; (b) in

1. According to the ‚History of the Almohads', *al-Mu'ǧib fī talḫīṣ aḫbār al-Maġrib*, by 'Abd-al-Wāḥid al-Marrākušī (ed. R. P. Dozy, Leiden 1881), p. 224, the cursory mention of the Almohad leader without adding his princely titles and epithets, was used against Averroes as a pretext to backbite him to the sultan Abū Yūsuf Ya'qūb in 590/1194. The source gives the commentary on Aristotle's *De animalibus* as the *locus* in

530/ 1135, at the age of nine, he saw an *idolum* (Arabic, *ṣanam*) destroyed on the Spanish side of the strait of Gibraltar: "in ultimo occidentis Hispanie, quod dicebatur Idolum Cadiz: et ego vidi ipsum elevatum, deinde fuit corruptum in prelio quod crevit inter Barbaros et homines istius insulæ in anno CCCCC et XXX anno a predicatione Machometi a quibusdam erigentibus se super quasdam regiones maris in illo tempore" (II, c. 111, commenting on *De cælo* II.14, 298a10 τὸν περὶ τὰς Ἡρακλείας στήλας τόπον). (c) According to the *Talḫīṣ* (*Commentarium medium*), he made observations on a journey to Marrakesh in 548/1153, mentioning "the mountain of Suhayl" in Andalusia ("where alone, of all Andalusia, Canopus [Arabic, *Suhayl*] is sighted"), and a mountain near Marrakesh where he saw a star not observed "in this country", said to be Canopus.[2] Also found in his commentary are various other observations regarding the visibility of certain stars in Marrakesh[3].

THE ARABIC TEXT OF AVERROES' *COMMENTARIUM MAGNUM*

Only a third of the original Arabic text of Ibn Rušd's *Tafsīr al-Samā' wa-l-'ālam* has been preserved in the unique manuscript of Tunis, National Library (al-Maktaba al-Waṭaniyya), no. 11821 (*olim* al-Madrasa al-Aḥmadiyya, no. 5538). It was endowed by *waqf* (*ḥabūs*) to the Aḥmadiyya library at the Great Mosque (al-Ǧāmi' al-A'ẓam) of Tunis in 1268/1851-2. There are no other clues for dating the codex; the quality of the paper, worm-eaten and frayed, and the neat Maġribī script seem to indicate a period not later than the 8th/14th century, the time when the scholarly traditions of both the Andalus and Morocco united in Ḥafṣid Ifrīqiyya.

The manuscript has 114 leaves. In a recent restoration, it was rebound in the old leather binding; the size of the written area is 14.5x22 cm on cahiers measuring 20.5x28 cm. The leaves are in a great disorder. The original volume, as it was presented to the library in 1268 H., can be reconstructed from the folio numbers which were

question, but the long commentary on *De caelo*, written within the reign of Abū Yūsuf (1184–99), is more likely to have served as a cause of offence.

2. *Talḫīṣ al-Samā' wa-l-'ālam* / li-Abī l-Walīd Ibn Rušd, ed. Ǧamāl-al-Dīn al-'Alawī (Fes 1984), 275.10–15.

3. F. J. Carmody, *Innovations in Averroes De Cælo* (Berkeley, Cal., 1982), 11-12.

added about that time: After the *waqf* record (fol. 1a), the text begins with book II, comm. 1–42, breaking off in the middle of comm. 42 (ff. 1b-47b); it goes on with book I, comm. 61 (truncated in the beginning) to the end of book I, comm. 140 (ff. 48a–117a). Thus the volume was given the outward appearance — presumably by a clever bookseller — of a complete copy of book II. At an earlier date already, two leaves after fol. 75, and one leaf after fol. 96, were lost.

The following is a table of the parts available in Arabic and the corresponding sections of the Latin version (where the section numbers in square brackets refer to the Juntine printing of 1550, *v. infra,* p. 37):

Book, textus/commentary	Arabic text, ms. Tunis	Latin version, Venice 1562
I: c. 61–c. 94 [95]	ff. 48a–75b	ff. 42vbK–63rbE
I: c. 96 [97]–c. 118 [119]	ff. 77a–96b	ff. 64vbL–80raC
I: c. 119 [120]–c. 140 [141]	ff. 97a–117a	ff. 81raA-95rbF
II: t. 1–c. 42	ff. 1b–47b	95vaG–125rbD

Facsimile edition: *Commentary on Aristotle's Book on the Heaven and the Universe. Sharḥ al-Samā' wa-l-'ālam,* by Ibn Rushd, Muḥammad ibn Aḥmad ibn Muḥammad al-Ḥafīd (d. 1198 A.D.), with an introduction by Gerhard Endress, Publications of the Institute for the History of Arabic-Islamic Sciences, ed. by Fuat Sezgin: series C. Facsimile editions, vol. 57 (Frankfurt am Main, 1994). — With a table of concordances between the folia of the Tunis manuscript, the sections of the principle Renaissance printings of the Latin version, and the Greek text of Aristotle's *De caelo* (pp. 4–12).

A recent (early 20th-century) copy of this manuscript in 110 folia is kept in the same library, no. 18117 (see 'Abd-al-Ḥafīẓ Manṣūr, *al-Fihris al-'āmm li-l-maḫṭūṭāt*; qism 1: Raṣīd Maktabat Ḥasan Ḥusnī 'Abd-al-Wahhāb. Tunis: Dār al-Kutub al-Waṭaniyya, Qism al-Maḫṭūṭāt, 1975, p. 397b).

The Arabic text of Aristotle's *De caelo,* quoted by Ibn Rušd in the lemmata of his commentary, is the vulgate translation by Yaḥyā (also: Yūḥannā, Johannes) ibn al-Biṭrīq, a *mawlā* (client) of the caliph al-Ma'mūn (d. 218/833) working with a circle of early translators who provided the source materials of the Arab scientist and philosopher al-Kindī. It is labelled as one of the 'translations of al-Kindī' by Ibn Rušd who deplores (*Comm. mag.* III, c. 35) not having one of the superior productions of Isḥāq ibn Ḥunayn (the translator of Aristotle's *Physics*) at his disposal. The Western tradition of this version is also represented by the Latin translation of Gerard of Cremona (*v.* infra, p. 13). The Eastern tradition exists in a family of several manuscripts, all coming from Isfahan of the late Ṣafavid period (11th/17th century) and going back to an exemplar (now lost) copied in Damascus in 580/1184.

A specific difficulty of this early translation of *De caelo* is its terminology, not consistent with the standard established in later translations and original writings of Aristotelian philosophy, and making use of transliterated Greek terms which, having fallen out of use, became prone to corruption in the manuscript tradition. Examples for transliterations are *abūfāsīs* for ἀπόφασις, *qāṭāfāsīs* for κατάφασις; in some cases, the transliterated term is combined with an Arabic word, as *šarḥ wa-ṭaqs* for τάξις. In some cases, Averroes himself was at a loss; for κύβος in *De caelo* III.8, he found the Persian *narda* 'die [in the game of *nard,* backgammon]' instead of the familiar *ka'b* or *muka"ab*, but due to wrong or missing punctuation, he read *barda* or *burūda* 'coldness' (*frigidum, grando* in the Latin of Gerard, *frigus* in Scotus' version; *v. Comm. mag. Cæl.* III, c. 71, f. 229F.).

Ibn Rušd had access to another version of *De caelo*, which he calls the translation of Abū l-Farağ (referred to three times in book III, *comm.* 52, 56, and 58). This is the version of Abū l-Farağ 'Abdallāh ibn al-Ṭayyib († A. D. 1043), physician and philosopher, who wrote a number of commentaries on Aristotle dependent on the late Alexandrian tradition. Only a large fragment of his commentary, for which he made his own translation of the lemmata from the Syriac, is extant in a manuscript of the Bibliothèque nationale at Paris.

For the Arabic versions of Aristotle's *De caelo* and their textual history, see G. Endress, *Die arabischen Übersetzungen von Aristoteles' Schrift De caelo* (Ph. D. thesis, Frankfurt am Main, 1966); also the same, 'The circle of al-Kindī: early Arabic translations from the Greek and the rise of Islamic philosophy', in: *The Ancient Tradition in Christian and Islamic Hellenism: studies on the transmission of Greek philosophy and sciences, dedicated to H. J. Drossaart Lulofs on his ninetieth birthday,* ed. by G. Endress and R. Kruk, CNWS Publications; vol. 50 (Leiden:

en 1215 al IV Concilio de Letran', *Hispania sacra*, 4 (1951): 335–55; Charles H. Haskins, 'Michael Scot and Frederick II', *Isis*, 4 (1921–22): 250–75, revised in: id., *Studies in the history of medieval science*. Cambridge, Mass., 1924, 219–27; [2]1927, 272–98; *id.*, 'Michael Scot in Spain', in: *Estudios eruditos in memoriam Adolfo Bonilla y San Martín*, 2 (Madrid 1930); Raoul Manselli, 'La corte di Federico II e Michele Scoto', in: *L'Averroismo in Italia: convegno internazionale (Roma, 18–20 aprile 1977)*, Atti dei Convegni Lincei, 40 (Roma: Accademia Nazionale dei Lincei, 1979), 63–80; R.-A. Gauthier, 'Notes sur les débuts (1225–1240) du premier «averroïsme»', *Revue des sciences philosophiques et théologiques*, 66 (1982), pp. 331-334, id., 'Préface', in: Thomas Aquinas, *Opera Omnia*, t. 45,1: *Sentencia libri De anima* (Roma: Leonina, 1984); P. Morpurgo, 'Le traduzioni di Michele Scoto e la circolazione di manoscritti scientifici in Italia meridionale: la dipendenza della Scuola Medica Salernitana da quella Parigina di Petit Pont', in: *La diffusione delle scienze islamiche nel medio evo europeo: convegno internazionale (Roma, 2–4 ottobre 1984)* (Roma: Accademia Nazionale dei Lincei, Fondazione Leone Caetani, 1987), 167–192; Charles Burnett, 'Michael Scot and the transmission of scientific culture from Toledo to Bologna via the court of Frederick II of Hohenstaufen', in: *Le scienze alla corte di Federico II — Sciences at the Court of Frederick II*, Micrologus, 2 (1994): 101–26; Wolfgang Stürner, *Friedrich II.* (Darmstadt: Wissenschaftliche Buchgesellschaft, 1992–2000), Teil 2: 'Der Kaiser 1220-1250', 400–22.

On individual works, see S[ibyl] D[ouglas] Wingate, *The Mediæval Latin versions of the Aristotelian scientific corpus, with special reference to the biological works*, London, Univ., phil. diss. 1930 (London 1931), 72–85, 94–96 and index; Palémon Glorieux, *La Faculté des Arts et ses maîtres au XIIIe siècle*, Études de philosophie médiévale, 59 (Paris: Vrin, 1971), 258–60; Al-Biṭrûjî, *De Motibus Celorum*: critical edition of the Latin translation of Michael Scot, ed. by Francis J[ames] Carmody (Berkeley and Los Angeles: University of California Press, 1952), 17–21; Aristotle, *De Animalibus: Michael Scot's Arabic-Latin Translation*, part 3: Books XV-XIX. Generation of Animals, ed. by Aafke M. I. van Oppenraaij, Aristoteles Semitico-Latinus, 5, Leiden, 1992 (pp. vii–xv: 'Preface', by H. J. Drossaart Lulofs); Aafke M. I. van Oppenraaij, 'Michael Scot's Arabic-Latin translation of Aristotle's *Book on Animals*: some remarks concerning the relation between the translation and its Arabic and Greek sources', in: *Aristotle's Animals in the Middle Ages and Renaissance* (Leuven: Leuven University Press, 1999), 31–43.

On his place in the reception and development of Arabic science in the West: Clemens Baeumker, *Petrus de Hibernia, der Jugendlehrer des Thomas von Aquin und seine Disputation vor König Manfred,* Sitzungsberichte der Bayerischen Akademie der Wissenschaften, Phil.-Hist. Klasse, Jhg. 1920 (München 1920); cf. Martin Grabmann, 'Magister Petrus von Hibernia etc.', in: Grabmann, *Mittelalterliches Geistesleben: Abhandlungen zur Geschichte der Scholastik und Mystik,* 1 (München: Hueber, 1926): 249–65, esp. p. 260; id., *Mittelalterliches Geistesleben,* 2, 1936: 103–37.

Michael Scot's translation of Aristotle's De caelo with the commentary of Averroes

Scot's translation of the Long Commentary on Aristotle's books 'On The Heavens', *Commentum Averrois super libro de celo et mundo,* was made after 1217[7] since it mentions the version of al-Biṭrūǧī's *De motibus celorum.* This appears from the dedication to Stephen of Provins, preceding the text in several codices:

> Tibi Stephane de Provino hoc opus quod ego Michael Scotus dedi latinitati ex dictis Aristotelis specialiter commendo. Et si aliquid Aristoteles incompletum dimisit de constitutione mundana in hoc libro, recipies eius supplementum ex libro Alpetrangii quem similiter dedi latinitati, et es in eo exercitatus[8].

Stephanus de Provino was a French ecclesiastic — canon of Reims and archdeacon of Paris — who is found in various documents from 1211, and in 1231 is named as a member of the abortive commission appointed by Pope Gregory IX to examine the newly translated works of Aristotle on natural science, and to revise them for use in the University[9]. According to a report of Roger Bacon, Michael Scot

7. Probably not before he joined the court of Frederick II in the late twenties; *v.* Grabmann, ‚Kaiser Friedrich II. und sein Verhältnis', 111f.

8. Cf. R. de Vaux, 'La première entrée d'Averroès chez les Latins', *Revue des sciences philosophiques et théologiques,* 22 (1933): p. 196 (from cod. Vat. lat. 2184, with n. 2, enumerating the mss. containing this dedication).

9. Wingate, *Mediæval Latin versions,* 95; Lynn Thorndike, *Michael Scot,* 27; id., *University records and life in the Middle Ages,* Records of Civilization, 38 (New York: Columbia University Press, 1944; repr. New York 1949 et sæpius), 34, 38–40. President of this commission was William of Auxerre, author of the *Summa aurea* of theology conceived as demonstrative science; his death on the 3rd November, 1231, put an early end to its activities.

where Scotus appears to rely on Gerard on the evidence of the Renaissance editions of the Giuntina press, this is due to "contamination effected by scribes and editors, notably of the edition of 1550".[16] One remarkable case of Scot's use of Gerard is not found here, but in his version of al-Biṭrūǧī's *De motibus celorum,* where (according to F. J. Carmody), "for no apparent reason, Scot borrows from Gerard the Latin of al-Ḥajjâj's *Almagest* [i.e., translated from al-Ḥaǧǧāǧ ibn Maṭar's Arabic version], wholly distinct in style and vocabulary from the remainder of the work"[17].

Gerard's translation is rarely given jointly with Michael's and the *Commentarium magnum* in the manuscripts (as in MS Paris 17155), whereas the early printed editions, beginning with the incunabula of 1489 and 1495, give the lemmata of Aristotle's text both in the Arabo-Latin version of Michael and the *translatio nova,* i. e., the Græco-Latin version probably made by William of Moerbeke († 1286).

Incipit of the *translatio vetus* (Gerardus Cremonensis):
Summa cognitionis naturæ et scientiæ ipsam significantis in corporibus existit ...
Incipit of the *lemmata* in Michael Scot's version:
Maxima cognitio naturæ et scientia demonstrans ipsam est in corporibus ...
Incipit of the Græco-Latin version (Guillelmus de Moerbeke):
Scientia de natura [v.l. De natura scientia] fere plurima videtur circa corpora et magnitudines et horum existens passiones et motus ...

A thorough study of Michael's translation technique has been made by Aafke M. I. van Oppenraaij in connection with her edition of his translation of Aristotle's *De animalibus,* to be published in the

16. See Carmody, *Innovations in Averroes De Cælo,* 7: „Contamination of Scot from Gerard includes the interpolation in the Juntine [sc. of 1550, *v. infra* p. 36*] alone of the long presentation of syllogisms of 5ra45–5va30 [= I, c. 11], and the substitution of some 50 words in [book] 1. [textus] 11. [line] 2–5 in the Great manuscript B [i.e. Paris, Bibl. nat., lat. 15453, dated 1243], and in [Y =] Bibl. nat. lat. 14385, effected therefore before 1243. The Juntine, and related manuscripts, also show frequent substitution of single words and, for example in 1.96, other editorial intervention in 30 interpolated words absent in all other versions here considered."

17. Carmody, *ibid.*, 7.

first volume of her critical edition[18]; see also her article: 'Quelques particularités de la méthode de traduction de Michel Scot', in: *Rencontres de cultures dans la philosophie médiévale: traductions et traducteurs de l'Antiquité tardive au XIVe siècle* (Louvain-la-Neuve & Cassino, 1990), 121–29. Based on the present version of Averroes' *De caelo* is a study of the editor, Francis J. Carmody, 'The Latin style of Michael Scot in *De Cælo*', in: *Humaniora: essays in literature, folklore, bibliography, honoring Archer Taylor on his seventieth birthday*, ed. by Wayland D. Hand, Gustave Otto Arlt (Locust Valley, N. Y.: Augustin, 1960), pp. 208–18.

Manuscripts of the Latin Version

Abbreviations:

AL = *Aristoteles Latinus,*: codices descripsit Georgius Lacombe [etc.], 1 (Romae 1939); 2, supplementis indicibusque instruxit L. Minio Paluello (Cantabrigiae 1955); supplementa altera, ed. L. Minio-Paluello (Bruges, Paris 1961).

d'Alverny, *Avicenna Latinus* = Marie-Thérèse d'Alverny, *Avicenna Latinus: codices*, Louvain-la-Neuve, Leiden 1994

The sigla appearing in the left margin are those used by Carmody in the *apparatus criticus* of his edition (*v. infra*, p. 40* and 46*).

a) Complete Commentary

Admont, Bibl. Monasterii, 480 sæc. xiv

AL I: 252f. no. 32. — Jakob Wichern, *Catalogus codicum manu scriptorum Admontensis* (1889) [manuscript, copy in München, Staatsbibliothek, shelf no. Hbh Ea 1620].

Parchment; 291 ff. in 2°, in double columns.

- ff. 1r-116: Aristotelis *De celo et mundo* cum Averrois Commentario magno.

  ¶ Cum epistola ad Stephanum de Provino nuncupatoria in capite præmissa [manu altera contemporanea exarata].

18. *V. supra*, p. 18*.

Assisi, Bibl. Communalis, 279 sæc. xiv

AL II: 869 no. 1255. — On Frater Urbanus Averroista (first half of 14th century) as a reader of this ms., *v.* Horst Schmieja, 'Urbanus Averroista und die mittelalterlichen Handschriften des Physikkommentars von Averroes', *Bulletin de philosophie médiévale*, 42 (2000): 133–43, esp. p. 144.

- ff. 102r-145r: Aristotelis *De celo et mundo* cum Averrois Commentario magno.

Basel [cod. Basileensis], Bibl. Universitatis, F. I. 31 sæc. xiii ex.–xiv in.

AL II: 808 no. 1134. — Carolus (Charles) Lohr, *Aristotelica Helvetica: catalogus codicum latinorum in bibliothecis Confederationis Helveticae asservatorum quibus versiones expositionesque operum Aristotelis continentur,* Scrinium Friburgense, Sonderband 6 (Freiburg [Switzerland]), 1994); P. O. Kristeller, *Iter italicum* (London, Leiden, 1963–92), V: 5: 65; K. Escher, *Die Miniaturen in den Basler Bibliotheken, Museen und Archiven* (Basel 1917), p. 70 no. 83.

Parchment; 21+123ff., 331×233 mm, 2 coll.

- ff. 1r-123v: Aristotelis *De celo et mundo* cum Averrois Commentario magno.

  ¶ Inc. mutil., cum commento Averrois mutil.

V Cambridge, Bibl. Collegii Gonvillæ et Caii, 486 (482) sæc. xiv in.

AL I: 348f. no. 235. — Montague Rhodes James, *A descriptive catalogue of the manuscripts in the library of Gonville and Caius College* (Cambridge 1907–8), 556f.

Parchment; 498 + 4 ff., 375×235 mm, double columns.

- ff. 235r-384r: Aristotelis *De celo et mundo* cum Averrois Commentario magno.

  ¶ Cum epistola ad Stephanum de Provino dedicatoria in capite praemissa.

Cambridge, Bibl. Domus Petri, 126 sæc. xiv

AL I: 357 no. 254. — Montague Rhodes James, *A descriptive catalogue of the manuscripts in the library of Peterhouse* (Cambridge 1899), p. 144.

Parchment; 146+60+13+8 ff., 337×216 mm, double columns.

- I, ff. 1r-146r: Aristotelis *De celo et mundo* cum Averrois Commentario magno.

  Contrary to the statement of the catalogue, this is the complete commentary, and not part I only.

CHICAGO, BIBL. SCIENTIARUM MEDICARUM
University of Illinois at Chicago, Library of Medical Sciences [now: Library of the Health Sciences].

AL, Suppl., p. 227f.; Seymour De Ricci & D. J. Wilson, *Census of medieval and Renaissance manuscripts in the United States and Canada, Supplement,* originated by C. U. Faye, continued and ed. by W. H. Bond (New York 1962), 227f.

- ff. 157r-188v: Aristotelis *De celo et mundo* cum Averrois Commentario magno.

CLERMONT-FERRAND, BIBLIOTHECA MUNICIPALIS, 168 sæc. xiii

Catalogue général des manuscrits des bibliothèques publiques de France: Départements, t.14: Clermont-Ferrand [etc.] (Paris 1890), p. 61; see also database of the Institut de recherche et d'histoire des textes (IRHT).

- ff. 73–103 [IRHT: 75–101v]: *Liber de celo et mundo* [dubium an sit Comm. mag. completum].

  Inc. textus: Summa cognitionis naturæ et scientia ipsam significantis ... [= translatio vetus Gerardi Cremonensis]. — *Inc. comm.* [IRHT]: Incipit primo narrare subiectum istius artis et dicit quod maxima cognitio nature et scientie demonstrans ipsam ... — *Expl.:* ... venimus super intentionem nostram in eo.

DURHAM [COD. DUNELMENSIS], BIBL. CAPITULI, C. I, 17 (B) sæc. xv in.

AL I: 365 no. 270.

- ff. 1r-147r: Aristotelis *De celo et mundo* cum Averrois Commentario magno.

  ¶ Expl. ex abrupto in libro quarto.— Cum epistola ad Stephanum de Provino dedicatoria.

Φ DURHAM [COD. DUNELMENSIS], BIBL. CAPITULI, C. I, 18 sæc. xiv

AL I: 365f. no. 271.

- ff. 1r-101r: Aristotelis *De celo et mundo* cum Averrois Commentario magno.

  ¶ Cum epistola ad Stephanum de Provino dedicatoria in capite præmissa.

ERFURT, AMPLONIANA [Wissenschaftliche Bibliothek der Stadt Erfurt], 2° 351 s. xiv in.

AL I: 662f. no. 886. — Wilhelm Schum, *Beschreibendes Verzeichnis der amplonianischen Handschriften-Sammlung zu Erfurt,* [...] bearb. u. hrsg. mit einem Vorwort über Amplonius und die Geschichte seiner Sammlung

(Berlin: Weidmann, 1887), 246f.; Miecislaus Markowski, *Repertorium commentariorum medii ævi in Aristotelem Latinorum quæ in Bibliotheca Amploniana Erffordiæ asservantur* (Wrocław [etc.] 1987), p. 59f., 186 (with reference to Schum's entries).

- ff. 191v-275r: [Commenta] super quattuor libros de celo et mundo Aristotelis.

  ¶ Cum epistola ad Stephanum de Provino dedicatoria in capite præmissa.

KYNŽVART, MS. CASTELLI (Zámeska knihovna), 22 D 2 (11 Phil. antiqu.) sæc. xiii ex.

d'Alverny, *Avicenna Latinus* (Addenda), p. 372-5; Franciscus Čáda, *Codices manu scripti Bibliothecæ Castelli Kynžvart in Bohemia Occidentali* (Pragæ, 1965), pp. 122-30 no. 80.

Parchment; 288 ff., 290×205 mm.

- ff. 81–205v: Aristotelis *De celo et mundo* cum Averrois Commentario magno.

  ¶ Cum epistola ad Stephanum de Provino dedicatoria in capite præmissa.

LEIPZIG, BIBL. UNIVERSITATIS 1433 sæc. xiv in.

AL I: 710f. no. 992.

- ff. 70r-203r: Aristotelis *De celo et mundo* cum Averrois Commentario magno.

  ¶ Cum epistola ad Stephanum de Provino dedicatoria in capite præmissa.

  Expl.: Explicit commentum Averroys super librum de Celo et Mundo scriptum completum per manus Roberti scolaris, Parisii anno domini MCCC quinto, XVI kalendas Augusti.

D MADRID, BIBL. PALATII 150, ff. 142r-203r:

v. Salamanca, Bibl. Universitatis, 2671.

MILANO [COD. MEDIOLANENSIS], BIBL. AMBROSIANA, E. 80 inf. sæc. xiv

AL II: 985f. no.1443 et p. 1280 (Errata). — *Inventario Ceruti dei manoscritti della Biblioteca Ambrosiana,* vols. 1-5 (facsim.) ed. A. Paredi, Fontes Ambrosiani vol. 50, 52, 57, 60, 63 (Trezzano 1973–9), vol. I, p. 740; Astrik L. Gabriel, *A summary catalogue of microfilms of one thousand scientific manuscripts in the Ambrosiana Library, Milan* (Notre Dame, Indiana: The Mediæval Institute, University of Notre Dame, 1968), p. 135 no. 288; Ezio Franceschii, 'L'«Aristotele Latino» nei codici dell'Ambrosiana', in: *Miscellanea Giovanni Galbiati*, 3 (Milano 1951), 227–47.

Parchment; 57 ff.

- Averrois commentaria in Aristotelis *De celo et mundo* [incomplete]. *Inc.:* [...] duorum elementorum ... — *Expl.:* ... quæ existunt [...].

MONTPELLIER [COD. MONTEPELLUSANUS], BIBL. FACULTATIS MEDICINÆ, 33 sæc. xiv

AL I: 477f. no. 493. — *Catalogue général des manuscrits des bibliothèques publiques des départements, publié sous les auspices du Ministère de l'Instruction publique,* t. 1. Paris 1849, pp. 279ff.: Manuscrits de la Bibliothèque de l'École de médecine de Montpellier [rédigé par G. Libri et H.-M. Kühnholtz], p. 298 no. 33; Charles Samaran & Robert Marichal, *Catalogue des manuscrits en écriture latine portant des indications de date, de lieu ou de copiste,* tome IV (Bourgogne, Paris 1968), p. 303.

Parchment; 140 folia in 2°. — Dated A. D. 1321.

- ff. 1r-82vb: Aristotelis *De celo et mundo* cum commentario [Adveroys].

σ NAPOLI, BIBL. NATIONALIS, VIII E. 36 sæc. xiv in.

AL II, 1011f. no. 1480.

- ff. 181r-284v: Aristotelis *De celo et mundo* cum Averrois Commentario magno.

OXFORD, BIBL. BODLEIANA, Canon. lat. class. 287 sæc. xiv

AL I: 392 no. 327.

- ff. 1r-147r: Aristotelis *De celo et mundo* cum Averrois Commentario magno.

OXFORD, BIBL. COLLEGII BALLIONENSIS 243 sæc. xiv in.

AL I: 404f. no. 351; Roger A. B. Mynors, *Catalogue of the Manuscripts of Balliol College, Oxford* (Oxford 1963), p. 265.

Parchment; 117 ff., 368×241 mm, 2 coll.

- ff. 2r-112v: Aristotelis *De celo et mundo* cum Averrois Commentario magno.

OXFORD, BIBL. COLLEGII BALLIONENSIS 244 sæc. xiv in.

AL I: 405 no. 352. — Roger A. B. Mynors, *Catalogue of the Manuscripts of Balliol College, Oxford* (Oxford 1963), p. 265f.

Parchment; 134 ff., 375×229 mm, 2 coll. — Dated A. D. 1308.

- ff. 2r-134r: Aristotelis *De celo et mundo* cum Averrois Commentario magno.

F PARIS, BIBL. MAZARINEA 3473 sæc. xiii ex.—xiv in.
(olim S. Iacobi Parisiensis)

AL I: 495f. no. 534. — Auguste Molinier, *Catalogue des manuscrits de la Bibliothèque Mazarine,* Catalogue général des bibliothèques publiques de France (Paris 1885–98), 3 (1890): 96f., no. 3473; Wladyslaw Senko, *Repertorium commentariorum medii ævi in Aristotelem Latinorum quæ in Bibliothecis publicis Parisiis asservantur,* fasc. 2, Opera philosophorum medii ævi. Textus et Studia, V, 2 (Warszawa 1982), p. 138f [with further bibliography]; d'Alverny, *Avicenna Latinus,* 26-8. — Siglum **J** in *Comm. mag. Metaph.* ed. Darms (p. 32, with further bibliography), ed. Bürke, ed. Ruggiero (p. 50*f.) and Ponzalli (p. 45f.); siglum **F** in *Comp. Parva Naturalia* ed. Shields & Blumberg (descr. p. xx), *Comm. med. De gen. et corr.* ed Fobes & Kurland (p. xiv), *Comm. mag. De an.* ed. Crawford (p. xiii).

Parchment; I + 194 folia 400×277 mm, double columns.

- ff. 69r–100r: Aristotelis *De celo et mundo* cum Averrois Commentario magno.

  ¶ Cum epistola ad Stephanum de Provino dedicatoria in capite præmissa.

γ PARIS, BIBL. NATIONALIS, lat. 6504 sæc. xiv

AL I: 519f. no. 585. — Siglum **D** in *Comm. mag. Metaph.* edd. Bürke (p. 17), Ponzalli (p. 47), Ruggiero (p. 51*). — François Avril & Marie-Thérèse Gousset, *Manuscrits enluminés d'origine italienne,* vol. 2: XIIIe siècle (Paris 1984), p. 91 (no. 113) + ill.

- ff. 151r-222v: *Comm. mag. de Cælo.*

Y PARIS, BIBL. NATIONALIS, lat. 14385 sæc. xiii ex.

AL I: 541ff. no. 634. — Siglum **V** in *Comm. mag. Metaph.* edd. Darms (p. 35f.), Ponzalli (p. 46), and Ruggiero (p. 52*f.).

- ff. 1r-44v: Aristotelis *De celo et mundo* cum Averrois Commentario magno.

  ¶ Cum epistola ad Stephanum de Provino dedicatoria in capite præmissa.

B PARIS, BIBL. NATIONALIS, lat. 15453 (olim Sorbonne 943). sæc. xiii

AL I: 551f. no. 654. — Charles Samaran & Robert Marichal, *Catalogue des manuscrits en écriture latine portant des indications de date, de lieu ou de copiste,* t. 3: Bibl. Nat. 8001 à 18613 (Paris 1974), p. 429. — On the significance of the codex as one of the first — still incomplete — collections of the commentaries of Averroes, *v.* Fernand Van Steenberghen, 'Le problème de l'entrée d'Averroès en Occident', in: *L'Averroismo in Italia (Roma,*

18-20 aprile 1977), Atti dei Convegni Lincei, 40 (Roma: Accademia Nazionale dei Lincei, 1979), 81–9. — Siglum **B** in *Comm. mag. De an.* ed. Crawford (pp. xiv, xvii), *Comm. med. De gen. et corr.* ed. Fobes & Kurland (p. xii), *Comp. Parva Naturalia* ed. Shields & Blumberg (p. xiv); siglum **C** in *Comm. mag. Metaph.* edd. Darms (p. 31f.), Ruggiero (p. 48*f.) and Ponzalli (p. 44f.).

Parchment; 414 ff., 395×275 mm, double columns. — Dated A. D. 1243, of Italian origin.

- ff. 121r-202r: Aristotelis *De celo et mundo* cum Averrois Commentario magno.

ϰ [?] PARIS, BIBL. NATIONALIS, lat. 16155 sæc. xiii

AL I: 567f. no. 681.

- ff. 2r-140v: Aristotelis *De celo et mundo* cum Averrois Commentario magno.
  ¶ Cum epistola ad Stephanum de Provino dedicatoria in capite præmissa.

C PARIS, BIBL. NATIONALIS, lat. 16156 sæc. xiii

AL I: 568 no. 682. — Siglum **C** in *Comm. mag. De an.* ed. Crawford (pp. xiv, xvii).

- ff. 3r-187r: Aristotelis *De celo et mundo* cum Averrois Commentario magno.
  ¶ Cum epistola ad Stephanum de Provino dedicatoria in capite præmissa.

ν [?] PARIS, BIBL. NATIONALIS, lat. 16159 sæc. xiii

AL I: 568f. no. 683. — Cf. *Comm. mag. Metaph.* edd. Darms (p. 34), Ruggiero (p. 53*), and Ponzalli (p. 47).

- ff. 158r-223v: Aristotelis *De celo et mundo* cum Averrois Commentario magno.

Z PARIS, BIBL. NATIONALIS, lat. 17155 sæc. xiii

AL I: 582 no. 708. — François Avril & Patricia Danz Stirnemann, *Manuscrits enluminés d'origine insulaire VII*[e] *— XX*[e] *siècle* (Paris 1987), p. 141f. (no. 177); planche LXXI.

- ff. 225r-344v: *De Celo* veteris translat. et Averroes *in De Celo*.
  ¶ Cum epistola ad Stephanum de Provino nuncupatoria in capite præmissa.
  ¶ Also giving the translatio vetus of Aristotle's text made by Gerard of Cremona.

ω PARIS, BIBL. UNIVERSITATIS, 601 sæc. xiii

AL I: 592 no. 731. — *Catalogue général des manuscrits des bibliothèques publiques de France: Université de Paris et universités des départements* (Paris 1918), 1–368: Bibliothèque de l'Université de Paris / Ch. Beaulieux, no. 601.

Parchment; 150 ff., 290×220 mm, 2 coll.

- ff. 104r-150r: Aristotelis *De celo et mundo* cum Averrois Commentario magno.
  ¶ Cum epistola ad Stephanum de Provino nuncupatoria in capite præmissa.

D SALAMANCA, BIBL. UNIVERSITATIS, 2671 sæc. xiii ex.

(*olim* Collegii S. Bartholomei 176; Palatii Regii Matritensis VII G 2, 2 c. 2, 150)

AL II: 840f. no. 1207. — Siglum **D** in *Comm. mag. De an.* ed. Crawford (descr. p. xvii); *Comm. med. De gen. et corr.* ed. Fobes & Kurland (p. xxv f.). — G. Beaujouan, *Manuscrits scientifiques médiévaux de l'Université de Salamanque et de ses "Colegios mayores"* (1962), pp. 178–81, table p. 193; d'Alverny, *Avicenna Latinus,* 229–31.

Parchment; 250 ff., 340×215 mm., double columns.

- ff. 142r–203r: Aristotelis *De celo et mundo* cum Averrois Commentario magno.
  ¶ Cum epistola ad Stephanum de Provino nuncupatoria in capite præmissa.

W TOLEDO, BIBL. CAPITULI, 98.22 sæc. xiii

AL II: 860 no. 1246. — J. M. Millás Vallicrosa, *Las Traducciones orientales en los manuscritos de la Biblioteca Catedral de Toledo* (Madrid 1942), pp. 202-6 (cf. P. O. Kristeller, *Iter Italicum* [London, Leiden 1963–92], IV: 645).

- ff. 2r-65v: Aristotelis *De celo et mundo* cum Averrois Commentario magno.
  ¶ Cum epistola ad Stephanum de Provino dedicatoria in capite præmissa.

A VATICAN, COD. OTTOBONIANUS, lat. 2215 sæc. xiii

AL II: 1186f. no. 1768. — Siglum **A** in *Comm. mag. De an.* ed. Crawford (descr. p. xvi); *v.* R.-A. Gauthier, 'Quelques questions à propos du commentaire de S. Thomas sur le De anima', *Angelicum*, 51 (1974): p. 421.

- ff. 174r-244r: Aristotelis *De celo et mundo* cum Averrois Commentario magno.

Ψ VATICAN, COD. URBINAS, lat. 221 sæc. xiv med.–ex.

AL II: 1207f. no. 1814. — Cosimus Stornajolo, *Codices Urbinates latini,* Bibliothecæ Apostolicæ Vaticanæ codices manu scripti recensiti (Romæ 1902–21), 1: p. 213 no. 221; Roland Hissette, 'Le *corpus averroicum* des manuscrits vaticans Urbinates latins 220 et 221 et Nicoleto Vernia', in: *Miscellanea Bibliothecæ Apostolicæ Vaticanæ III,* Studi e Testi, 333 (Vatican, 1989), pp. 257–356; id. [ed.], *Commentum medium super libro Peri Hermeneias Aristotelis,* Averrois Opera, series B: Averroes Latinus, 12 (Lovanii 1996), pp. 15*–17* (p. 15*, n. 44: "Dans ma récente étude 'Le *corpus averroicum* [...] (1989), p. 257–258 et passim, j'ai pu établir que les deux manuscrits Vat. urb. lat. 220 et 221 sont complémentaires et conservent le *corpus* pratiquement complet des commentaires aristotéliciennes d'Averroès connus par le moyen âge latin").

Parchment; 257 ff., double columns of 21 lines.

- ff. 2r-107r: Aristotelis *De celo et mundo* cum Averrois Commentario magno.

E VATICAN, CODD. VATICANI, lat. 2089 sæc. xiii ex.

AL II: 1221 no. 1846. — Claudius Leonardi, Maria Magdalena Lebreton, *Codices Vaticani Latini: Codices 2060–2117,* Bibliothecæ Apostolicæ Vaticanæ codices manu scripti recensiti (Vatican 1987), pp. 124-9; d'Alverny, *Avicenna Latinus,* 89-91; Antonio Manfredi, *I codici latini di Niccolò V,* Studi e Testi, 359 (Città del Vaticano: Biblioteca Apostolica Vaticana, 1994), p. 365f. no. 584; Richard C. Taylor, 'The «Liber de Causis»: a preliminary list of manuscripts', *Bulletin de philosophie médiévale,* 25 (1983): p. 78. — Siglum E in *Comm. med. De gen. et corr.* ed. Fobes & Kurland (descr. p. xii–xiv).

Parchment; 359 ff. (358 + 79bis), 315×207 [210] mm, 2 coll.

- ff. 6ra-195rb: Aristotelis *De celo et mundo* cum Averrois Commentario magno.

X VATICAN, CODD. VATICANI, lat. 2184 sæc. xiii–xiv

AL II: 1227 no. 1862. — Anneliese Maier, *Codices vaticani latini: Codices 2118-2192,* Bibliothecæ Apostolicæ Vaticanæ codices manu scripti recensiti (Vatican 1961), pp. 189-91.

Parchment; 98 ff., 330×215 mm, 2 coll.

- Aristotelis *De celo et mundo* cum Averrois Commentario magno.
  ¶ Cum epistola ad Stephanum de Provino nuncupatoria in capite præmissa.

η VATICAN, CODD. VATICANI, lat. 2436 — sæc. xiv

AL II: 1229 no. 1866.

- ff. 1r-55r: Aristotelis *De celo et mundo* cum Averrois Commentario magno.

  Expl.: Explicit comm. Averroys super libro de Celo et mundo Aristotilis quod complevi in scribendo Bon[onie] 1345 in vigilia Ascensionis Domini.

λ VENEZIA, BIBL. S. MARCI, lat. VI, 52 — sæc. xiv

AL II: 1101f. no. 1610.

- ff. 1r-186v: Aristotelis *De celo et mundo* cum Averrois Commentario magno.

b) Fragments and Excerpts

ASSISI, BIBL. COMMUNALIS 281 — sæc. xiv

AL II: 870f. no. 1255.

- ff. 44v-67v: Aristotelis *De celo et mundo*; in notulis marginalibus textus commentarii Averrois additi.

BASEL, BIBLIOTHECA UNIVERSITATIS, D. I.9 — sæc. xiv

Carolus [Charles] Lohr, *Aristotelica Helvetica,* Scrinium Friburgense, Sonderband 6 (Freiburg [Switzerland] 1994), p. 28.

Parchment; 337×228 mm, 2 coll.

- ff. 1r–2v; 59r–62v: Aristoteles, *De celo et mundo,* II-III, fragmentum cum Averrois Commento.

BOLOGNA [COD. BONONIENSIS], BIBL. UNIVERSITATIS, lat. 1180 (2344) — sæc. xiii

AL II: 889f. no. 1286.

- ff. 127r-175v: Aristotelis *De celo*, in margine alia translatio, translatio commentatoris et commentator.

CAMBRIDGE, BIBL. AULÆ VALENCIÆ MARIÆ, ms. 312 — sæc. xiv

AL II: 1260 no. 1938.

- B. 22-23 (2 ff.): Excerptum ex Averrois comm. in *De celo* III.

ERFURT, AMPLONIANA [Wissenschaftliche Bibliothek der Stadt Erfurt], 2° 337 — sæc. xiv med.

AL II: 661 no. 882. — Wilhelm Schum, *Beschreibendes Verzeichnis der amplonianischen Handschriften-Sammlung zu Erfurt,* [...] bearb. u. hrsg. mit einem Vorwort über Amplonius und die Geschichte seiner Sammlung (Berlin: Weidmann, 1887), p. 232f.; Miecislaus Markowski, *Repertorium commentariorum medii ævi in Aristotelem Latinorum quæ in Bibliotheca Amploniana Erffordiæ asservantur* (Wrocław [etc.] 1987), pp. 62, 183.

Parchment; 85 ff.

- ff. 45r-50v: Excerpta ex commentario in Aristotelis librum cui *de cœlo et mundo* inscribitur.

  Inc.: Maxima cognitio nature et sciencia demonstrans ipsam ... — *Expl.:* ... propter quantitatem et similiter intelligendum est de terra. Expl. exscerptum de commento sup. de cel. et mun.

FIRENZE, BIBL. LAURENTIANA, S. CRUCIS, PLUT. XV Sin. 1.

AL II: 946 no. 1376.

- ff. 60r-109v: *De celo* (corpus vetustius); in marg. et in interlin. glossæ et notulæ ex Averrois comm.

GRAZ, BIBL. UNIVERSITATIS II, 482 sæc. xiii ex.

AL I: 263–5 no. 57. — d'Alverny, *Avicenna Latinus,* 173-80 (p. 177 no. 35).

Parchment; 242 ff., 296×210 mm., double columns.

- ff. 214r-216v. Averrois *Commentarium magnum* in Aristotelis *De celo et mundo* <excerpta>.

  Inc.: Corpora que sunt partes tocius ..., *Expl.:* ... diuerse huismodi diuersitate, etc. hoc significat.

LONDON, MUS. BRIT. ADD. 19582

AL, Suppl., 75 no. 2045.

- ff. 60r-82v: *De celo* (corpus recentius), cum glossis multis in *De celo,* partim ex Averroe.

PARIS, BIBL. NATIONALIS, lat. 16110 sæc. xiv

AL I: 561–3 no. 671. — Charles Samaran & Robert Marichal, *Catalogue des manuscrits en écriture latine portant des indications de date, de lieu ou de copiste.* Tome III (Bibl. Nat. 8001 à 18613) (Paris 1974), p. 699; Wladyslaw Senko, *Repertorium commentariorum medii ævi in Aristotelem Latinorum quæ in Bibliothecis publicis Parisiis asservantur,* fasc. 2 = Opera philosophorum medii ævi. Textus et Studia, V,2 (Warszawa 1982), p. 7-14.

- ff. 115r-118r: Aristotelis *De celo et mundo* cum Averrois Commentario magno, in librum II, textus et commentaria 43-59.

  Inc.: Dixit Aristotiles quod stelle cum videntur... — *Expl.:* ... et non est sicut diximus.

PARIS, BIBL. NATIONALIS, lat. 17837

AL I: 584f. no. 715.

- ff. 63r-96v: *De celo* (corpus recentius) in marginibus fragmenta commenti.... Averrois.

VENEZIA, BIBL. S. MARCI, lat. XIV, 259 sæc. xiv

AL II: 1114 no. 1632. — Pietro Zorzanello, *Catalogo dei codici latini della Biblioteca Nazionale Marciana di Venezia non compresi nel calatogo di G. Valentinelli*; vol. III, class XIV (Trezzano 1985), 457-9 (cf. P. O. Kristeller, *Iter Italicum* [London, Leiden, 1963–92], VI: 258).

- f. 118v: *De celo* veteris translat. fragmentum perbreve cum Averrois comm. in idem.

WÜRZBURG [COD. HERBOPOLITANUS], BIBL. UNIVERSITATIS, Mp. Med. f. 3 sæc. xiii ex.

AL I: 191f. no. 952. — Hans Thurn, *Die Handschriften der Universitätsbibliothek Würzburg*, B. 1: Die Ebracher Handschriften (Wiesbaden 1970), pp. 8–10.

- ff. 214va-218vb: Averroes, Comm. mag. in *De celo* II, c. 43–59.
  Inscr.: Sermo ubi de comparatione motu<s> stellarum. —*Inc.:* Dixit Aristotiles quod stelle cum videantur....
- ff. 227vb-228rb: In *De celo* I,1-2
  Inscr.: Differentia de Trino. — *Inc.:* Dixit magister: Continuum quidem est ...

EARLY PRINTED EDITIONS

a) Incunabula

Gesamtkatalog der Wiegendrucke, vol. 2: Alfarabius—Arznei (Leipzig: Hiersemann, 1926), coll. 567–89 [abbrev. *GW*].

Aristoteles: *Opera.* [With commentaries of Averroes.] Ed. Nicoletus Vernia. — T. I,1–III,2. — Venetiis: Andreas Torresanus & Bartholomæus de Blavis, 1483 [–1484]. 2°

GW 2: 563–67 no. 2337. — On the manuscript basis of Nicoleto Vernia, *v.* R. Hissette, 'Le corpus averroicum' [*v. supra,* p. 31*], and esp. p. 299 n. 179 on the dating of the edition.

- T. 2, no. 1, ff. 2ra–101rb: *De celo et mundo.* — Dated VI. Kal. Jun. [27th May], 1483.
  Lemmata of Aristotle are given both in the Græco-Latin and in Scotus' Arabo-Latin translation.

Aristoteles: *Opera.* [With commentaries of Averroes; also Leonardus Brunus Aretinus, Comm. in libros ethicos.] — T. 1–5. — Venetiis: Bernardinus Stagninus (de Tridino), 1489. 2°

GW 2: 568–72 no. 2339, col. 569.

- T. 2, ff. 2ra–71vb: Aristotelis Stagyrite peripateticorum principis *de celo & mundo* liber primus: cum Averrois cordubensis expositione. — Dated pridie kal. feb. [January 31], 1489.
  Lemmata of Aristotle are given both in the Græco-Latin and in Scotus' Arabo-Latin translation.

Aristoteles: *Opera.* [With commentaries of Averroes and Leonardus Brunus Aretinus, Comm. in libros ethicos. Ed.: Augustinus Niffus. Supplement ed. by Nicolaus Grassetus.] — T. 1; 2. — Venetiis: [Iohannes et Gregorius de Gregoriis, for:] Octavianus Scotus, 1495–96. 2°

GW 2: 572–74 no. 2340, col. 572f.

- T. 1, no. 2, ff. 161ra–262rb: Aristotelis Stagyrite peripateticorum principis *de celo & mundo* liber primus: cum Averois cordubensis expositione. — Dated 28 November, 1495.
  Lemmata of Aristotle are given both in the Græco-Latin and in Scotus' Arabo-Latin translation.

b) 16th Century Editions

F. Edward Cranz, *A bibliography of Aristotle editions, 1501-1600*, Bibliotheca Bibliographica Aureliana, 38 (Baden-Baden: Koerner, 1971) [based upon, and following the numbering sequence of the *Index Aureliensis*]; see also the same: 'Editions of the Latin Aristotle accompanied by the commentaries of Averroes', in: *Philosophy and Humanism: Renaissance essays in honor of Paul Oskar Kristeller*, ed. by E. P. Mahoney. Leiden 1976, pp. 116-28; on the editions of the Giuntina press, see P. Camerini, *Annali dei Giunti* (Firenze, 1962). — The first part of the following short list of editions of the works of Aristotle, containing *De caelo* with Averroes' *Commentarium magnum* in the translation of Michael Scot, is based on the data provided by Cranz.

Aristotelis Physica. De coelo et mundo [etc]. — Venetiis: Paganinus de Paganinis / Ioa. & Gregorius de Gregorii, 1501. 2°

Index Aureliensis, no. 107.693.

Aristotelis Opera. [With the commentaries of Averroes and of Leonardo Bruni.] [Ed.:] Marcus Antonius Zimara. — Venetiis: Bonatus Locatellus, 1507. 2°

Index Aureliensis, no. 107.753a. Based on the edition Venetiis 1495–6, *GW* 2: 2340.

Aristotelis opera. [With the commentaries of Averroes and of Leonardo Bruni.] [Ed.:] Augustinus Niphus. — Venetiis: Octavianus Scotus, 1516. 2°

Index Aureliensis, no. 107.838. Based on the edition Venetiis 1459–6, *GW* 2: 2340, cf. *GK* VI: 5957.

Aristotelis ex physiologia libri duodetriginta. [...] *De coelo quattuor.* [...]. — Parisiis: Henricus Stephanus, 1518. 4°

Index Aureliensis, no. 107.850.

Aristo. De celo & mundo, cum com. Aver.: Aristote. Stagyrite Lib. iiij de celo et mundo : subnexis eius duobis illis de generatione et corrup. cum singulorum Epitomis hactenus non impressis: ac Averro. fidiss. interprete: necnon eiusdem opusculum de substantia orbis: in calce operis apposito: cum apostillis M. Ant. Z[imaræ]. — Lugduni [Lyon]: apud Scipionem de Gabiano, per Iacobum Myt., 1529. 271 ff. 8°

Index Aureliensis, no. 107.909.

Aristotelis Stagyritæ libri quatuor de coelo, et mundo, subnexis eius duobis illis de generatione & corrup. cum singulorum Epitomis hactenus non impressis, ac Averro. fidiss. interprete: necnon eiusdem opusculum de substantia orbis, in calce operis appositum, cum apostillis M. Ant. Z[imaræ] quibus nil inesse vitij deprehendes. — Lugduni [Lyon]: apud Iacob Giunctam, 1542. 324 ff. 8°

Index Aureliensis, no. 108.039.

J *Aristotelis quæ extant opera,* nunc primum selectis translationibus, collatisque cum Græcis emendatissimis exemplaribus, margineis scholiis illustrata: et in novum ordinem digesta: additis etiam nonnullis libris *nunquam* antea Latinitate donatis: Averrois Cordubensis in ea opera omnes qui ad nos pervenere commentarii Græcorum, Arabum et Latinorum monumenta quædam ad hoc opus spectantia. — Vol. 1–10. — Venetiis: Apud Iuntas, 1550–1552. 2°

Index Aureliensis, no. 108.193. — This edition was recommended as a reference basis for the *Corpus Commentariorum Averrois in Aristotelem*, ed. H. A. Wolfson et al. Its folia are given in the margins of Crawford's edition of the *Commentarium magnum* on Aristotle's *De anima* (1953). — On the Giuntine edition of the works of Aristotle with the commentaries of Averroes, *v.* Charles B. Schmitt, 'Renaissance Averroism studied through the Venetian edition of Aristotle–Averroes (with particular reference to the Giunta edition of 1550–2)', dans *L'Averroismo in Italia (Roma, 18-20 aprile 1977)*, Atti dei Convegni Lincei, 40 (Roma: Accademia Nazionale dei Lincei, 1979), 121–42; on the methods of emendation and correction (*castigatio, purgatio*) of the available translations, see Charles Burnett, 'The second revelation of Arabic philosophy and science: 1492–1562', in: *Islam and the Italian Renaissance,* ed. by Ch. Burnett and Anna Contadini, Warburg Institute Colloquia, 5 (London: Warburg Institute, 1999), 185–98.

- Vol. V: Aristotelis De Coelo, De Generatione & Corruptione, Meteorologicorum, De Plantis, cum Averrois Cordubensis variis in eosdem commentariis. — Folia 1-271: Aristotelis De Coelo.
  The *textus* of Aristotle are given first in the Græco-Latin translation, probably made by William of Moerbeke[19], followed by the Arabo-Latin version of Michael Scot.

Aristotelis Stagiritæ omnia quæ extant opera, nunc primum selectis translationibus emendationibus ex collatione græcorum exemplarium, scholiis in margine illustr., nova etiam ordine digesta: add. præterea non nullis libris nunquam antea latinitate donatis. — Averrois Cordubensis in ea opera omnes, qui ad nos pervenere, commentarii. Nonnulla super add. dubia, figuræ, notationes, nunquam antea ed., ut Averrois media in libros metaphys. commentatio: eiusdem de spermate libellus. Græcorum, Arabum, & latinorum monumenta quædam, ad hoc opus spectantia. Marci Antonii Zimaræ in Arist. et Aver. dicta contradictionum solutiones, quibus nunc addidimus doctissimorum virorum solutiones 100. Acc. autem omnia tum ex præfatione, tum ex indice librorum clarius innotescunt. — Vol. 1–11. — Venetiis: apud Cominum de Tridino Montisferrati, 1560. 8°

Index Aureliensis II, 225: no. 108.423.

J_2 *Aristotelis omnia quæ extant opera.* Selectis translationibus, collatisque cum græcis emendatissimis, ac vetustissimis exemplaribus, illustr., prȩstantissimorumque ætatis nostrȩ philosophorum industria diligentissime recogn. — Averrois Cordubensis in ea opera omnes qui ad hæc usque tempora pervenere, commentarii. Nonnulli etiam ipsius in logica, philosophia, & medicina libri, cum Levi Gersonidis in libros logicos annotationibus, quorum plurimi sunt, à Iacob Mantino in lat. conversi. Græcorum, Arabum & Latinorum lucubrationes quædam, ed hoc opus pertinentes. Marcantonii Zimaræ in Aristotelis [...]. — Vol. 1–11. — Venetiis: apud Iunctas, 1562. 4°

Reprinted as: *Aristotelis opera cum Averrois Cordubensis commentariis.* Frankfurt a. M., 1961-2 [in 14 vol.].

19. *Cf. Aristoteles Latinus,* pars prior, 53–54; M. Grabmann, *Mittelalterliches Geistesleben,* 3, p. 85.

Index Aureliensis, no. 108.456. — On the original titles and contents of the single volumes, *v.* Paolo Camerini, *Annali dei Giunti*, vol. 1 (Firenze, 1962), parte 1: p. 426 no. 658; p. 430 no. 663.

- Vol. V: Aristotelis De Coelo, De Generatione & Corruptione, Meteorologicorum, De Plantis, cum Averrois Cordubensis variis in eosdem commentariis. — Folia 1-271: Aristotelis De Coelo.

  The *textus* of Aristotle's *De caelo* are given first in the Græco-Latin translation, probably made by William of Moerbeke, followed by the Arabo-Latin version of Michael Scot (as in J, Venice 1550–52, *v. supra*). — The folia of this edition are given in the left margin of the present edition of *De caelo.*

Aristotelis omnia quæ extant opera. Selectis translationibus, collatisque cum græcis emendatissimis, ac vetustissimis exemplaribus, illustr., pręstantissimorumque ętatis nostrę philosophorum industria diligentissime recogn. — Averrois Cordubensis in ea opera, omnes qui ad hæc usque tempora pervenere, commentarii. Nonnulli etiam [...]. — Vol. 1–10. — Venetiis 1675: Iuntæ.

Index Aureliensis, no. 108.599; cf. *GW* 6: 5973.

Francis J. Carmody's Edition

A critical edition of Michael Scot's Latin translation of Averroes' *Commentarium magnum* was undertaken by Francis James Carmody (1907–1983), professor of French at the Univerity of California, Berkeley, as a sequel to his edition of al-Biṭrūǧī's *De motibus celorum*[20], on the suggestion of Francis H. Fobes[21]. On completion, in 1962, this was offered to Harry Austryn Wolfson, who was general editor of the *Averrois Commentaria in Opera Aristotelis* on behalf of the Mediæval Academy of America[22], for publication in the series. After a lengthy correspondence, and prolonged evasion and stalling on the

20. *V. supra*, p. 17* with note 4.

21. Editor of the Middle Commentary of *De gen. et corr.*: *Commentarium medium in Aristotelis de Generatione et corruptione,* recensuit Franciscus Howard Fobes adiuvante Samuele Kurland, Corpus Commentariorum Averrois in Aristotelem, versionum latinarum vol. IV, 1 (Cambridge, Mass., 1956).

22. For the general outline of this project, see H. A. Wolfson, 'Plan for the publication of a *Corpus Commentariorum Averrois in Aristotelem*', *Speculum,* 6 (1931): 412–22; the same, 'Revised plan for the publication of a *Corpus Commentariorum Averrois in Aristotelem*', *Speculum,* 38 (1963): 88–104; repr. in Wolfson, *Studies in the history of philosophy and religion,* vol. 1 (Cambridge, Mass.: Harvard University Press, 1973): 430–54.

side of the Academy, the proposal was finally declined in 1962, (a) because, for lack of funds for the publication, they could not promise publication within a reasonably time, (b) because, since the Juntine edition of 1562 had been reprinted by offset in 1962, Wolfson felt that for those works that did not have Arabic and/or Hebrew counterparts, it was not worth-while to work on the Latin text alone. Carmody contined working on the text, and finding out from the dissertation of the present writer that part of the Arabic had been discovered, procured himself a photocopy of the Tunis manuscript, and started on a comparison between the Arabic and the Latin. But giving up hope of a publication in print, he put together a booklet *Innovations in Averroes De Caelo* (Berkeley, Cal., 1982 [privately printed]), "an adaptation of the traditional»prolegomena«of my critical edition" which, rather than presenting the philological basis of his edition, expounded the sources and substance of Averroes' cosmological doctrine. Shortly before his death in 1983, he provided a carbon copy of the typescript which he had been using as a working copy for entering variants and emendations gleaned from the Arabic and other documentation for his apparatus criticus, to Albert Zimmermann, then director of the Thomas-Institut at Cologne and general editor of the *Averrois Opera* editorial project. After his demise in 1983, his widow, Carole Carmody, made photostats of the clean copy of his edition and sent them to the present writer who had started work on an edition of the part preserved in Arabic.

Apart from two pages of remarks in his *Innovations in Averroes De Caelo,* on the "critical method" he followed, and fragments of a typescript found among his papers, Carmody did not leave any *Prolegomena* to his edition to be consulted for a listing of the available manuscripts, an explanation in detail of his *recensio codicum*, and a key to the sigla and abbreviations used in his apparatus. Most of the sigla could be identified only through comparison of the variants listed with copies of the manuscripts available at the Thomas-Institut of the University of Cologne; some of them, rarely referred to, could not be identified at all and have been omitted from the revised apparatus presented here in print. The evidence of the original Arabic of Averroes's commentary was not adduced in Carmody's edition as submitted for publication, and consequently, has not been considered for the present printing.

For the preparation of his edition, Carmody adhered to certain guidelines prescribed by the Mediæval Academy, and taking the edition of the *Commentarium medium in Aristotelis de Generatione et corruptione,* published by F. H. Fobes in 1956, as a model for his general layout. The alineation was to be that of the Giuntina of 1550; in the present edition, we have indicated the folio and column breaks of the 1562 edition because it is now easily accessible in a modern reprint (v. supra, p. 37*, siglum J_2). Also imposed was the choice of the Parisinus 15453, dated in 1243, as a basic reference text (siglum B). Four other manuscripts were to be selected as representatives of four sub-families, presupposing a stemma to be determined by the collation of specimen passages from a larger number of codices. Resulting from this procedure, and noting all variants from 24 manuscripts for the procemium, book I, c. 1–5, and book IV, c. 40–45, Carmody obtained four such families, viz.

b ABJXγδστ
w EWκλυ
y Yημνω
z αβεζοπρφχ

The capital letters used as sigla represent the following manuscripts, reported regularly in the apparatus of variants for the whole work (adding D, not figuring in one of the "four families"):

A Vaticanus, cod. Ottobonianus lat. 2215, ff. 174r-244r
B Parisinus, Bibl. Nationalis lat. 15453, ff. 121r-202r
C Parisinus, Bibl. Nationalis lat. 16156, ff. 3r-187r
D Salmanticensis, Bibl. Universitatis, 2671 (*olim* Matritensis, Bibl. Palatii VII G 2, 2 c 2, 150), ff. 143–204
E Vaticanus, lat. 2089, ff. 6r-195r
F Parisinus, Bibl. Mazarinea 3473, ff. 69r-100r
V Cantabrigiensis, Bibl. Collegii Gonvillæ et Caii 486/482, ff. 235r-384r
W Toletanus, Bibl. Capituli 98.22, ff. 2r-65v
X Vaticanus, lat. 2184, ff. 1r-98r
Y Parisinus, Bibl. Nationalis lat. 14385, ff. 1r-44v
Z Parisinus, Bibl. Nationalis lat. 17155, ff. 225r-344v

Further quoted are one early printed edition, and — in regard of its influence on the reading of the *lemmata* in some of the mss. — the *translatio vetus* of the text of Aristotle by Gerard of Cremona:

J impr. Venetiis, apud Iunctas, 1550-52, t. V, ff. 2r–125r
G Aristotelis *De Celo et mundo* in versione Gerardis Cremonensis [from a collation of MS Paris, Bibl. Nat. lat. 17155, and four other manuscripts]

The Greek sigla denote those manuscripts which had been collated only for the sample passages. Only the following codices could be identified, occurring, however, rarely in the apparatus criticus to support specific editorial decisions in doubtful cases:

γ Parisinus, Bibl. Nationalis lat. 6504, ff. 151r-222v
η Vaticanus, lat. 2436, ff. 1r-55r
κ [?], presumably Parisinus, Bibl. Nationalis lat. 16155, ff. 2r-140v
λ Venetus, Bibl. S. Marci, lat. VI, 52, ff. 1r-186v
ν [?], presumably Parisinus, Bibl. Nationalis lat. 16159, ff. 158r-223v
σ Neapolitanus, Bibl. Nationalis VIII E. 36, ff. 181r-284v
φ Dunelmensis (Durham, Bibl. Capituli), C. I, 18, ff. 1r-101r
ψ Vaticanus, cod. Urbinas, lat. 221, ff. 2r-107v
ω Parisinus, Bibl. Universitatis 601, ff. 104r-150r

The following are some further remarks gleaned from Carmody's *Innovations in Averroes De Caelo* (pp. 63f.) and from his typescript notes, in lieu of a *recensio* which would have to be re-established on the basis of the manuscripts and other witnesses:

"The sigla C, E, F and A[23] were set [...] by Crawford[24] and Fobes: it was obvious that in these manuscripts *De Cælo* belonged to subfamilies other than those formed by these researchers. Before rejecting them I made further extensive samples. It was also clear that A was better than B in not being so extensively infected by Gerard (*v. infra*).

The relative reliability of BJVWXYZ, and of my groups bwyz, may be measured by the percentage of error in 1, 2 or 3 copies: in one the basic quality; in two according to accord or disaccord with the groups; in three with a possible doubt regarding my solution — in 2ra–4ra [= *procemium* and book I, c. 1–5] 1% in WZ, JW and JWZ; omitting the remaining combinations, I note 5.5 for W, 10 for Z, 11.5 for B, J and X.

23. Carmody, *Innovations,* 63, line 14: "C, E, F and *L*" — this is clearly a lapsus calami.

24. Averrois Cordubensis *Commentarium magnum in Aristotelis De anima libros,* rec. F. Stuart Crawford, Corpus Commentariorum Averrois in Aristotelem, Versionum latinarum vol. VI, 1 (Cambridge, Mass.: The Mediæval Academy of America, 1953).

Another measure of reliability is the presence of forms borrowed from the *translatio vetus* of Gerard of Cremona (*v. supra*, p. 13), particularly important in JZ, BJ, Z and especially J.

> W: No decisive trace of Gerard's translation, in contrast to the interlineary cribs W_1, written on erasures, the original being quite readable, and W_2 (probably by the same hand) in the margins, almost all present also in Z and λ, probably transmitted by D.
>
> B: The most striking example of contamination in B, in common with γ, and hence present in their model copy, but otherwise absent from family b, is the substitution of almost one entire *textus* [viz. book I, c. 11] from Gerard's translation (*Innovations*, p. 11; v. supra, note 16). A number of isolated words or expressions illustrate the same phenomenon, less significant than in J and Z (cf. a further borrowings from Gerard in BZ, book I, t. 96, line 1, and in BW_2, book I, t. 64, l. 6).
>
> J: The editors of 1550 "improved" Scot's text by numerous adjustments from Gerard and from the Greek, perhaps the latter through William of Moerbeke. There are substantial interpolations from Gerard (I, t. 2, line 14; I, t. 4, line 23; I, t. 8, ll. 3–4; I, t. 20, l. 6, etc.), and from unidentified sources (2rb60–61, 4ra51, etc.). The most important is a massive application of syllogisms (5ra45–5rb56 [= I, c. 11]), in Gerard's style, with a formal commentary also in his style. Further, the editors reworded many *textus* to match the lemmata interpolated in the commentary, or vice-versa.
>
> X and Y: Although of little merit, I maintained these as pilot copies. They were hastily and carelessly prepared, showing many of the same lacunæ, and many incorrect spellings of common words.

The critical apparatus gathers all readings common to two or more copies, and all readings in B, W, Y, and Z. In order to relieve the apparatus of excessive and plausibly useless details, I omitted mention of variants peculiar to J, V, X, Y or Z, and those peculiar to XY. With few exceptions, the text is determined by three among BJWZ, and the dilemma BJ versus WZ is resolved by recourse to V, X or Y. In principle, BZ vs JW results from contamination. Finally, the sources and parallels may serve to resolve dilemmas. It remains true however that certain omitted readings are or might be of interest [*n.b.* quotations of such readings found in Grosseteste and Albertus Magnus]."

Gerhard ENDRESS

EDITORIAL NOTE

The following edition presents the text of Averroes' Commentary as intended for publication by Francis J. Carmody. Apart from some minor exceptions, we have refrained from altering the wording fixed by Carmody in his final typescript and confined ourselves instead to incorporating his final handwritten corrections. Although Carmody's decisions in favour of a particular variant reading might be disputable in some cases, any substantial change of the text would have been unjustifiable from a philological point of view in the absence of a new, complete and systematic *recensio codicum*. Such a project, however, having the possible result of a different critical evaluation of the filiation and its consequences for the *constitutio textus*, would have exceeded the scope, aim and timetable of the present series of Latin Averroes editions.

Alterations of Carmody's text have been limited to (1) obvious typographical errors in the typescript, (2) cases of doubt left undecided by Carmody in his clean copy and (3) a very few instances of evident mistakes in the Latin tradition emendable on the basis of the unique Arabic manuscript. In several cases it seemed reasonable to modify the punctuation, orthography or capitalization of letters for the sake of consistency, yet always in accordance with Carmody's general systematics and without impinging upon his classicizing standardizations. In addition, long commentary sections typed by Carmody throughout in a single paragraph (following the line breaking of the Giuntina of A.D. 1550) have been subdivided into paragraphs (with an independent line breaking beginning with line # 1 for each Aristotelian lemma) in order to facilitate the approach to the often quite intricate structure of argumentation.

Since the comparison of the Latin text with the original Arabic of the commentary, as far as it has been preserved in ms. Tunis, National Library 11821, has brought to light several significant divergences, we have indicated the folio numbers of the Arabic manuscript in the right margin of the edition in order to facilitate references. For the correspondence between the page numbers of the facsimile edition and these folio numbers, unfortunately cut off in this edition, see pp. 4-12 of the facsimile edition (v. supra, p. 14*). Most of the divergences between the Arabic translation of the Aristotelian text, quoted in the lemmata of the commentary, and the Greek original, caused by the paraphrastic and verbose style of the Arabic translator, are mirrored in the corresponding passages of the Latin translation. Those of passages lacking any equivalence in the Greek text have been marked in the present edition by ⌜superscript hooks⌝.

The critical apparatus has undergone by far the most significant modification. In place of Carmody's minimalist notation, which aimed at avoiding explanatory notes by using a system of diverse brackets and, in cases of variant readings or inversion of words, even omitting the word(s) of the edited text referred to in the apparatus, we have employed a more explicit notation adopting standard abbreviations and sigla. Generally, we cite the word(s) to which the variant refers at the beginning of the note followed by the variant reading(s) or an abbreviated explanation of how the text edited diverges from one or more manuscripts. In some difficult instances, we have given a 'positive' apparatus. Wherever variant readings of subsequent words are evidently attributable to one common error or problem in the manuscript tradition, we have combined Carmody's separate notes into a single note of the apparatus. Most of this revision could be carried out on the basis of Carmody's own working copy, in which he had entered numerous variants and emendations not included in the apparatus intended for publication. Nevertheless, in some cases it was necessary to consult a microfilm of the manuscript(s) in question. Moreover, we have incorporated several variants and discrepancies mentioned in Carmody's working copy but left out in his clean copy of the apparatus. In other cases, we have decided to document variants of the second Juntine edition (A. D. 1562) or of the original Arabic in support of a specific Latin reading not taken into consideration by Carmody. For reasons mentioned in the foregoing preface, other references of Carmody's original apparatus had to be omitted.

Finally, two indices have been added, in part based on preparatory notes by Carmody. For the purposes of further medievalist studies of the Latin reception or transmission of Greek and Arabic philosophical works, names and titles of or references to works appear in the way they are adduced in the edited text, which often does not correspond to contemporary orthographical and bibliographical habits (e. g., *Tholomeus* instead of *Ptolemeus*, *Posteriora analetica* instead of *Analytica posteriora* etc.). Where necessary, additional information has been supplied in square brackets or footnotes. Names and works mentioned by Aristotle but not taken up by Averroes for commentation have not been included in the indices.

We wish to express our warmest thanks to Dr. Roland Hissette, who made various helpful suggestions for the preparation of the critical apparatus, as well as to Andreas Buschmann, Dr. Heidrun Eichner, and Hermann Hastenteufel for their invaluable editorial assistance at various stages of the project.

Rüdiger ARNZEN

SIGLA

A Vatican, cod. Ottobonianus lat. 2215, ff. 174r-244r
B Paris, Bibl. Nationalis lat. 15453, ff. 121r-202r
B_2, B_3 = annotationes alterius, tertiae manus.
C Paris, Bibl. Nationalis lat. 16156, ff. 3r-187r
D Salamanca, Bibl. Universitatis, 2671 (*olim* Madrid, Bibl. Palatii Regii VII G 2, 2 c 2, 150), ff. 143–204
E Vatican, cod. Vat. lat. 2089, ff. 6r-195r
F Paris, Bibl. Mazarinea 3473, ff. 69r-100r
G Aristotelis *De Celo et mundo* in versione veteri Gerardis Cremonensis (v. *supra,* pp. 21*, 41*f.) secundum cod. Parisin., Bibl. nat. 17155 [= Z]
J impr. Venetiis, apud Iunctas, 1550-52, vol. V, ff. 2r–125r.
J_2 impr. Venetiis, apud Iunctas, 1562, vol. V, ff. 1r–271v.
V Cambridge, Bibl. Collegii Gonvillæ et Caii 486 (482), ff. 235r-384r
W Toledo, Bibl. Capituli 98.22, ff. 2r-65v
W_1 = annotationes adiectae inter lineas, W_2 = in margine, eaedem manus ut vid.
X Vatican, cod. Vaticani lat. 2184, ff. 1r-98r
Y Paris, Bibl. Nationalis lat. 14385, ff. 1r-44v
Z Paris, Bibl. Nationalis lat. 17155, ff. 225r-344v
Z_1, Z_2 = annotationes alterius, tertiae manus.

Raro citantur:

γ Paris, Bibl. Nationalis lat. 6504, ff. 151r-222v
η Vatican, cod. Vaticani lat. 2436, ff. 1r-55r
κ [?], presumably Paris, Bibl. Nationalis lat. 16155, ff. 2r-140v
λ Venice, Bibl. S. Marci, lat. VI, 52, ff. 1r-186v
ν [?], presumably Paris, Bibl. Nationalis lat. 16159, ff. 158r-223v
σ Naples, Bibl. Nationalis VIII E. 36, ff. 181r-284v
φ Durham, Bibl. Capituli C. I, 18, ff. 1r-101r
ψ Vatican, cod. Urbinas lat. 221, ff. 2r-107v
ω Paris, Bibl. Universitatis 601, ff. 104r-150r

Division and foliation:

The division and counting of the sections (*commentum,* with preceding *textus*) are taken from the Juntine edition of 1562. The segments of the Arabic, and of the Giuntina of 1560, are generally identical (without section numbers in the Arabic), with two exceptions:

t. / c. 64 is not a separate section in the Arabic, the *textus* being quoted only as a subordinate *lemma* within the commentary, fol. 49b19–23.

t. / c. 65 of the 1550 Juntine edition is not a separate *textus* and *commentum* in the Arabic, nor in the Juntine edition of 1562 (following some of the Latin manuscripts). In the latter, the first *lemma* belonging to this section, quoted within *c.* 64, starts fol. 45raA, line 1ff. From here to the end of book I, the count of the *commata* in the two Juntine editions differs by one. In the present edition, the count of the 1550 Juntine edition appears in square brackets.

¶ In the left margins, folio references are given to the ed. Venice, apud Iunctas, 1562, vol. 5, ff. 1r–271r (*v. supra,* p. 37*).

¶ In the right margins, numerals preceded by the letter T indicate folios of the original Arabic text, as far as extant in ms. Tunis (=T, *v. supra,*, pp. 13*-15*)

¶ The numerals printed in the right margins beside the lemmata of Aristotle key the Latin translation to Immanuel Bekker's standard edition of the Greek text of 1831 (*De caelo,* 268a1–313b23).

ABBREVIATIONES

app. crit.	apparatus criticus
c.	commentum
cf.	conferatur
cod. cett.	codices ceteri
coni.	coniecit
corr.	correxit
del.	delevit
dittogr.	dittographia
exp.	expunxit
fol.	folium, folia
haplogr.	haplographia
homoiarch.	homoiarchon
homoiotel.	homoioteleuton
inv.	invertit
iter.	iteravit
lac.	lacuna
leg.	legendum
marg.	margine
obsc.	lectio obscura
om.	omisit
praem.	praemisit
spat. vac.	spatium vacuum
suppl.	supplevit
t.	textus
transp.	transposuit
v.	versus

AVERROIS CORDVBENSIS COMMENTVM MAGNVM SVPER LIBRO DE CELO ET MVNDO ARISTOTELIS

LIBER PRIMVS

etiis apud *ctas 1562* Tibi Stephane de Provino hoc opus quod ego Michael Scotus dedi latinitati ex dictis Aristotelis specialiter commendo. Et si aliquid Aristoteles incompletum dimisit de constitutione mundana in hoc libro, recipies eius supplementum ex libro Alpetrangii quem similiter dedi latinitati, et es in eo exercitatus. *De Caelo ed. Bekker*

1r7 In primo tractatu istius libri continentur decem summe magne: prima de substantiis istius artis; secunda de diffinitione corporis naturalis et quod ipsum solum inter omnia corpora est completum; tertia in demonstratione quod mundus est perfectus; quarta in demonstratione eius quod est corpus esse simplex rotundum consimilis nature cuilibet corporum simplicium motorum motu recto; quinta in demonstratione quod hoc corpus neque est grave neque leve; sexta in demonstratione quod non accipit generationem neque corruptionem neque aliquam specierum motuum nisi localem tantum, et quod numerus simplicium est quinque; septima quod totum

---

Ad titulum operis: Averrois—primus *scripsi* : Incipit commentum Averrois yspani cordubensis super libro De celo et mundo Aris. greci stragelite *B* : Incipit commentum Averrois super libro primo De celo et mundo et Michael Scotus transtulit in latinum *EJ*λ : Incipit commentum Averrois super libro celi et mundi *X*γ

Ad dedicationem: 1-5 Tibi—exercitatus *CDFVWYZ*$\lambda_2\varphi\psi_2\omega$] *om. cod. cett.* 2 commendo] continendo *Y* 3 incompletum] *om. Z* : imple- *Y* 4 recipies] incip- *Y* | Alpetrangii] -trandi *Y*ω

Ad tabulam: 1-8 In—demonstratione *ABJX*$\gamma\sigma$] *om. cod. cett.* 1 magne] *om. B*γ 2 istius artis] *inv. X* 3 et] eo *J*σ | omnia] alia *A* : omnis alia *J* 5 consimilis] dissim- *J* 6 cuilibet] cuius- *J* 7 quod] + est *BX* | hoc corpus] *inv. B* : corpus *X* | neque[1]] non *J* 8-16 quod—corruptibile *J*σ] *om. cod. cett.*

est finite quantitatis; octava quod unus mundus est numero tantum; nona quod non est extra mundum neque vacuum neque plenum; decima in demonstratione quod mundus neque generabilis est neque corruptibilis, neque habet potentiam corruptionis, et quod impossibile est aliquid esse generabile incorruptibile neque aliquid non generabile esse corruptibile.

1. *Maxima cognitio nature et scientia demonstrans ipsam est in corpori-* *268a1*
bus et in aliis magnitudinibus et in passionibus et in motibus eorum, et in principiis cuiuslibet quod assimilatur isti. Nature, etiam naturalium rerum quedam sunt corpus et magnitudo, et quedam habent corpus et magnitudinem, et quedam sunt principia habentium corpora et magnitudinem. *268a5*

1va Quia iste liber primus, in quo loquitur de rebus sensibilibus naturalibus, scilicet in singulis partibus istius mundi quarum quelibet continetur in uno libro de suis libris in hac scientia, incepit primo narrare vel dicere subiecta istius artis, et dixit *Quod maxima cognitio nature et scientie demonstrans ipsam* etc., vult dicere quod scientia naturalis maxime est in corporibus et in magnitudinibus et in consequentibus istorum et in motibus eorum et causa omnium illorum. Et dixit *Maxima cognitio* quoniam artifex in hac facultate considerat etiam in rebus extra istas sicut de vacuo et de infinito et de loco et sibi similibus ex eis que dicta sunt in libro predicto; et intendebat dicere per cognitionem cognitionem certam, et per naturam naturalia, et intendebat per scientiam demonstrantem ipsam scientiam certificatam que acquiritur per demonstrationem. Et videtur nobis quod

11 est[1]—quantitatis] finitum est secundum magnitudinem *σ* | unus] *post* est[2] *J* | tantum] 3 *σ* 12 non est] *om. σ* | mundum] + nihil est *σ* 13 generabilis] -ratus *σ* 14 neque—potentiam] et quod non est in eo potentia *σ* 15 esse—incorruptibile] esse generatum non corruptibile *σ* 15-6 neque—corruptibile] aut aliquid non generatum corruptibile *σ*

1. 2 eorum] ea- *Y*: ipso- *BX* 3 etiam] entium enim *J* 9 incepit] -cipit *BJY* 10 subiecta] -tum *JZ* | et] vel *Y* | Quod] *om. JZ* 11 scientie] -tia *JZ* 12 maxime est] *inv. W* : maxime *BXY* | in[2]] *om. JXZ* 15 istas] ipsas *Z* | et[3]] *om. B* 16 eis] hiis *W* 17-8 dicere—intendebat] *om. homoiotel. WY*

nomen cognitionis et scientie non dicebantur sinonima, cum doctrina demonstrativa non utatur talibus nominibus: intendebat igitur per cognitionem intelligere imaginationem, et per scientiam fidem, aut intendebat per cognitionem primam cognitionem et per scientiam cognitionem provenientem a prima cognitione, scilicet acquisitam per rationem; quapropter inquit *Scientia demonstrans ipsam*, et non dixit Scientia eius.

Et intendebat dicere cum dixit *Erit in corporibus* corpus divisibile in tres dimensiones; et quia magnitudo dividitur in corpus, superficiem et lineam, dixit *Et in aliis magnitudi*|*nibus*, scilicet quod ista ars *1vb* considerat in corpore quod est unus modorum magnitudinis, et in aliis magnitudinibus. Et bene potes scire ex aliis libris quomodo Naturalis considerat in magnitudinibus et quomodo Mathematicus: Mathematicus enim considerat in eis que sunt abstracte a materia, Naturalis autem secundum quod sunt ultima materie, ut determinatum est in aliis locis. Et cum narravit quod maxima consideratio istius artis est in tribus magnitudinibus, inquit *Et in passionibus et motibus eorum*, et intendebat per passiones qualitates essentiales eis que sunt naturalia, et intendebat per motus eorum vias ducentes ad generationes istorum: omnia enim existentia aut sunt substantie aut accidentia, aut sunt vie ad esse istorum et sunt universaliter transmutatio. Themistius autem inquit quod magister intendebat per motus motus locales quibus non mutatur substantia rei mote, et intendebat per passiones mutationes quibus mutatur substantia rei; et complementum istarum transmutationum sunt qualitates sensibiles; et quocumque sit, isti sermones propinqui sunt. Deinde dixit *Et in principiis cuiuscumque assimilati isti nature*, idest et Naturales etiam considerant in sua consideratione in corporibus naturalibus et in accidentibus eorum in causis omnium istorum scilicet in causis corporum et accidentium: scientia enim modorum naturalium et consequentium eorum non completur nisi per scientiam causarum. Et

22 et] *om.* *W*   27 Erit] erant *B* : est *JW*   28 et] *om.* *W* | corpus] + et *W*   31 bene] hoc *WZ* : habere *Y* : unde *X*   32 considerat] communicat *Z* | Mathematicus] + considerat *B*   33 que] secundum quod *BJX*   37 et] *om.* *B*   39 existentia] *om.* *Y* : entia *BJX*   41 magister] Aristoteles *J*W_2 *Z*   44 sunt] scilicet *BWY*   45 sit] sint *WYZ*   46 cuiuscumque] cuiuslibet *J* : cuiusque *X* : uniuscuiusque *Y*   46-7 etiam considerant] *inv.* *JY*   49 et[1]] + in *B* | modorum] + corporum *BJX*

cum dixit *In principiis cuiuscumque assimilati nature*, intendebat quod considerat Naturalis in causis istorum predictorum et in causis omnium generatorum ab eis; quapropter dixit Aristoteles *Naturalium enim quedam sunt corpus et magnitudo, et quedam habent corpus et magnitudinem*; et intendebat cum dixit *Quedam sunt corpus et magni-*
2ra *tudo* corpora simplicia, et precipue habentia | animas. Deinde dixit *Et quedam sunt principia habentium corpora et magnitudinem*, intendendo quod causarum quedam sunt cause simplicium, ut materia et forma, et quedam cause compositorum ut elementa ex quibus componuntur res consimilium partium; et intendebat monstrare per hoc quod Naturalis considerat in duobus modis corporum naturalium scilicet simplicium et compositorum, et dat causas eorum et causas accidentium eorum.

2rb 2. *Continuum quidem est quod est divisibile in recipientia divisionem* *268a6*
successive, corpus autem est divisibile in omnibus mensuris: magnitudinis vero quod est unius mensure est linea, et quod duarum superficies,
trium autem corpus, et post ista nulla mensura est; omnia enim sunt tria *268a10*
et divisa in tres mensuras. Et similiter inquierunt Pitagorici quod omne et res terminantur tribus mensuris, fine, medio et principio. Et est hic numerus cuiuslibet, et est demonstrans trinitatem rerum; et non invenimus istum numerum nisi ex natura, et sustinemus ipsum quasi legem *268a15*
nobis; et secundum istum numerum tenemur magnificare Deum creatorum remotum a modis creaturarum. Et etiam appellamus istum numerum secundum hunc modum: dico quod nominamus duos numeros duo et duos viros duos viros; et non dicimus omnes, sed hoc omne non dicitur nisi de tribus, et per ipsum nominantur tria primo. Et hoc fuit dictum quoniam natura naturata ita fecit; et nos sequimur ita suum opus, sicut prius narravimus. *268a20*

---

51 dixit] -cit *BX* | cuiuscumque] -libet *JZ* | assimilati] + isti *JZ* 57 intendendo] -tendo *B* 60 monstrare] *om. BZ*: demonstrare *J* 61 considerat] est *Z*

2. 1 Continuum] Et *praem. WZ* 2 est divisibile] *inv. W* | in] secundum *BJX* 4 ista] illa *Z* | mensura est] *inv. BX* 5 inquierunt] -quiunt *Z* 6 fine medio] *inv. B* | est hic] *inv. JZ*: quod est *B* 8 sustinemus] + per *W* 9 istum] illum *BX* 10 appellamus] nomina- *JG* 11 numeros] viros *W* 12 duos viros] *haplogr. B* | omnes] eque *B* 13 nominantur] -namus *BJ* 14 fecit] facit *B*

Cum intendebat describere magnitudinem que est corpus cum est subiectum universale istius artis, quoniam oportet quemlibet artifi-
2va cem prius describere genus in quo considerat; in | diffinitione autem corporis accipitur continuum: cum sit quasi genus ipsius, incepit diffinire continuum dicendo *Et continuum est quod est divisibile in recipientia divisionem successive*, idest quod continuum est illud quod dividitur in res divisibiles semper et numquam terminabitur divisio, sicut declaratum est in sexto de Auditu Naturali; et hec est differentia inter quantitatem discretam et continuam. Continuum autem est quasi genus corporis quoniam continet in se corpus et locum et tempus et superficiem et lineam et motum; quapropter incepit diffinire ipsum antequam diffiniret corpus. Deinde incepit diffinire corpus, dicendo *Et corpus est divisibile secundum omnes mensuras*, idest et corpus dividitur in tres partes, in longitudinem, latitudinem et spissitudinem, scilicet quod corpus est illud quod possibile est in eo ponere tres diametros secantes se; dixit autem *Secundum omnes mensuras* in loco trium mensurarum quoniam non est mensura quarta in qua dividitur magnitudo.

Et cum diffinit corpus, diffinit superficiem et latitudinem: ut demonstret quod quantitates sunt tres, dixit quod linea non dividitur nisi in una parte scilicet in longitudinem, et quod non est in ea nisi mensura que dicitur longitudo. Deinde dixit *Quod est duarum mensurarum est superficies*, scilicet secundum longitudinem et latitudinem. Deinde dixit *Trium autem mensurarum est corpus*, scilicet quod corpus dividitur in tres mensuras: magnitudo enim aut longitudinis erit tantum, aut longitudinis et latitudinis, aut longitudinis et latitudinis et spissitudinis, et quia manifestum est per se ex hac divisione; et quia manifestum est per se quod nulla mensura est quarta, dixit *Et post istas nulla quantitas est*. Deinde reddidit causam illius dicendo

---

18 prius describere] *inv.* *B*: prius declarare *Z*: prius *WY* 19 incepit] -cipit *BX* 21 idest] idem W_2XY 22 res] tres *W* | terminabitur] -natur *W* 24 discretam ... continuam] *inv.* *B* | autem] *om.* *W* | est] *om.* *WY* 26 superficiem] -ficies *B* 29 longitudinem] + et *W* 30 in eo] *post* tres *(31) JZ* 31 diametros] -metras *W* 34 diffinit^2] *om.* *B* | latitudinem] lineam *JZ* 35 demonstret] -straret *JZ* | dixit] -cit *WY*: + enim *Z* 36 ea] eo *BWY* 37 dixit] + et *JZ* 38 scilicet] *om.* *WY* | secundum] *om.* *B* 40 in] + istas *BX* 41 aut^2—latitudinis^2] *om.* *homoiotel.* *WY* 42 quia] hoc *Z*: quod *W* 43 et—se] *om.* *JX* 44 reddidit] recludit *B*

Omnia enim sunt tria et divisibilia in tres quantitates, et est intelligere *2vb* sic: necesse est ut non | sit magnitudo quarta, quoniam omnes dimensiones sunt tres quoniam dividuntur in tria scilicet in longitudinem et latitudinem et spissitudinem, et non in plura; et ista ratio non est accepta per locum a divisione quo utimur ad declarandum ea que sunt latentiora primis cognitionibus et apertiora primis questionibus; et similiter inductione non utimur nisi ad declarandum talia occulta.

Et cum dedit in hoc rationem inducentem ad certitudinem, dedit rationem famosam secundum suam consuetudinem in dando rationes famosas post rationes certificantes, et dixit *Et similiter dixerunt Pitagorici quod omne et res terminantur tribus mensuris, fine et principio et medio, et hic est numerus cuiuslibet rei*, et est sic intelligere (et testatur hoc sermo Pitagoris) quod tria terminant numerum totius et omnis: totum enim et omnia habent principium et finem et medium, et tria tantum habent istos tres modos; et ideo dixit *Hic est numerus cuiuslibet*, scilicet tria. Deinde dixit *Et est demonstrans trinitatem rerum*, scilicet quod hoc significat quod natura omnium rerum est trina; deinde dedit etiam aliam rationem famosam quod hoc fuit causa ut lex apud eos preciperet ut orationes eorum essent tres aut sacrificia; et dixit *Et non invenit legislator hominibus istum numerum nisi ut sequeretur naturam ut ista lex sit quasi sequens naturam*. Deinde dixit *Et secundum istum numerum tenemur magnificare creatorem remotum a modis creaturarum* in orationibus et sacrificiis, nam omnia ista non fiunt nisi ad magnificandum creatorem. Deinde dixit *Et etiam nominamus numerum secundum hunc modum, dico quod duos numeros numeramus duo et duos viros duos viros, et non dicimus omnes*: omnis enim non dicitur nisi de tribus, et per ipsum nominantur *3ra* prius tria. Et ista etiam est ratio famosa grammatice: vide|mus enim

---

45 est] *om. B* | intelligere] -gendum *Z* 46 omnes] + divisiones aut *Z* 49 non] *om. JWY* | per] secundum *W* | utimur] -tur *BJX* 51 utimur] -tur *BJX* 53 certitudinem] -ficationem *Z* 55 Et^{2}] *om. B* 56 terminantur] + in *B* | et^{2}] *om. JZ* 58 testatur—sermo] hic sermo testatur *W*: testatur hic sermo *BY*: testatur sermo *X* | Pitagoris] -goricus *JX*: -goricis *W* | quod] quia *YZ* 60 ideo] cum *BXY* 60-1 est numerus] *inv. X*: numerus *B* 63 trina] tria *B*: eterna *Z*: forma *X* 67 istum] hunc *BX* 69 magnificandum] glorifi- *B* 70 etiam] *om. B* | nominamus] numera- *BX* 71 duos viros] *haplogr. B* 72 enim] nomen enim omnis *BJX* 73 etiam] *om. J*: enim *BX* | est ratio] *inv. J*: ratio *WY*

idiomata ponere singulari unum nomen scilicet Sor vel Plato et duobus nomen proprium et pluribus nomen proprium; et paucior numerus quem poterimus nominare numero plurali est tres; hoc igitur quod dixit *Quod nominantur duo numeri duo* est intelligere quod due res non dicuntur numerus sed tantum duo, et duo viri duo viri non viri simpliciter. Deinde dixit *Et hoc fuit ita quoniam natura naturata ita fecit, et nos sequimur suum opus*, scilicet quod in rebus voluntariis invenitur trinitas, quoniam natura ita fecit in rebus naturalibus; quapropter voluntas sequitur ipsam in lege et in grammatica, scilicet quod nos sequimur opus eius in eis que ponimus per legem et institutionem: lex enim et institutio posite sunt voluntarie, et res voluntarie que vere ponuntur assimilantur naturalibus et consequuntur opus eius.

3. *Dicimus etiam quod res et omne et complementum non differunt secundum formam sed secundum naturam; et cum ita est, manifestum* 268a20
3rb *est igitur quod solum corpus est per|fectum et perfectius omnibus magnitudinibus. Corpus enim terminatur per tres mensuras, quapropter dividitur in omnes mensuras; aliarum autem quedam sunt duarum quantitatum et quedam unius quantitatis.* 268a25 *Et sicut de eis dicitur quantitas, ita dicitur continuatio de eis aut discretio: habens enim unam mensuram dicitur continuum, habentia autem duas dicuntur discreta; et corpus recipit duo nomina.* 268a28

Vult dicere quod magis generalia nomina pluralium, que sunt *Res et omne et perfectum*, significant idem, et non diversantur secundum formam sed secundum naturam; subiectum enim istius rei si fuerit quantitas discreta dicitur omne, et si quantitas continua dicitur totum, et si in forma et qualitatibus dicitur perfectum. Deinde dixit *Manifestum est quod solum corpus est perfectum* etc., et intellige quod,

74 idiomata] -mate *X*: predicta *Y* | singulari] -larum *B*: significari *Z* | Sor vel Plato] Sortem vel Platonem *Z* 74-5 scilicet—nomen[2]] *om. homoiotel. B* 76 hoc] hic *BX* 78 viri[1]] + et *Z* 79 viri] numerus *J*W_2 80 sequimur] subsequi- *W* | scilicet] secundum *B* 85 consequuntur] sequu- *BJX*

3. 1 et[2]] *om. BX* 2 naturam] materiam *JZ* 5 aliarum] alia *WYZ* | quedam sunt] *inv. JWYZ* 10 pluralium] -lia *Z* 11 diversantur] -sificantur *Z* 12 subiectum] + est *B* 13 dicitur[1]] + origo vel *Z* 15 intellige] -gere *WY*: -gitur *J*

cum declaratum est quod numerus perfectus est ternarius, manifestum est quod corpus est perfectum: corpus enim solum dividitur in tres mensuras, alie autem magnitudines non sunt perfecte cum non dividantur in tria, sed quedam dividuntur in duas mensuras scilicet superficies, et quedam in unam mensuram scilicet linea.

Deinde dixit *Et secundum quod de eis dicitur mensura, sic dicitur de eis continuatio aut discretio*: potest ergo hoc intelligi quod illud quod ex eis habet unam mensuram dicitur esse continuum secundum unam partem et discretum secundum aliam, ut linea: linea enim continua est secundum longitudinem et discreta secundum divisibilitatem. Quod autem habet duas mensuras dicitur esse continua secundum duas partes et discretum secundum duas, ut superficies *3va* que est continua in longitudi|ne et latitudine et divisibile in eas, quapropter describitur hiis duabus descriptionibus; corpus autem discretum est in tribus partibus et continuum in tribus. Et potest intelligi per descriptionem predictam descriptio unius eorum est in diversas mensuras; et secundum hoc erit linea continua tantum non discreta, cum non dividatur in aliam magnitudinem, et erit superficies discreta secundum duo scilicet longum et latum; corpus autem erit discretum in tres magnitudines, et hoc est apertius. Dixit enim post hoc *Habens unam mensuram dicitur continuum et quod habet duas dicitur discretum, corpus autem recipit duo nomina*, scilicet quod linea, inquantum non dividitur in aliam magnitudinem, dicitur tantum continua, et superficies inquantum dividitur in aliam dimensionem dicitur discreta tantum, cum hec divisio sit opposita continuationi linee inquantum non dividitur in aliam dimensionem. Corpus autem recipit duo nomina, idest quia corpus dividitur in superficiem et spissitudinem (spissitudo autem est linea), contingit quod totum corpus producitur ex corpore et linea; inquantum igitur continet in se lineam, dicitur continuum non discretum, secundum quod in linea

16 cum] *om. B* 17 quod] + solum *BX* 19 quedam] qui- *B* 20 quedam] qui- *B* | scilicet] ut *J*: ut sunt *Y* 22 ergo hoc] *inv. BX*: hoc *J* 24 linea] *haplogr. BX* | enim] est *B* 25 continua] -nuum *Z* | secundum divisibilitatem] scilicet divisibilis *B* 26 continua] -nuum *BX* 27 secundum[2]] + que *B* 29 descriptionibus] *om. Y*: discretio- *Z* 31 descriptionem] discretio- JW_2 | diversas] duas *YZ* 38 non dividitur] *om. B* 41 linee] nature *Y*: sed *X* 42 et] + in *B* 44 corpore] superficie JW_2

dicitur continuum, et inquantum continet in se superficiem, dicitur discretum non continuum, secundum quod dicitur de ipsa superficie, dicitur discretum.

3vb20 4. *Et dicimus etiam quod omnis magnitudo discreta necessario etiam est* 268a28 *continua. Utrum autem continuum sit discretum aut non adhuc non est manifestum. Quod vero est notum et manifestum est hoc quod impossibile est ferri in aliud genus sicut linea fertur in planitiem, deinde planities in corpus: si autem possibile est ferri in aliud genus, tunc non erit corpus magnitudo perfecta sed imperfecta. Talis autem latio non existit necessario nisi in imperfecto; perfectum igitur impossibile est esse imperfectum diminutum; discretum enim est in omnibus mensuris. Unum-* 268b5 *quodque autem aliorum corporum que sunt sicut corpus in formis suis est* 4ra *perfectum sicut diximus de diffini|tione perfecti. Quodlibet enim illorum habet tres mensuras, tamen habet terminum quo tangit illud quod est post ipsum ex corporibus, quapropter unumquodque corporum est plura. Totum autem cuius hec sunt corpora partes necessario est perfectum et non diminutum; et est sicut nomen eius significans totum significatione perpetua, et non est quibusdam rebus totum et quibusdam non totum. Et* 268b10 *nos dicemus post naturam totius, et utrum habeat finem.* 268b11

Dixit *Dicamus etiam* etc.: dicit quod manifestum est per se quod omnis magnitudo recipiens divisionem est continua; sed si omne continuum recipiat discretionem non est adhuc manifestum scilicet conversive; quapropter non vere diffinitur corpus hac diffinitione, scilicet quod sit divisibile secundum omnes dimensiones, donec declaretur quod nullum corpus sit indivisibile, quapropter visum est quod quamvis geometria ponat hanc diffinitionem, tamen Naturalis experitur probare. Quoniam autem nulla pars est indivisibilis decla-

46 continet] + igitur *B* 48 discretum] *om. BXY*

4. 1 etiam est] *inv. B* 2 non adhuc] *inv. Y*: aliud ergo hoc *X* 3 vero] autem *JZG* | hoc] + scilicet *BJX* 4 genus] *om. XY* 7 esse] *om. BX* | imperfectum] + vel *Z* 8 diminutum] *om. BX* 8-10 diminutum—perfectum] *om. Y* 8 enim est] *inv. BX* | est—mensuris] est secundum omnes quantitates *Z* 9 aliorum] *om. Z* | corpus] pars *Z* 10 diximus] -cimus *JX* 13-4 et non diminutum] *om. WXY* 15 quibusdam[2]] + rebus *JX* 16 finem] + aut non in magnitudine *JG* 21 scilicet] *om. B* | dimensiones] divisio- *Z* 22 sit] est *BXY*

ratum est in sexto de Naturali Auditu, unde continuum diffiniebatur superius quod est continuum illud quod est divisibile in semper recipientia divisionem; deinde diffinivit corpus et alias mensuras.

Et cum narravit quid sit manifestum de natura continua et quid non, incepit dicere quod voluit intendere de diffinitione corporis; et est quod est perfectum, cum sit manifestum quod impossibile est ferri in aliam naturam sicut fertur superficies in corpus et linea in superfi-
4rb ciem, et dixit *Quod autem* | *est notum est hoc* etc., scilicet impossibile est ut corpus feratur in quartam magnitudinem addita quarta mensura, cum nulla quarta mensura sit per quam corpus feratur sicut linea fertur in superficiem per mensuram additam et superficies in corpus per aliam mensuram additam; et voluit demonstrare per hoc quod solum corpus est perfectum: perfectum enim est quod non recipit additionem nec fertur in aliud esse. Deinde dixit *Si autem impossibile est ferri in aliud* etc., scilicet si impossibile est ut corpus feratur in aliam magnitudinem sicut est possibile in linea et superficie, tunc non erit corpus perfectum sed diminutum. Deinde dixit *Latio autem non existit nisi imperfecto necessario*, scilicet quod necessarium est ut corpus sit diminutum si possibile est ipsum ferri: omne enim latibile est diminutum quoniam possibilis est in eo additio. Deinde dixit *Impossibile est igitur ut perfectum sit diminutum, discretum enim est secundum omnes quantitates*, et hec est quasi corroboratio cause propter quam perfectum non est diminutum. Deinde dixit *Unumquodque autem aliorum corporum que sunt sicut pars in suis formis perfectum est sicut diximus in dispositione perfecti.*

Et cum declaratum est de diffinitione corporis quod est perfectum, et quedam corporum non continentur in aliis et est corpus totius mundi, et quedam continentur et sunt corpora que sunt partes corporis totius mundi aut sunt in suis formis sicut partes eius, manifestum est quod in corporibus que sunt partes mundi est perfectio que videtur esse in corpore inquantum est corpus, et est quod impos-

---

25 unde] nam *WYZ* 26 semper] *om. BWY* 27 diffinivit] -niunt *Z*: -nitum *B* | et] + secundum *Z*: + in *Y* 29 incepit] -cipit *B* 30 ferri] servari *B* 32 scilicet] + quod JW_2 39 impossibile^2] poss- *JX* 40 in^2—superficie] lineam et superficiem *Z* 42 Latio autem] *om. Y*: *inv. J* 43-4 si—diminutum] *om. homoiotel. B* 44 est^2] + diminutum *B* 46 enim est] *inv. JZ* | quasi] quod *BX*: cum *Y* 48 autem] *om. Y*: aut *X* | aliorum] *om. WYZ* | suis formis] *inv. BXZ*

sibile est in eo ipso ferri; unde dixit *Quodlibet enim eorum habet tres*
4va *mensuras.* | Deinde dixit *Tamen habet terminum contingentem illud quod est post,* idest sed tamen iste partes que sunt partes totius habent ultima scilicet superficies tangentes superficies corporum continentium, et est intelligere quoniam secundum hunc modum sic sunt diminuta et possunt recipere additionem. Deinde dixit *Quapropter unumquodque corporum per se multum est,* idest quod corpora que sunt partes totius non sunt multa nisi quia sunt partes continentis; et quod invenitur ex eis plusquam unum est diminutum, quia si esset perfectum sufficeret esse unum sicut est in mundo; et hec est causa esse multorum individuorum unius speciei, scilicet quod propter diminutionem non fuit natura contenta esse unius.

Deinde dixit *Totum autem cuius hec corpora sunt partes est perfectum necessario non diminutum,* idest quod totus mundus est perfectus omnibus modis perfectionis: est enim perfectus secundum quod partes eius sunt perfecte, et est etiam perfectus quoniam nichil continet ipsum; et non est plus uno individuo, sive illud totum ponatur finitum sive infinitum, necessario contingit ut nichil contineat ipsum. Si enim ponatur infinitum, manifestum est quod nichil continet ipsum, et si ponatur quod aliquid continet ipsum, tunc erit pars non totum; sed positum est totum scilicet quod hoc nomen totum quod dicitur de toto mundo significat hanc intentionem vere, cum numquam contingat ei diminutio que contigit toti in partibus inquantum continentur in aliquo et corrumpuntur; unde dixit *Significatione perpetua,* scilicet quod non est aliquando totum et aliquando non totum, sicut est totalitas in rebus generabilibus et corruptibilibus. Et quia non manifestum est adhuc, de toto quod est mundus,
4vb utrum sit finitus aut infinitus, promisit declarare quod | finitus est, et dixit *Et nos dicemus post naturam totius et utrum habet finem aut non in magnitudine,* idest quod ipse perscrutabitur utrum mundus sit finitus aut infinitus.

---

56 enim] *om.* *Z* 59 corporum] eorum *BX* 60 sic sunt] *inv.* *BX* 61 Quapropter] + dixit *Z* 62 idest] + patet W_2Z 64 est] quia est in eis *Z* | quia] quoniam *BX* 65 hec est] *inv.* *B* 66 esse] *om.* *Z* 69 idest] + unum est W_2Z 71 est etiam] *inv.* *Z*: etiam *BX* | quoniam] quod *B* 72 illud totum] *inv.* *BX*: illud *Z* 74-5 Si—ipsum[1]] *om. homoiotel.* *WYZ* 76 positum] com- *YZ* 77 vere] necessario *Z* 82 manifestum est] *inv.* *YZ* 83 finitus est] *inv.* *BX* 84 post] prius *XY* | habet] -beat *BX* 85 magnitudine] imagine *WXY* | idest] *om.* *Z* 86 infinitus] non finitus *B*

5. *Primo modo autem volo dicere partes totius diversarum formarum, et* 268b13
incipiam sic: Omnia corpora naturalia et magnitudines moventur per se 268b15
motu locali: natura enim est principium motus in eis. Et omnis quidem
motus localis aut est equalis aut circularis aut compositus ex istis duobus;
5ra *et isti duo motus sunt tantum simplices, | et causa illius est quod corpora*
simplicia sunt ista, longum et rotundum. Motus autem circularis est 268b20
circa medium, motus autem equalis aut est superius aut inferius; et est
dicere ad superius motum a medio, et ad inferius motum ad medium. 268b22

Cum dixit et diffinivit corpus quod est universalius subiectorum eius, et declaravit ipsum esse perfectum et quod mundus qui est totum est perfectum duobus modis scilicet secundum quod est corpus et secundum quod nichil continet ipsum, incepit loqui de suis simplicibus et primo de nobiliori eorum, et est corpus quod movetur circulariter; et incepit proponere ea ex quibus apparet natura istius corporis: esse enim eius sensibile est per se. Et dixit *Omnia corpora naturalia* etc., idest quod manifestum est per se quod omnia corpora naturalia moventur secundum locum per principium existens in eis quod dicitur natura, et corpora dicuntur naturalia ista per ipsum; et hoc fundatum est super hoc scilicet quod naturam esse in rebus naturalibus est manifestum per se, sed in hoc est questio non parva, natura enim ut dicitur in sua diffinitione est principium in aliquo per quod movetur et quiescit res ipsa ex se, et videtur quod huiusmodi primo dicatur de forma et secundo de materia.

Si ergo naturam esse in omnibus rebus naturalibus notum est per se, et quod movens in istis corporibus aliud est a moto (ut estimatur quod hoc declaratum est in principio octavi de Naturali Auditu cum declaravit quod omne motum per se habet movens in se), tunc corpus celeste, cum movetur per se aut pervenit ad motum per se,
5rb necesse est ut sit in eo principium quo movetur per se et quod | illud
sit aliud a moto; sed cum hoc ponitur, continget quod corpus celeste

---

5. 1 modo autem] *inv. B*: autem *X* | totius diversarum] *inv. B* 2 magnitudines] imagines *WZ*: imaginationis *Y* 3 quidem] *om. JX* 4 ex] *om. BXY* 7 aut est] *inv. BY*: est autem aut *X* 8 ad[1]] quod *WY* 9 subiectorum] -iectum *WY* 12-3 suis simplicibus] *inv. BX* 15 eius] + esse *Z* 16 quod[1]] + est *Z* 18 ista] *om. J*: istum *WXY* | ipsum] ipsam *JZ* 19 scilicet] secundum *Z* | quod] + secundum *Z* 21 natura enim] *inv. B*: non enim *Y* 23 et] *om. B* 25 moto] motibus *B* 26 octavi] septimi *BX*: li *Y* 29 necesse—per se] *om. homoiotel. B*

sit compositum ex materia et forma, quapropter continget ut sit generabile et corruptibile. Et si posuerimus hanc formam in eo non consistentem per naturam, scilicet formam que declarabitur in fine octavi Phisicorum non esse per naturam, quomodo ergo contingit ex diversitate motus ut diversetur natura moti? Natura enim moti non diversatur nisi propter diversitatem sue forme, et hoc non invenitur nisi in forma existente per naturam, verbi gratia quia ignis non differt a terra in motu nisi propter diversitatem suarum formarum, scilicet quod terra est gravis et ignis levis. Non est ergo intelligendum hoc quod dicitur quod natura in corpore celesti est causa motus esse formam: forma enim abstracta non significat naturam corporis, neque in corpore celesti est forma naturalis que sit causa motus ipsius in quo Aristoteles dicit neque esse grave neque esse leve; queret igitur aliquis quod si ita esset, necessario corrumperetur corpus celeste aut in eo est potentia corruptionis; sed nos post declarabimus hoc esse impossibile. Et cum ita sit, forte intelligendum est in hoc loco de natura que est causa motus subiectum simplex forme corporis abstracte corporis celestis, et hoc ipsum corpus non habet in se formam corporalem; et iste locus esset fortior locorum facientium errare opinantes quod corpus celeste componitur ex materia et forma qua dicitur neque esse grave neque leve, sicut ex forma ignis dicitur ignis levis et terra ex sua forma gravis: secundum hoc igitur possumus intelligere hoc nomen natura que dicitur de corpore celesti et de aliis corporibus simplicibus gravibus scilicet et levibus; et erit hoc equivoce. *5va* | Aut dicemus quod forme abstracte, secundum quod sunt forme abstracte, innate sunt ut quod illud quod ab eis movetur sit corpus neque grave neque leve, sicut innate sunt ut motum ab eis sit rotundum et sicut innata est forma ignis ut habeat materiam propriam. Dicitur igitur hoc nomen natura secundum hoc de forma, sicut dicitur de formis aliorum corporum simplicium; et post perscrutabimur

33 naturam] materiam *BJ* | scilicet] + per *W* 34 octavi] *om. B* | naturam] materiam *BJ* | ergo] igitur *JX* 36 propter] per *BW* 39 est[2]] *post* intelligendum *Z* 40 quod[2]] *om. W* 43 dicit] -xit *JW* | esse[2]] *om. B* 49 locus esset] *inv. Y*: locus *W* 51 esse grave] *inv. W* | leve] + esse *W* | ignis dicitur] *om. BX* | ignis[2]] *om. B* 52 forma] materia *Z* 54 scilicet] *post* levibus *Z*: *om. X* 55 sunt forme] *inv. Z* 56-7 quod[1]—ut] *om. homoiotel. X* 56 illud quod] *om. B* | ab eis] *om. Y* | movetur] + de eis *Y* 58 sicut] sic *YZ* 59 igitur] ergo *Y*: prius *X*

plus de hoc; et quocumque modo sit, non debemus intelligere de natura que est neque gravis neque levis nisi naturam corporis simplicis, non naturam compositam ex corpore et forma; et forte hoc est manifestius, et iste due explanationes sunt propinque.

Et cum posuit hanc propositionem dicentem quod natura est principium motus et quietis in omnibus rebus mobilibus secundum locum, ad probandum quod causa diversitatis motuum secundum genus aut speciem est diversitas naturarum agentium, et quod impossibile est unam eandem naturam habere diversos motus secundum speciem, incepit narrare motus diversos ad demonstrandum quod numerus naturarum debet esse secundum numerum motuum secundum locum, et dixit *Et omnes motus locales aut sunt equales aut circulares* etc., idest recti. Deinde dixit *Et isti tantum simplices*, scilicet aut rectus aut circularis. Deinde reddit causam propter quam necessario motus simplices dividuntur in ista duo, et dixit *Quod causa illius est quod corpora simplicia sunt ista longum scilicet et rotundum*, scilicet quod necessario dividuntur in ista duo, quia magnitudo continua que est causa motus continuationis dividitur in ista duo scilicet in rectam et rotundam. Et non est estimandum quod propter divisionem magnitudinis super quam est motus in ista duo, dividitur motus in ista duo, *5vb* | sed propter divisionem ipsius magnitudinis mote secundum ista duo, scilicet quod omne motum necessario aut est corpus rectarum dimensionum aut rotundarum, cum nulla dimensio sit tertia; et iam declaratum est in sexto Phisicorum quod omne motum est corpus.

Et cum ostendit modos motuum simplicium et reddit causam propter quam sunt tot, incepit describere motum circularem et rectum secundum quod sunt motus naturales sensibiles et in corporibus naturalibus motis natura non violentia, et dixit *Motus autem circularis est circa medium* etc., idest manifestum est sensui quod motus circularis naturalis est eorum que circa medium moventur; dixit enim

---

61 de—modo] quomodo de hoc modo *B* | et] in *X* | quocumque] -libet *X*: -modo *Z* | modo] *om. Z* 62 naturam] materiam *B* 64 explanationes] expositio- *X*: propositio- *W*: compositio- *Y* 66 rebus] corporibus *B* 67 locum] motum *B* 69 unam] + et *JZ* 70 incepit] -cipit *WY* | motus diversos] *inv. JWY* 71 naturarum] -raliter *W* | debet] *om. B* 73 aut] ut *B* 74 reddit] dedit *YZ* 75 illius] istius *B* 76 longum scilicet] *inv. BJX* 77 quod] de *B* | dividuntur] -ditur *BY* 78 in[2]] *om. YZ* | rectam] -tum *JWZ* 79 et[1]] *om. B* | rotundam] -dum *BJ* 80 duo[2]] + scilicet *Z* 90 moventur] –vetur *Y*

hoc quia motus circularis est terminatus natura propter terminationem medii: motus autem naturales debent esse terminati, alii autem motus circulares qui sentiuntur non sunt naturales cum non habeant centrum terminatum; et ideo videmus quod epicicli quos ponunt astrologi falsi sunt, et forte similiter est de ecentricis magis. Motus autem equalis est superius aut inferius scilicet motus duorum elementorum levium et gravium: motus enim levium est rectus et ad superius scilicet aeris et ignis, et gravium ad inferius scilicet aque et terre; et quia superius est superius naturale et inferius similiter, descripsit ea et dixit *Et dicere superius est motus a medio*, scilicet a medio mundi quod est centrum ad suam circumferentiam, et ad medium scilicet a circumferentia mundi ad eius medium quod est centrum terre deinde aque.

6ra3 6. *Manifestum est igitur quod omnis motus simplex aut a medio aut ad* *268b22*
medium aut circa medium est. Et hoc sequitur hoc quod diximus supe- *268b25*
rius quod corpus componitur ex tribus mensuris, quapropter et motus
corporis sunt tres. *268b26*

Hoc quod dixit manifestum est ex positis: cum enim posuerimus motus simplices esse duos scilicet rectum et circularem, et manifestum est sensui quod motus rectus naturaliter dividitur in duos motus contrarios; hoc autem non videtur in circulari: circularis enim non contrariatur circulari. Manifestum est ex hiis quod motus simplices sunt tres; et post declarabitur certius cum declaraverimus quod motus circularis non contrariatur circulari. Deinde testificat super hoc et dixit *Et hoc sequitur hoc quod dixit superius quod corpus componitur ex tribus mensuris*, idest ostendit motus simplices esse tot hoc quod prediximus, scilicet corpus componi ex tribus mensuris, scilicet quod trinarius est numerus perfectus, quasi igitur diceret quod isti motus sunt tres quia trinarius est numerus perfectus ut corpus com-

95 similiter] simile *WX* 96 aut] ad *Z* 99 superius naturale] *inv. Z* 100 dicere] + ad *JWZ* 101 medium] comfontiam (*!*) *Z*

6. 4 tres] + etiam *BZ* 5 Hoc] Ex *praem. Z* 7 quod] quia *WY* | motus[2]] modos *W* 8 videtur] dicitur *Z* 9 contrariatur] -riat *XY* 11 contrariatur] -riat *XY* | testificat] -catur *JW* 15 igitur] *om. Z*

ponitur ex tribus mensuris, scilicet quod causa eorum eadem est et est quod trinarius numerus est perfectus, non quia mensure sunt tres: inquantum sunt mensure sunt motus tres; et sic intelligendum est.

6rb6 7. *Et cum corporum hoc quidem est simplex hoc autem compositum ex* 268b26
simplici, et dicere est corpus simplex omne corpus in quo est principium
motus naturalis, verbi gratia ignis et terra et similia formis eorum et
generi. Et etiam necessario motuum hic quidem simplex hic autem com- 268b30
ponitur, et motus corporum compositorum erunt diversi, sed tamen
moventur secundum motum corporis simplicis in eis dominantis. 269a2

Cum posuit motus naturales simplices sensui manifestos esse tres, voluit demonstrare quod causa esse motuum simplicium est simplicitas corporum moventium, et dixit *Et cum corporum* etc., et cum quedam corporum sunt simplicia quedam composita, necesse est ut quidam motus sint simplices et quidam compositi, et quod dicere est *Simplex omne corpus* etc., idest omne habens in se aliquod principium principiorum naturalium moventium secundum locum; et non est questio quod descriptio compositi est intellecta ex hoc quod est quod in se habet principia plura uno. Et cum manifestum est quod corpora sunt aliqua in quibus non est nisi unum principium tantum principiorum moventium ut ignis et terra et sibi similia, scilicet quod est sicut dixit secundum formas eorum, manifestum est quod si aliqua corpora sunt
6va simplicia, et cum sint corpora simplicia, sunt motus simplices; | motus
enim simplices sunt corporum simplicium et compositi compositorum. Et cum quedam composita non moventur naturaliter nisi a simplici dominanti, et si in ipso sint diversa quamvis estimetur per sensum in simplicibus eorum esse equalem et unum, et potest aliquis dicere fingendo quod ita est in motibus elementorum, dixit *Et motus corpo-*

19-20 intelligendum est] *inv. WXY*

7. 1 cum] + quedam *XG* 2 omne corpus] *om. BX* 3 et[1]—similia] et forma terra *B* 4 Et] + similia *B* | componitur] *om. BY*: -positus *JZ* 9 et cum] *om. WY*: idest *X* 10 simplicia] + et *BJ* 11 dicere] *om. J*: -citur *B* 12 principium principiorum] *inv. B* 14 hoc] eo *BJX* 17 et[1]] *om. WY* | dixit] -cit *WZ*: -cimus *J* 19 et—simplicia] *om. homoiotel. WYZ* | sunt motus] motus eorum *Z* 21 quedam] + corpora sunt *B* 22 estimetur] -matur *BX* 23 simplicibus] pluribus *BJX*

rum compositorum erunt diversi sed tamen moventur secundum motum corporis simplicis dominantis, idest quod motus eorum erunt secundum naturam corporis simplicis dominantis: impossibile est enim in corporibus compositis aliqua componi equaliter; et nos post declarabimus.

Et si esset contingeret quod aliquod corpus non moveretur omnino sed staret in quocumque loco poneretur sed staret aut superius aut inferius aut in medio duorum contrariorum et moveretur in ceteris, quod non invenitur; et dixit totum hoc ne aliquis fingeret quod ista elementa quatuor non sunt simplicia. Et cum motus eorum sit secundum partem dominantem, erit rectus secundum dominium abundantis, aut erit incurvatus, aut ad inferius ut turbines descendentes ex aere ad terram, aut ad superius ut turbines ascendentes, aut circa medium ut motus ventorum, et que sunt septentrionales et dextri et sinistri et ad ante et ad retro: omnes enim isti preter ad superius et ad inferius sunt compositi.

6vb4 **8.** *Et etiam cum motus equalis sit simplex et motus circularis sit simplex,* *269a2*
corporis simplicis est motus simplex, et motus simplex habet corpus sim-
plex; si ergo corpus compositum habuit motum simplicem, erit ille motus *269a5*
corporis simplicis dominantis in illo. Manifestum est igitur ex hiis que
diximus quod necessario est corpus simplex quod per naturam eius move-
tur motu circulari. *269a7*

Quia eadem natura est principium motus unius, necesse est ut eadem natura habeat eundem motum et econverso; et quia corpus simplex est habens eandem naturam, oportet ut corpus simplex habeat eundem motum; et quia nature plures una necesse est ut habeant plus uno motu, corpora in quibus sunt plures una natura

---

27 est enim] *inv. BXZ* 28 componi] -positionem *Y* 34 partem] + eorum *Z* 35 abundantis] -davit *W*: duorum *X* 37 que] etiam *BJX* 38 preter] prestent *B*: principaliter *WZ* 39 compositi] oppositi *B*

8. 1 etiam] *om. BJYZ* | equalis] *om. BY*: rectus *J* | sit[1]] *om. BXY* | sit[2]] *om. Z* 2 et—simplex[2]] *om. homoiotel. Z* | et—simplex[3]] *om. homoiotel. B* | simplex[2]] -pliciter *Y* 3 corpus] *om. Z*: -pora *J*: -pori *(obsc.) B* | habuit] -buerit *Y*: inveniatur *JG* 4 illo] ipso *W* 5 est] *om. B* | simplex] + et *W* | quod[2]] quia *Z* 7 Quia] Et *praem. Z* 8 et[2]] *om. WY*: ergo *Z* 9 corpus simplex] *om. B* 10-1 eundem—habeant] *om. homoiotel. B* 11 motu] + in *WYZ* 11-2 corpora—motu] *om. homoiotel. B* 11 plures] plus *XY* oporteret ut habeant plus uno motu; et si forte habuerint unum motum erit secundum corpus dominans simplex, non secundum quod est compositum. Quoniam si esset ita, contingeret ut compositum sit simplex: motus enim unus non provenit nisi ab una natura, et habens unam naturam necessario est simplex; et cum hoc positum est sic, et etiam quod motus sint simplices, oportet ut corpora simplicia sint secundum numerum modorum motuum simplicium; et cum positum est quod motus circularis circa medium est simplex, necesse est quod corpus quod movetur in illo motu sit simplex necessario.

Aristoteles igitur, cum posuit quod natura est principium motuum
7ra corporum scilicet motuum localium, et posuit | quod corpora simplicia habent principium unius motus (composita autem habent plus una natura), ex quibus declaratum fuit quod corpora simplicia non habent nisi motus simplices; posuit etiam quod causa motuum simplicium est esse corporum simplicium, et cum hoc dictum sit et descripti sint modi motuum simplicium, posuit etiam quod sequitur ex hiis, et est quod si inventus est motus simplex, oportet ut habeat corpus simplex, idest manifestum est ex hiis que diximus cum motus simplices sint, et unus eorum est motus circularis, et omne corpus simplex habet motum simplicem, et omnis motus simplex habet corpus simplex, necesse est ut motus circularis habeat corpus simplex. Sed tamen tacuit consequens istius propositionis, quoniam sequitur a consequenti parte quod sequitur ab hoc quod posuit precedens; et posuit consequens hoc quod corpus simplex habet motum simplicem et motus simplex habet corpus simplex; et hoc quod sequitur ab hoc quod posuit precedens et potest accipi pars precedentis sicut nos fecimus, quoniam quando acceptum fuit cum precedenti quasi precedens, concludet conclusionem quesitam.

---

12 oporteret] -tet *WXY* 14 esset ita] *inv. X*: est ita *Y*: ita essent *Z* 15 unus] *om. B*: unius *W* 16 est[1]] *om. B* | positum] com- *B* 18 sint] *om. WXY* 19 medium] *post* simplex *Z* 20 quod[1]] ut *WXY* | in] *om. JX* 21 Aristoteles igitur] *om. Z* | igitur] modo *W* 23 principium] -pia *W* 25-6 posuit—corporum simplicium] *om. homoiotel. J* 25-7 posuit—simplicium] *om. homoiotel. B* 27 sint] sunt *X* 29 idest] et *W* 30 simplices sint] *inv. W* | unus] unius *BX* 31-2 motum—simplex[1]] *om. B* 33 tamen] + non *WYZ* | istius] huius *W* 34 quod[1]] que *Z* 36 hoc[1]] *om. WY* | quod—hoc] *om. homoiotel. BX* 38 quando] *om. Z*: + que *Y* 39 concludet] -deretur *B*: -deret *J* | quesitam] finitam *B*

Et quia propositio dicens quod motus simplex est corporis simplicis que est pars precedentis, potest contradici ab aliquo dicente quod motus simplex est corporis compositi, dixit respondendo *Si ergo in corpore composito inveniatur motus simplex, ille motus erit corporis simplicis dominantis in eo*, idest et quocumque modo sit, necesse est quod omnis motus simplex habeat corpus simplex aut singulare per se aut pars compositi. Et voluit dicere quod cum inveniatur pars compositi, necesse est ut inveniatur per se; et cum ita est, necesse est ut numerus corporum simplicium sit secundum numerum motuum *7rb* simplicium. Et quia conclu|sio sicut diximus collocatur in sillogismo potentia propinqua narravit illud quod declarabitur ex hoc quasi diceret ipsam actu, et si non dixit, dicendo manifestum est igitur ex hiis que diximus quod necesse est corpus simplex quod per naturam eius movetur motu circulari; et iam apparuit ex sermone nostro quomodo hoc sequitur necessario ex propositionibus positis. Et laboravit Iohannes Grammaticus secundum quod narravit Alfarabius in contradicendo propositioni dicenti quod omnis motus simplex idem secundum speciem est corporis simplicis eiusdem secundum speciem ex hoc quod cum in corporibus diversis secundum speciem invenitur unus motus specie, verbi gratia quia in aere et igne invenitur unus motus in specie, similiter fingit de terra et aqua; quapropter non sequitur ut omnis unus motus specie habeat unum corpus in specie proprium. Et hoc quidem quod dixit non est ita: motus enim ignis esset idem cum motu aeris si moverentur ad eundem locum specie, quia iam dictum est quod motus unus specie est qui ad unum locum specie est: locus autem ignis necessario differt a loco aeris; et ideo si aer poneretur in loco ignis, descenderet ad inferius, similiter de motu aque et terre scilicet quod specie differunt in se propter diversitatem suorum locorum.

---

43 motus[1]] corpus *B* 46-7 Et—compositi] *om. homoiotel. E* 50 quasi] quod *B* 51 igitur] autem *B* 52 est] + quod sit *Z* 55 Alfarabius] + secundum quod narravit *WY* 57 est—simplicis] habet corpus simplex *BX* | est—speciem] *om. homoiotel. W* | simplicis—speciem] *om. Y* | eiusdem] idem *B* : *om. X* 61 omnis unus] *inv. W*: unus *B* | specie[1]] non *Z* 64 est[2]] *om. WY* 66 in] *om. WY*

impossibile; erit enim aut motus ad superius, et tunc erit aut ignis aut aeris, aut motus ad inferius, et tunc aut aque aut terre. 269a18

Cum declaravit quod impossibile est ut iste motus circularis sit alicuius corporum simplicium recti motus, neque natura neque accidente, et potest aliquis dicere quod est accidentalis alii corpori preter ista corpora mota motu recto, et omnino non erit naturalis corpori moto secundum circulationem. Hoc enim possibile est poni in loco 9rb propositionis minoris dicentis in prima demonstratione | quod iste motus idest circularis est naturalis corpori naturali, quoniam potest aliquis dicere quod est ei accidentalis, sive sit unum elementorum sive aliud, aut naturalis ei ita quod sit unum elementorum quatuor.

Et cum posuit primam demonstrationem, incepit destruere istos tres modos confirmando suam demonstrationem, et dedit tres demonstrationes contradicentes istis modis; et quedam eorum sicut diximus auferunt dubitationem contingentem in minori propositione et quedam in maiori prime demonstrationis per quam probavit corpus quintum esse; secundum hoc ergo intelligendum est hunc locum non quod quelibet istarum demonstrationum sit sufficiens per se; prima enim demonstratio dat quod iste motus circularis debet esse corporis circularis natura; et omnia que diximus post ex tribus sermonibus sunt confirmantia primum sermonem. Cum ergo dixit *Si ergo corpus aliud preter ista* etc., vult dicere si aliquis dixit quod iste motus potest esse accidentalis vel alii corpori preter ista que recte moventur, necessario continget ut hoc corpus habeat motum naturalem preter motum istorum quatuor corporum; sed hoc est impossibile. Non enim est motus rectus naturalis si non confiteatur a nobis quod motus circularis sit naturalis nisi ad superius aut ad inferius: si ad superius, erit motum necessario aut ignis aut aer, si ad inferius necessario erit aut terra aut aqua; si ergo iste motus erit accidentalis alicui corpori, impossibile est ut sit aliud ab uno istorum elementorum.

---

3 et tunc] *om. X* | aut[2]] ut *WY* 4 aeris] aer *Z* | aut[3]] et *W* 5 est] *om. XY* 7 alii] alicui *BY* 8 omnino] ideo *Z* 9 circulationem] -cumlocutionem *BX* 11 idest] scilicet *BX* 12 sive sit] cum sint *Y* 13 ei] *om. WY* 18-9 corpus quintum] *inv. JZ* 20 quod] *om. WY* 22 circularis] naturalis *WY* 23 confirmantia] -manda *XYZ* : + post *WYZ* 25 vel] *om. JZ* 27 quatuor corporum] *inv. W* 28 confiteatur] -tetur *J* : -terentur *Y* 29 nisi] + aut *WXY* | aut—superius] *om. homoiotel. B* 30 motum] -tus *JW* 31 terra … aqua] *inv. JZ* 32 aliud] *om. WY*

9va18 **12.** *Et etiam motus circularis est primus motus motuum necessario: perfectum enim et completum est ante diminutum, et est prius eo prioritate naturali, et quod circulus est perfectior figurarum, sed linearum equalium nulla est perfecta, sive infinita sive finita. Si autem infinita, erit diminuta non perfecta; non enim habet complementum neque finem, perfectum enim habet complementum et finem. Si finita, etiam erit diminuta, cum habeat continens, quapropter possibile est eam augmentari in omnem partem et crescere. Et etiam revertar et dicamus quod, si primus motus est naturalis primo corpori, et motus circularis est principium motus equalis, et motus equalis simplicibus corporibus ut* 9vb | *ignis vadens ad superius equaliter et terra descendens ad inferius scilicet ad medium, necessario erit motus circularis quorundam corporum simplicium. Et iam diximus superius quod corpora composita sunt secundum corpus simplex existens in eis et dominans in eis.*

269a18 269a20 269a25 269a29

Hec est alia demonstratio a predicta, et vult declarare quod motus circularis est naturalis corpori moto hoc motu; et incepit dicere *Et etiam motus circularis* etc., idest quod motus circularis est prior secundum naturam omnibus motibus naturalibus. Deinde dedit causam per quam declarabitur hec propositio, dicendo *Perfectum est et completum* etc., idest quod motus circularis est primus omnium motuum, nam est perfectus et alius motus diminutus; et quod est perfectum est prius diminuto, ergo motus circularis est prior recto. Et cum propositio, dicens quod motus circularis est perfectus et rectus diminutus, non est manifesta per se, incepit declarare eam et dixit *Quod circularis est perfectior figurarum*, idest quod motus circularis est perfectior motuum, proprius enim est corpori circulari et quod est circulare est perfectius figurarum; ex quibus contingit quod motus circularis est in eo quod est perfectioris figure; et quod est perfectioris est perfectius, ergo motus circularis est perfectissimus motuum.

---

12. 2 ante] aut *WY* 3 quod] quia *W* | circulus] -laris *BX* 4 erit] et *Y* 6 perfectum] *dittogr. WY* 7 augmentari] -mentum *W*: -menta *X*: augeri *JY* 8 revertar] -tamur *J*: -tatur *B* | dicamus] -cam *BX* 10 et] + est *J* | simplicibus corporibus] *inv. BX* 14 et] *om. Y* | et—eis] *om. homoiotel. W* 19 causam] terminum *BX* | quam] quod *XY* 22 diminuto] -tum *W* | est prior] *inv. WZ* 23 cum] + hec *JZ* 25 circularis[1]] -culus *JZ* 26 motuum] + quia W_2Z | enim est] *inv. X*: est *WYZ*

Et cum propositio dicens quod circulare est perfectius secundum figuram recte, incepit declarare eam dicendo *Linearum autem equalium nulla est perfecta* etc., idest quod corpus circulare differt a recto quia est finitum et impossibilis est in eo additio; corpus autem rectarum *10ra* dimensionum | est diminutum, sive finitum sive non. Deinde incepit declarare hoc, dicendo *Quoniam si infinitum, erit diminutum non perfectum* etc., et compositio istius sermonis est in secunda figura ita: omne corpus infinitum non habet complementum neque finem; et omne corpus perfectum habet complementum et finem; ergo omne corpus infinitum non est perfectum. Deinde dixit *Si finitum erit etiam diminutum* etc., idest quod si posuerimus corpus rectarum dimensionum finitum, non erit finitum nisi propter corpus continens ipsum extrinsecus, ergo possibilis est in eo additio inquantum corpus, et omne in quo possibilis est additio est diminutum, et quod est diminutum non est perfectum; corpus igitur rectum est imperfectum. Et omne hoc quod affirmavit in corpore recto intelligendum est per negationem in circulari: corpus enim circulare videtur nullo modo posse recipere additionem et maxime ex sua concavitate, quare est perfectum. Et difficile est imaginari corpus circulare infinitum nisi ex parte superficiei extrinsece; et ipse declarabit quod corpus circulare impossibile est ut sit infinitum ex parte superficiei extrinsece, et universaliter quod hic apparet principaliter est quod corpus rectum est diminutum ex omnibus modis, et quod rotundum est perfectum ex parte concava; et est dubium in gibbositate.

Et cum declaravit quod motus circularis est primus motuum in natura, incepit declarare ex hoc illud quod intendebat declarare, et est quod motus circularis existit secundum naturam in corpore quod movetur eo, et dixit *Etiam revertamur et dicamus* etc., iste sermo est quasi sillogismus hipotheticus. Sed tamen posuit in eius precedenti *10rb* maiorem propositionem | que potest esse per se in sillogismo categorico et ab ea concluditur questio que propositio maior est, motus

31 circulare] -laris *B* 37 ita] + quod *Z* 39 omne[2]] *om. JZ* 40 non est] *inv. B* | finitum] in- *B* 40-1 erit etiam] *inv. JY* 42 corpus] *om. WYZ* 44 possibilis est] *inv. WY* 45 est[1]] + additio *B* 50 extrinsece] + et naturaliter quod hic apparet principaliter est *W* | quod—circulare] *om. B* 53 ex] + hiis *WYZ* | est[2]] + et *B* 57 corpore] + et *WY* 58 revertamur] -tar *XY* | dicamus] -cam *XY* 59 in eius] motus *Y* 61 propositio maior] *inv. W* | est] + si *WY*

primus est naturalis primo corpori, quoniam quando huic propositioni fuerit adiuncta hec alia, scilicet quod motus circularis est primus, erit conclusio quod motus circularis est naturalis. Alia autem accepta in precedente faciunt perfectum precedentem per se in sillogismo hipothetico facto et confirmato per locum a digniori, scilicet si motus circularis est prior naturaliter motu recto, et motus recti existunt naturaliter in corporibus simplicibus, tunc multo fortius motus circularis existit in corpore simplici; et hoc non contingit posteriori motui scilicet recto nisi quia invenitur in eo ex precedente, et ideo locus iste est necessarius in precedente et posteriori naturaliter. Et cum dixit *Iam diximus superius quod corpora composita* etc., dixit muniendo se ab aliquo dicente quod motus simplices inveniuntur in corporibus compositis; et hoc iam dicebatur superius, sicut dixit.

13. *Quoniam igitur hoc corpus aliud est, et alia substantia preter ista* 269a30
corpora que nobis vicinantur et est substantia et corpus nobile et gloriosum, et est primum istorum omnium corporum manifestum est. 269a32

Manifestum est ex demonstrationibus predictis quod hoc corpus celeste neque est unum elementorum quatuor neque factum est et
10va compositum ex eis, | sed est alia natura preter istas naturas et est corpus divinum. Non enim generatur neque corrumpitur, ut declaratum est in Phisicis et ut declarabitur hic, quapropter est primum corporum simplicium et compositorum cum sit ingenerabile et incorruptibile, et secundum partem et secundum totum: alia autem que sunt sub eo sunt generabilia aut secundum totum aut secundum partem.

14. *Et si aliquis dixerit quod omnis motus aut est naturalis aut acciden-* 269a32
talis, motus autem naturalis quibusdam corporibus erit accidentalis alicui alii corpori verbi gratia ascensus et descensus, motus enim ad supe- 269a35

---

67 si] + locus *B*   68 existunt] -tant *WY*   70 nisi] *om. WY*   74 iam] *om. WY*

13. 2 nobis vicinantur] *inv. B*   2-3 gloriosum et] *om. homoiotel. Z*   5 et] neque *J*   9 incorruptibile] corr- *VZ*

14. 1 dixerit] *om. BWXY* | quod] *om. BWY* | omnis] *om. B* : causas *X*   3 alii] *om. JZ* | enim] autem *Z*

levitate idest omnia levia; deinde dixit *Quod elevatur super omnia ascendentia superius*, et forte intendebat compositionem unam ex duabus ut esset manifestior, non ex utraque proprietas coniuncta gravi ita quod faceret unam descriptionem.

Deinde dixit *Si igitur ita est, omne corpus vadens ad superius* etc., et hoc quod dixit manifestum est, quoniam hec descriptio convertitur super se sicut innatum est descriptionibus; quoniam si grave est illud quod descendit ad inferius, necessario omne descendens ad inferius est grave, et similiter si leve est illud quod ascendit ad superius, necessario omne ascendens ad superius est leve, et si simpliciter simpliciter et si in respectu in respectu; et similiter de gravi. Quapropter dixit *Non eodem modo sed in respectu* etc., idest cum dixerimus quod omne descendens est grave, non est intelligendum eodem modo scilicet grave simpliciter, sed intelligendum est ex eo illud quod intelligendum est ex descensu; si igitur descensus simpliciter grave simpliciter, et si descensus in respectu grave in respectu, et similiter de ascensu et levi.

13ra19 **18.** *Manifestum est ergo modo etiam quod corpus rotundum non habet* *269b30* *levitatem neque gravitatem; impossibile est enim moveri ad medium neque a medio motu naturali neque motu accidentali. Motus enim equalis non est ei naturalis, quoniam non est proprius nisi corporibus simplicibus; si ergo moveretur eo motu, esset cum illo corpore idem, et si corpus rotundum moveretur motu accidentali, aut esset motus eius ad* *269b35* *inferius motus accidentalis et motus eius ad superius motus naturalis, aut moveretur ad superius motu accidentali et ad inferius motu naturali. Et iam diximus quod cum alter motuum contrariorum fuerit naturalis,*

65 idest] *om.* *Y*: et *X* 67 ex] quod *B* | proprietas] + esset *B* 72 descendit] *post* inferius[1] *JZ* 74 et] sed *W* | si] *om.* *BWZ* | simpliciter[1]] -plex *BXY* | simpliciter[2]] -plex *B* 75 in respectu] *haplogr.* *X* | et[2]] *om.* *J* 75-6 et[2]—respectu] *om. homoiotel.* *B* 75 similiter] + est *W* 76 sed] si *Y* 77 grave] *om.* *Z* 78 simpliciter] -plex *B* | est] *om.* *Z* 79 ex] de *JZ* | igitur] ergo *B* | simpliciter[1]] -plex *B*: sit *Y* | simpliciter[2]] -plex *WZ* 80 si] *om.* *JX*: est *Z* | in respectu[1]] *om.* *WXY*: + et *BJ* | in respectu[2]] *om.* *Z* 81 et] *om.* *W*

18. 2 est enim] *inv.* *Z* 3 motu[2]] *om.* *B* 5 esset] erit JW_2: erunt *Y* 6 moveretur] -vetur *Z*: -ventur *Y* | motu] *om.* *Z* 7 naturalis] accidenta- *BXY* 8 moveretur] -vetur *Z*: -ventur *Y* 9 iam] ideo *W* | diximus] pre- *BJX* | cum] *om.* *B*

13rb *alter erit accidentalis necessario; et cor|pus rotundum non habet aliquod contrarium omnino.* 270a3

Cum declaravit quod omne leve aut grave movetur aut a medio aut ad medium, et iam declaratum erat quod corpus rotundum non movetur neque ex medio neque ad medium neque naturaliter neque accidentaliter, concluditur ex hoc quod corpus rotundum neque est grave neque est leve, et dixit *Manifestum est ergo modo etiam* etc. Deinde propalavit maiorem minorem propositionem cum maior esset iam declarata; deinde incepit determinare eam, et dixit *Motus enim equalis* etc., idest causa in hoc est quod motus rectus non est ei naturalis, quoniam iste motus non est naturalis nisi alicui corporum simplicium scilicet elementorum, idest quoniam iste motus non est nisi nature corporum elementarium, non nature corporis rotundi. Deinde dedit rationem dicendo *Si ergo moveretur hoc motu esset idem cum illo corpore*, idest quod si hoc corpus rotundum moveretur motu recto natura, esset corpus rotundum idem cum illo corpore quod naturaliter movetur illo motu; esset ergo corpus rotundum unum elementorum quatuor, quod est impossibile; unus enim motus in specie non est nisi unius corporis in specie.

Et cum declaravit quod impossibile est corpus equale rotundum moveri motu naturali recto, incepit destruere aliam partem, et est quod etiam impossibile est moveri motu recto extra naturam scilicet violenter, et dixit *Si ergo corpus rotundum movetur motu accidentali* etc., idest si hoc corpus rotundum movetur recte extra naturam recte et simpliciter extra naturam, aut erit iste motus ad inferius extra naturam, quapropter necessario contingit ut motus eius naturalis
13va esset ad superius, et tunc esset ne|cessario aut ignis aut aer, quod est impossibile. Deinde dixit *Aut moveretur ad superius* etc., idest neces-

12 aut[1]] et *B* 12-3 a—medium] ad medium aut a medio *B* 13 et] *om. Z* | iam] + erit *Y*: + erat *BX* | erat] *om. BXY* | non] *om. WZ* 15 hoc] eo *B* 16 est[1]] *om. JZ* | ergo] *om. WXY* | modo] *om. B* | etiam] *om. BJ* 17 maiorem] *om. JZ*: + et *X* 20 corporum] -pori *BJX* 22 elementarium] -tatium *Y*: -torum tantum *X* | non] nec *W* 23 dedit] reddit *W* | moveretur] -vetur *B* 29 equale] *om. JZ* 30 naturali] + et *Y*: + ex natura *X* | aliam] secundam *JZ* 32 violenter] -tum *BX*: -tam *Y*: -tem *W* 33 movetur] -veretur *B* 35 contingit] -get *JW*: + et *WY* | motus eius] *inv. BX* 36-7 quod est impossibile] *om. Z* 37-8 necessarium est] *om. JW*

sarium est si motus eius ad superius esset accidentalis, ut esset motus eius ad inferius naturalis, quapropter erit tunc aqua aut terra.

Et cum posuit quod cum poneretur quod haberet unum motum istorum motuum accidentaliter, contingeret ut oppositus motus esset naturalis, fecit rememorare illud ex quo apparet hoc, et dixit *Et iam prediximus quod cum alter motuum* etc., idest et iam prediximus et apparet ex potentia nostri sermonis quod quando posuerimus alterum duorum motuum contrariorum simplicium in aliquo corpore naturaliter, continget quod alter sit accidentalis in eo, et si non contingeret duos motus esse in eodem corpore naturaliter. Et iam posuit in sermone predicto quod impossibile est in corpore simplici esse duos motus naturales; et intentio eius ex ista propositione non est nisi eius conversa dicens quod cum in aliquo corpore posuerimus alterum duorum contrariorum esse accidentalem, necesse est ut alter sit ei naturalis; et hec conversio magis est occulta quam prima, potest enim aliquis dicere non est remotum ut in corpore rotundo sint duo motus recti insimul accidentaliter scilicet motus ad inferius et motus ad superius, et naturalis sit ei circularis. Et cum subintellexit hoc Aristoteles, dixit *Et corpus rotundum nullum contrarium habet omnino*, idest si aliquis duorum motuum rectorum esset in eo accidentaliter, oporteret ut corpus in quo est naturaliter esset ei contrarium, et nos declarabimus quod corpus rotundum non habet contrarium.

13vb15 **19.** *Et etiam locus motus totius rei idem est cum loco motus partium* 270a4 *eius, verbi gratia tota terra et una gleba moventur ad eundem locum; et cum sint secundum quod diximus, non est dubium quod corpus rotundum neque habet gravitatem neque levitatem omnino, nam si haberet gravitatem aut levitatem, possibile esset ipsum moveri a medio vel ad medium. Manifestum est etiam quod corpus primum rotundum impossibile est moveri motu locali, neque ad superius neque ad inferius, neque* 270a10

38 ut] aut *X*: necessario *JW* 39 erit tunc] *inv. J*: tunc *W* 40 poneretur quod] *om. B* 42 rememorare] removere *YZ* | et[1]] + iam *B* 44 posuerimus] -suimus *B* 46 continget] -tingi *J*: -tingeret *Z* 47 duos motus] *inv. Z* 48 posuit] -suimus *Z* 52 magis est] *inv. B* 53 est] esse *X* 55 sit ei] *inv. W* 56 Et] quod *Z*

19. 1 rei] *om. JZ* 2 una] tota *W* 5 aut] et *YZ* 7 est] *om. B* *alio motu accidentali neque secundum totum neque secundum partes; sermo enim de toto et parte est idem, sicut diximus superius.* *270a12*

Quia fundamentum demonstrationis est quod corpus rotundum non habet gravitatem neque levitatem est super duas propositiones quarum altera est quod corpus rotundum non movetur neque a *14ra* medio neque ad medium, alte|ra autem est quod ita movetur est aut grave aut leve, et iste sunt iam declarate, iam ratione vult declarare minorem sensu et ratione. Et fundatum est super duas propositiones, quarum altera est intellectus, idest quod locus totius corporis gravis aut levis idem est et partis. Quomodo enim movetur grave cum fuerit extra centrum illuc moveretur terra tota si esset extra suum locum? Natura enim partis corporis similium eadem est cum natura totius diffinitione, et cum sit eadem diffinitione, eadem est forma, et cum forma eadem actione proveniente ab illa forma: si igitur actiones provenientes a gravi et levi sint isti duo motus, necesse est ut sic proveniant a parte sicut a toto. Secunda autem propositio est accepta a sensu, et est quod non apparet in partibus corporis rotundi motus neque a medio neque ad medium, ergo neque levitas neque gravitas; et cum non in partibus, ergo neque in toto; *Et cum sit secundum quod diximus, non est dubium* etc., idest cum sit corpus celeste secundum hoc scilicet quod partes eius non videntur moveri a medio neque ad medium, manifestum est quod neque est grave neque leve.

Deinde dixit *Quoniam si haberet gravitatem aut levitatem, possibile esset* etc., idest possibile esset in isto longo tempore ut partes eius viderentur moveri vel a medio vel ad medium, quapropter non habent gravitatem neque levitatem; et stelle cadentes non sunt celi, ut estimant quidam: stelle enim celi note sunt non transmutabiles; et si stelle caderent a celo, diminuerentur in isto longo tempore, et post dicetur de causa calefactionis solis quia non sequitur ut omne calefa-

9 est—sicut] idem est sicut iam *JZ*   10 est] *om. JZ*   14 iam^{2}] *om. J*: nunc *X*   16 intellectus] -ligitur *WY*: -ligibilis *J*: -ligi *Z* | quod locus] *inv. B* | gravis] + est *W*   18 illuc] *om. WYZ* | terra tota] *inv. XY*   22 sint] sunt *JXZ* | proveniant] -niat *XY*   24 est] + secundum *Z*   26 Et^{2}] ergo *Z*   27 dubium] nobis *B* | idest] *om. BXZ*: + et *JXZ*   30 Quoniam] quod *WY* | haberet] -bet *WYZ*   31 ut] aut *B*   32 vel^{2}] *om. B*   34 celi] *om. WZ* | note] nate *X*: tote *YZ* | non] in *W* | non transmutabiles] *om. YZ*   35 post] possumus W_2Z 36 dicetur] -cent *Y*: -cere *Z* | quia] quoniam *WX*

ciens sit ignis. Et Alexander secundum quod Themistius narrat de eo *14rb* dicit quod Aristoteles | dixit hunc sermonem opponendo et contradicendo dicentibus celum delatum super totum, et quod si hoc non esset, caderet, respondendo quod si ita esset, oporteret quod in isto longo tempore caderent eius partes; quoniam igitur corpora celestia non sunt unum quatuor elementorum bene manifestum est ex isto sermone. Quod autem non est ut composita ex eis, ut plures antiquorum putabant, manifestum est ex hoc quoniam si essent, essent aut ad superius ignis aut ad inferius scilicet terre: omne enim compositum tendit ad locum partis dominantis; et etiam si esset in loco, necessario haberet continens ipsum extrinsecum, cum hoc sit natura loci.

14va **20.** *Et oportet nos dicere etiam quod corpus gloriosum non est factum* *270a12* *neque generatum ab aliquo, neque cadit sub corruptione, neque recipit augmentum neque diminutionem neque alterationem neque aliquam transmutationem. Omne enim generabile non generatur nisi a subiecto* *270a15* *illi contrario, neque corrumpitur nisi in suum contrarium. Et iam diximus quod motus contrariorum corporum sunt contrarii etiam; et cum ita sit, nullum contrarium habet corpus primum, nullum enim contra-* *270a20* *rium motum habet motus circularis. Consulta igitur bene fuit natura cum fecit hoc corpus gloriosum non generabile et removit ipsum a contrariis; generatio enim et corruptio necessario existunt in contrariis.* *270a22*

Cum declaravit quod corpus quod circulariter movetur est nature quinte, quoniam natura movens circulariter est alia a natura motus equalis, et declaravit ipsum neque grave neque leve; vult etiam ostendere quod non est generabile neque corruptibile neque augmentabile neque diminuibile neque transmutabile neque alterabile alteratione

---

39 si hoc] *inv. Z* 40 ita esset] *inv. JZ* | oporteret] -tet *Z* 42 bene] unde *BYZ* 43 Quod] quia *YZ* | non] + sunt *BZ* | ut[1]] *om. JVW* 44 essent[2]] + motus eorum *JZ* 45 aut[1]] *om. B* 46 partis] *om. Z* 47 ipsum] ipse *B* | hoc] *om. B*

20. 1 etiam] *om. YZ* 4 generatur] *om. Y*: -rantur *X* | a subiecto] ab *Y*: ab altero *X*: ab fabricato *J*: a fabricato *G* 5 contrario] + quod *YZ* 8 motum] *om. JW* | circularis] -lares *JZ* | igitur] *post* fuit *BX* | bene] *om. BXY* 12 circulariter] -laris *B* 13 equalis] *om. BXY* | declaravit] -rat *Z* 14 neque[1]] vel *XY* 15 neque alterabile] *om. B* passiva, et dixit *Neque generabile ab aliquo*, et non dixit neque generabile simpliciter quoniam hec est differentia inter opinantes quod hoc corpus est creatum a prima causa et opinantes ipsum esse primam causam scilicet quod dicentes ipsum esse primam causam opinantur ipsum esse non generabile simpliciter, et dicentes quod est causatum *14vb* a prima causa opinantur quod non est | generabile ab aliquo, et quod si diceretur esse generabile a nullo et in nullo tempore, bene posset hoc institui sed invoce secundum transmutationem: generabile vero est terminatum tempore et generabile ab aliquo. Et hoc nomen generationis est vulgariter quod Aristoteles vult removere a celo; et intelligit per transmutationem quod est generabile tribus modis, et intelligit per alterationem alterationem quam sequitur generatio et corruptio; alteratio enim que est illuminatio non est impossibilis.

Et cum dixit hoc, incepit ponere propositiones ex quibus apparet: omne enim generabile non generatur nisi ab aliquo ente ei contrario, et hec propositio est vera in generatione simplici et composita scilicet in generatione quatuor corporum simplicium abinvicem et in generatione compositorum ab eis; sed compositum est ei contrarium ex quo componitur, sicut medium dicitur contrarium extremis contrariis cum fuerit prima compositio ex quibus componitur, et hoc erit quando compositum a primis contrariis fuerit prima compositio, non mediante aliquo composito ut compositio mineralium et plantarum que non fiunt a semine sed ab elementis, et generatio animalium per putrefactionem generatorum. Compositum autem a primis contrariis mediante secundo composito, ut generatio sanguinis a pane et membrorum a sanguine, est ei contrarium ex quo fit, secundum quod media que sunt inter extrema contraria contrariant se adinvicem, cum inter extrema fuerint plura uno medio sicut contrarietas colorum qui sunt inter album et nigrum adinvicem. Manifestum est ergo

---

18 opinantes] + quod hoc corpus est *B* | ipsum] ipsam *Z* 19 scilicet—causam] *om. homoiotel.* *B* | dicentes] videntes *XY* 20 esse non] *inv.* *JZ* | est] esse *W* 21 aliquo] alio *XY* 22 diceretur] -ret *X* 23 sed] secundum *BJX* | invoce] vocem *BJ* 23-4 vero est] *inv.* *W* 24 hoc] + est *B* 25 est] *om.* *BWXY* 26 modis] motibus *BX* 27 quam] qua *W* 28 illuminatio] lumi- *BY*: quelibet modo *X* 30 ei] sibi *W* 31 et[2]] + contraria *B* 32 corporum simplicium] *inv.* *JZ* 33 ei] eis *JW* | ei contrarium] *inv.* *BX* 34 contrarium] + in *WY* 35 cum] *om.* *JY* 37 mineralium] numerabi- *Y*: anima- *X* 38 fiunt] fuerint *BY* | animalium] + que *JZ* 42 contrariant] -antur *JW* 43 colorum] oculo- *B* 44 ergo] igitur *JZ*

sive in simplicibus sive in compositis quod omne generatum generatur ab aliquo contrario, et hoc firmabitur per inductionem et ratione; *15ra* generans enim simile est generato et contra|rium ei ex quo fit generatum, non enim agit in materia generati nisi inquantum est contrarium; si ergo illud ex quo est generatio esset simile generato, non posset generans agere in materiam generati; nihil enim agit in suum simile sed in contrario.

Deinde posuit secundam propositionem et dixit *Et iam diximus quod motus contrariorum corporum sunt contrarii*, idest quod motus corporum contrariorum simplicium est contrarius; et non est dubitandum quin concludatur ex istis duabus propositionibus, quamvis iste non propalavit hoc quod motus omnis generati simplicis est contrarius motui eius ex quo generatum est; si ergo motus alicuius corporis non habet contrarium, ergo neque illud corpus habet contrarium; et cum non habet contrarium, ergo neque est generabile neque corruptibile. Quapropter dixit post *Et cum ita sit, corpus primum non habet contrarium omnino, non enim motus circularis habet motum contrarium*, idest ut declaratum est ex suo sermone, non quod hoc sit manifestum per se.

Deinde dixit *Bene consulta fuit natura cum posuit* etc., et innuit per hoc quod impossibile esset hic generatio et corruptio si non esset corpus non generabile vel corruptibile: hoc est consilium primum ex operatione nature; consilium autem secundum est quod cum voluit facere neque generabile neque corruptibile, non fecit ei contrarium neque generavit ipsum ex contrario; quoniam si haberet contrarium, corrumperet ipsum omnino, quapropter necesse est ut omne generabile sit corruptibile. Et ex hoc apparet bene quod corpus celeste non est compositum ex materia et forma sicut sunt corpora quatuor simplicium, et quod est simplex; forme enim que sunt in materia contrarie sunt; *15rb* et | si esset forma in materia sine contrario, tunc natura

46 aliquo] *om.* *JYZ* | firmabitur] con- *JZ* | et[2]] + in *WY* 47 generato] -ratio *BX* 48 generati] -ranti *YZ* : -randi *J* 50 generans] -rari *X* : agens *B* | materiam] naturam *BXY* | generati] -randi *J* : -ranti *Y* 51 contrario] -rium *JZ* 52 Deinde] + dixit *B* 54 corporum contrariorum] *inv.* *JY* | contrariorum simplicium] *inv.* *X* | dubitandum] dubium *BX* 55 istis] hiis *JW* 56 iste] ipse *BJ* : ipsum *X* 59 ergo] *om.* *W* | est] *om.* *WX* 60 neque] *om.* *WX* 64 innuit] + et *B* 68 fecit] *om.* *Z* 69 generavit] + igitur *Z* 70 corrumperet] -rumpet *Y* : -rumpi *X* | ipsum] eum *W* 72 simplicium] -plicia *JZ*

otiose ageret cum in ista materia non sit aliqua potentia omnino; in materia enim non est potentia nisi apud separationem eius a forma. Si igitur substantia materie est in potentia, et fuerit aliqua materia cuius potentia numquam exit ad actum neque in aliquo tempore, ista potentia est otiosa et superflua; et si non esset generatio et corruptio, non staret operatio istius virtutis scilicet potentie, operatio enim istius potentie que est ex ea que est abscissio ex acumine gladii est generatio et corruptio; si ergo ista potentia esset sine generatione et corruptione, contingeret ut aliquod ens esset sine actione propter quam fuit ens; quod est impossibile.

21. *Et etiam omne quod augmentatur et omne quod corrumpitur non* 270a22
augetur aut corrumpitur nisi ex aliquo homogeneo quod fertur in ipsum;
15va *diminuitur ergo | quod preparatum est ad corruptionem in sua materia.*
Corpus autem gloriosum non est ex aliquo omnino; et si non recipit aug- 270a25
mentationem et additionem, propter hoc etiam non recipit transmutatio-
nem nec lationem: transmutatio autem est motus qualitatis, et qualitates
habent dispositiones et formas que non sunt sine alteratione activa, verbi
gratia sanitas et egritudo. Et visum est quod omne corporum naturalium 270a30
alteratur alteratione per quam recipit additionem et diminutionem, sicut
corpora animalium et partes eorum, et corpora plantarum et partes
eorum; similiter est alteratio elementorum mutatorum adinvicem. 270a33

Cum declaravit quod corpus celeste non recipit generationem neque corruptionem, quia non habet materiam neque contrarium, vult declarare quod neque augmentum neque diminutionem neque iuventutem neque senectutem, et dixit quod *Omne quod augetur* etc., idest omne quod augetur et diminuitur non augetur neque diminuitur nisi propter nutrimentum: nutrimentum autem oportet esse homogeneum cum nutriendo cum in ipsum alteratur, quapropter

---

75 in[1]] *om. BX* | materia] *om. Z* 78 exit] *post* actum *WZ* 79 otiosa ... superflua] *inv. Z*

21. 1 Et[1]] *om. BWXY* 2 augetur] augmentatur *BX* 3 quod] *om. Z* | est] *om. YZ* 4 autem] + primum *BJ* | non est] *inv. W* 7 et] in *B* | que] et *Z* 8 omne] + corpus *BJX* | corporum] *om. BX* 10 eorum] ea- *Z* 11 est] *om. JZ* | mutatorum] -tabilis *BX*: -tabit *Y*: -tationem *W*: + et *B* 13 corruptionem] corpora *XY* | materiam] naturam *B* | neque[2]] + habet *JZ* 15 quod[1]] *om. X*: et est *JZ*

nutrimentum ultimum rei debet esse equale in specie cum materia eius ex qua fit propinqua, verbi gratia quia sanguis est materia animalium, contingit ut sit nutrimentum propinquum; diminuitur ergo quod preparatum est corruptioni, quod si nutriendum non fuerit *15vb* preparatum corruptioni, augmentabi|tur propter nutrimentum adveniens; si autem fuerit preparatum corruptioni, diminuitur per ipsa elementa donec dissolvitur in naturam ex qua componitur aut in materias. Deinde declaravit quod corpus celeste non habet homogeneum, et dixit *Corpus autem primum gloriosum non est ex aliquo omnino*, idest et omne homogeneum alicui, inquantum eius alteratio et transmutatio in substantia alterius est, non est homogeneum nisi propter naturam communem illi, sed corpus celeste non habet materiam; et potest etiam intelligi quod nutrimentum aliquo modo debet esse simile et aliquo modo contrarium, sed corpus celeste non habet contrarium.

Et cum declaravit quod non recipit augmentationem neque diminutionem quam intendebat per corruptionem nisi sit error scripture, incepit narrare quod non recipit transmutationem, et dixit *Et si non recipit augmentationem et additionem, propter hoc non recipit transmutationem neque lationem*, et intelligit per transmutationem mutationem qualitatum existentium in eo, quapropter dixit *Neque lationem*, et non quia non recipit augmentationem non recipit mutationem, sed propter causam propter quam non recipit augmentum non recipit transmutationem in qualitate, quasi dicat propter illam causam non recipit transmutationem. Deinde incepit ostendere modum ex quo apparet quod non recipit transmutationem, quia transmutatio est motus qualitatis, idest non recipit transmutationem quia transmutatio est motus in predicamento qualitatis; genera autem prima qualitatum non sunt sine alteratione, ut genus quod est animatum in *16ra* eo quod est animatum, et genus quod | dicitur potentia naturalis, et

19 rei] *om. B* | debet] oportet *W* 24 per] propter *WY* 27 ex] ab *Z* 28 eius] + est *JWZ* 29 est[1]] *om. JW* : et *YZ* 30 illi] illis *BX* 30-2 materiam—habet] *om. homoiotel. YZ* 31 etiam] *om. Z* 32 et] ex *BX* 35 quam] que *B* : quod *J* | sit] esset *B* 37 additionem] diminutio- *J* : diminutionem additionem *Y* 41 sed—non[1]] sed causam quapropter non *B* 42 illam] istam *JZ* 44 quia] et *BJ* | transmutatio] + autem *BJ* 45-6 qualitatis—motus] *om. homoiotel. WXYZ* 46 predicamento] -dicto *B* : generatione *Z* | genera] -ratio *YZ* 47 sunt] fit *Z* 48 et[1]] vel *Z* : vel et *Y*

genus quod est in quantitate inquantum quantitate non transmutatur nisi alteratione: alteratio autem non est nisi in habente subiectum. Corpus autem celeste non transmutatur omnino in qualitatibus que fiunt per alterationem illuminatio autem et simile sibi ex qualitatibus que non fiunt per alterationem non est impossibile. Et ideo dixit in qualitatibus tamen sunt dispositiones que non sunt sine alteratione idest quorundam scilicet quatuor generum notorum; quapropter non est intelligendum per transmutationem in hoc loco transmutationem que fit sine tempore que intrat in capitulum relationis vel illuminationis.

Deinde dixit *Et visum est quod omne corpus corporum naturalium* etc., idest et testatur quod alteratio est ex conditionibus augmenti et diminutionis quod nos videmus quod omne quod augetur ex corporibus animalium et partibus eorum et corporibus plantarum et partibus earum non augetur neque diminuitur nisi per alterationem; et hoc non est tantum in compositis sed etiam in simplicibus, quod apparet ex eis quod non transmutantur adinvicem nisi per alterationem. Concluditur ergo ex omnibus istis quod omne generatum et omne augmentatum alteratur alteratione passiva, scilicet a qualitatibus activis que sunt calor et frigus et humidum et siccum, et sunt unum quator generum qualitatis, et hoc ergo genus impossibile est esse in corpore celesti, et similiter quod sequitur hoc ut sanitas, infirmitas, iuventus et senectus et omnia accidentia animalium provenientia ab istis qualitatibus; et etiam impossibile est esse in eo mutatio aliarum qualitatum que fit per alterationem, et non est *16rb* impossibile ut in eo sit semper ut figura; qua|litates autem quarum alteratio non est contradictio nisi secundum suum esse non est impossibile ut sint in eo secundum esse et secundum mutationem.

16va12 22. *Et etiam si corpus rotundum non recipit additionem neque diminutionem, dignum est ut non recipiat neque transmutationem neque alte-* *270a33*

52 simile] -liter *XY* 53-4 que—qualitatibus] *om. homoiotel. WYZ* 57 capitulum] -tulo *B* | relationis] *om. WYZ* 58 vel] *om. Z*: et *Y* 63 earum] eo- *BX* 63-5 et—alterationem] *om. homoiotel. B* 64 quod] quia *YZ*: quoniam *X* 65 quod] quia *J* | transmutantur] -tatur *XZ* 66 omnibus] *om. JZ* 67 a] *om. Y*: in *X* 68 et[1, 2]] *om. W* | humidum] -ditas *JW*: hiis *Z* | et[4]] *om. Z* 69 generum] genus ergo *B* 74 sit semper] *inv. JZ* 75 nisi] *om. BXY* | suum] *om. Z*

rationem. Cum igitur hoc corpus gloriosum est semper permanens, prop- 270b1
ter hoc etiam non recipit additionem neque diminutionem, neque intrat
ipsum senectus neque recipit passiones. Et hoc bene manifestum est ex hiis
quod diximus eis qui voluerunt credere et suscipere et considerare in pro-
positionibus et sillogismis predictis et intelligere eas. Et etiam ratio testa-
tur visui et visus rationi. Omnes enim homines conveniunt in hoc quod 270b5
hoc corpus primum gloriosum est locus spirituum, scilicet Greci et alii
primarum gentium qui cognoverunt Deum et eius divinum; et dixerunt
hoc quoniam illud super quod non cadit corruptio neque latio debet esse
in loco qui non corrumpitur neque diminuitur, neque transmutatur
neque alteratur. Et oportet esse, sicut est in loco qui non diminuitur
16vb *neque corrumpitur neque transmutatur | neque alteratur neque fertur,*
creare additionem, diminutionem, corruptionem, transmutationem, alte-
rationem et lationem, et creare quodcumque recipit et patitur additio-
nem et diminutionem corruptionem, transmutationem, lationem et alte-
rationem. Et impossibile est ut sit contrarium eius quod diximus, et recte
dixerunt in hoc, quoniam hoc corpus gloriosum est perpetuum. Iam 270b10
ergo declaravimus et demonstravimus sufficienter et perfecte in prima
substantia corporum, et inducemus testimonium de visu, et sufficiemus
hominibus per ipsum sufficientia perfecta. Non enim cessabit hoc corpus
scilicet celeste per secula preterita, secundum quod hereditabunt homines
unus post alterum permanere in suo esse, neque transmutatur neque alte- 270b15
ratur neque secundum totum neque secundum partem omnino. Antiqui
autem videbant et in hoc conveniebant secundum opinionem nostram, et
dixerunt in eo quod nos diximus non semel nec bis sed pluries, et perve- 270b20
niunt ad nos. Videbant igitur et dicebant quod primum corpus est aliud
a terra et ab igne et ab aere et ab aqua, et appellaverunt istum locum

22. 3 permanens] -net *JZ* 3-4 propter—additionem] *om. Z* 4 intrat] curat *Y*: mutat *Z*: inferentem *X* 6 quod] que *XY* | voluerunt] -lunt *BJX* | et[1]] que *Z* 7 et[2]] *om. JZ* | intelligere] *om. JZ*: -lige *WY* | eas] *om. JZ*: causas *B* | Et[3]] dicitur *WY* | etiam] *om. WYZ*: + super ea *JZ* 8 in hoc] *om. WXY* 10 eius divinum] *inv. JZ* 14 neque alteratur] *om. Z* 15 corruptionem] + et *JWY* 17 diminutionem] *om. BJ*: *post* corruptionem *Z* | corruptionem] + et *JY* 18 contrarium] -rius *BX* 19 hoc[2]] est *BX* | est perpetuum] *inv. J*: perpetuum *BX* 20 et demonstravimus] *om. WXY*: et declara- *Z* 22 ipsum] ipsa *B* | cessabit] -vit *JZ* 23 preterita] predicta *B* | homines] + vel heredes *Z* 24 permanere] -nendo *JZ* 25 neque[1]—partem] neque secundum partem neque secundum totum *B* 26 autem] + hoc *XY* 27 nec] neque *JZ* | perveniunt] -nerunt *JXYZ* 28 igitur] enim *WX* 29 istum] illum *XY*

17ra *octavum ethereum, nomine derivatum a sua actione; est enim in motu eterno velocissimo permanens | sine diminutione in se neque cum transmutatione neque cum alteratione in se, neque corruptione neque latione accidente in se. Anaximandras autem non utebatur hoc nomine secundum suam veritatem, quia vocavit ignem ethereum, et peccavit in sermone suo.* 270b25

Vult ostendere in hoc capitulo convenientiam inter sensum et rationem de isto corpore, et convenientiam testimoniorum antiquorum, et verba eorum obscura in antiquo tempore dicta; et incepit dicere *Et etiam si corpus rotundum* etc., idest omne quod non recipit augmentum neque diminutionem necesse est ut non recipiat transmutationem neque alterationem: omne enim augmentabile est alterabile et non convertitur. Et est intelligere quod si non recipit additionem et diminutionem propter causam quam diximus, scilicet quod caret subiecto, necesse est ut propter illam causam non recipiat transmutationem neque alterationem nisi large intelligantur additio et diminutio, scilicet additio in quantitate et diminutio in ea. Quod est tribus modis: aut secundum augmentum verum et diminutionem veram, et hoc tantum est in plantis et animalibus, aut secundum alterationem ut accidit aque et aeri quod cum inspissantur fiunt minoris quantitatis, et hoc multum invenitur in corporibus consimilis nature, ut accidit in musto; tertius est additio secundum accumulationem, et hoc invenitur in corporibus consimilium partium compositis ut in mineris, et invenitur etiam in elementis, et est ut super rem addantur aliqua extrinseca et continuentur cum prima, et secundum hoc dici-
17rb tur quod ignis augetur a pre|dictis. Omne ergo quod non recipit augmentum aut diminutionem non recipit alterationem neque transmutationem, omne enim transmutabile et alterabile recipit additionem

30 octavum] *om.* *W* | a] in *YZ* 31 velocissimo] -mum *JWYZ* 33 accidente] + ei *JZ* 34 quia] quoniam *BX* 36 hoc] isto *WY* 38 eorum] eius *WYZ* 42 est] + etiam *Z* | intelligere] -lige *Z* 44 subiecto] + et motu *Z* | non] *om.* *BJWZ* 49 quod] et *Z* 51 accumulationem] accidens *X*: + vel accumulatum extensionem W_2Z 52 hoc] *om.* *W* | corporibus] animalibus *Z* | compositis] *om.* *B* 53 rem] res *Z* 53-4 addantur aliqua] *inv.* *Z* 54 aliqua] alia *BX*: + alia *JY* | continuentur] -nentur *Z* 56 aut] neque *JZ* | alterationem … transmutationem] *inv.* *WY* 57 enim] igitur *B*

transmutationem et ceteras alias mutationes, et non tantum agit has transmutationes sed etiam recipientia has transmutationes. Deinde dixit *Impossibile est ut sit econtra,* et innuit per hoc ligamenta inter generabilia et corruptibilia, et corpus ingenerabile et incorruptibile, et ligamentum istius corporis cum ente eterno non corporeo; necessarium est enim ut sit corpus ingenerabile et incorruptibile quod numquam in sua substantia transmutatur nisi tantum in loco ita quod ipso transmutato semper in loco remanente substantia non cesset agere alias mutationes predictas in corporibus que sunt sub eo, et etiam non cesset generare recipientia ista; necessarium est ut inter principia eterna et res generabiles sit ens talis modi, et si non ita, impossibile esset quod a principiis eternis proveniret aliqua actio temporalis: actio enim eterna est necessario, et hoc innuebat cum dixit *Impossibile est ut sit econtrario*; hoc est ergo quod dixerunt antiqui, et verum est in se. Et ideo via loquentium de lege nostra ad probandum Deum esse non inducit eas nisi ad corpus eternum necessario, quoniam cum posuerit mundum esse novum, continget necessario ut inquantum est novum sit generatum; et quia omne generatum habet generans, et omne generatum non generatur nisi *18ra* | postquam primo alteratur in qualitate ab agente, necessario est quod alterans ipsum sit corpus, quapropter necesse est ut ante mundum esset corpus alterans ipsum; et si ponatur illud corpus generatum, contingeret ut prius esset aliud corpus, et sic in infinitum; necessario ergo oportet ut sit corpus alterans non alterabile neque generabile omnino, et perscrutabimur de hoc corpore.

Apparet autem quod impossibile est quod semper mobile per se nisi per principium extrinsecum non corporeum neque quod sit virtus in corpore; revertamur ergo ad sermonem nostrum: *Iam ergo declara-*

116 etiam] *om.* *JWZ* : et *Y* 117 econtra] contrarium *JZ* 119 ligamentum] -mento *J* : -menta *X* | corporeo] corrupto *WZ* | necessarium] -ario *WYZ* 120 sit] + hoc *B* 121 transmutatur] *om.* *XY* | ita] ipso *WYZ* 122 remanente] perma- *B* 123 eo] ipso *JW* 124 est] + enim *JZ* 125 ita] + esset *Z* 126 eternis] eterni *B* 127 est necessario] *inv.* *Z* 130 eas] eos *BJ* | nisi] *om.* *B* 131 posuerit] -suerint *BX* | continget] -geret *BX* 134 ab] *om.* *XY* | est] *om.* *BW* 135 alterans—corpus] *om.* *B* 136 ipsum] ipse *B* | corpus generatum] *inv.* *B* : + esse *Z* 137 contingeret] -get *BXY* | aliud] illud *B* 138 neque] *om.* *X* : aut *B* 139 perscrutabimur] -bitur *J* : -buntur *Y* 140 autem] *om.* *WXY* | quod^{2}] + sit JW_2 142 ergo^{2}] igitur *Z*

vimus in prima substantia illius corporis, idest declaratum est ex hoc quod primum corpus oportet esse huiusmodi scilicet ut non recipiat permutationem neque alia, quod si non esset ita, non esset primum alterans. Deinde dixit *Et inducemus testimonia ad illud* etc., idest ad hoc quod apparet ratione ex hoc quod apparet sensui; et dixit *Sufficiemus per ipsum*, quoniam sufficere per illud est facere quod sensui videtur; et non apparet in aliqua hora aliquod contradicens isti sufficientie, sed semper permanet eodem modo in sufficientia perfectissima; et talis sufficientia, quando non fuerit sustentata per aliquid, est de modis propinquis veritati certe: cum autem fuerit sustentata per demonstrationes certas, erit complementum illorum sermonum, quoniam ex conditione sermonum demonstrabilium est convenientia sensus, quoniam quando different, sunt demonstrationes quarum alteri testatur sensus et alteri non, aut quia non est innata sentiri aut quia non est possibilis ut sentiatur, fides prime demonstrationis erit magis perfecta; et appellavit hanc sufficientiam, quoniam hic modus credendi compositus est ex sensu testimonio, quapropter est sufficien- *18rb* tior sermonibus probabilibus, et inquantum | intrat ipsum testimonium, descendit a sermonibus demonstrabilibus. Quapropter dixit *Et inducemus testimonium in hoc per visum* etc., et hoc quod dixit est verum et conveniunt in hoc omnes gentes, quoniam numquam latum fuit ab hominibus alicuius seculi quod ipsi viderunt transmutationem in celo, neque secundum totum neque secundum partem. Et quia ipse fuit primo dicens ex Phisicis grecis quod celum esset ingenerabile, dedit testimonium ex Caldeis: dicitur enim quod philosophia perfecta erat tempore eorum sicut tempore Aristotelis, et dixit *Et hoc videbant Antiqui et in eo convenerunt* etc., idest et illud quod latum est ab Antiquis in natura istius corporis convenit ei quod in eo diximus scilicet quod est natura quinta: dicebant enim quod

---

143 illius] istius *BX* 145 ita] utrum *Z* 146 alterans] -rius *W* | testimonia] -nium *J* 147 Sufficiemus] + facere *BX* 148 est] *om. Y* | est facere] *om. BX*: + per illud *J* 151 aliquid] -quod *B* 153 erit] ut *B* 154 conditione] -tradictione *Z*: dicto *X* | sermonum] + nam *WX* | demonstrabilium] -bilis *W* 155 quando] *om. Z*: ante *Y*: autem quando *JZ* | different] -rens *Y*: -rentes *J*: -rere *W*: due *BX* | sunt] *om. BWY sed suppl.* W_2 | alteri] -ra *YZ* 156 testatur] attest- *BX* | sensus] *om. Y*: senso *X* 157 est] *om. WY* | possibilis] *om. Z* 159 sensu] + et *WZ* 160 et] *om. B* 162 testimonium] *post* in hoc *JZ* 164 latum] rela- *JWZ* | viderunt] -bant *Z* 166 Phisicis] philosophis *B* 169 in eo] ideo *WZ* 170 latum] re- *JZ*

aliud est a quatuor elementis, et derivaverunt ei nomen a velocitate motus et eius permanentia scilicet ethereum, quapropter credimus quod illi dixerunt ipsum esse materiam quintam, et cum sit natura quinta, non habet contrarium; si enim haberet contrarium, esset natura sexta: contraria enim debent esse paria sicut in corporibus simplicibus quatuor, et cum non habet contrarium, non habet naturam neque est generabile neque corruptibile. Et quia quidam antiquorum peccaverunt in intellectu huius nominis scilicet ether, ostendit Aristoteles peccatum eius dicendo *Anaximandras non utebatur* etc., idest quod non utebatur hoc nomine secundum veram interpretationem cum estimabat quod hoc nomen non dicitur simpliciter nisi de igne, et credidit propter hoc quod celum esset ignis, dixit *Et est error in idiomate* scilicet quod hoc nomen non dicitur de igne sed de velociter moto; et cum ignis est levis et velocis motus, estimavit quod hoc nomen diceretur de eo.

18va11 23. *Manifestum est igitur ex hoc quod diximus quod hec corpora simpli-* *270b26*
cia impossibile est ut essent plura. Et iam diximus quod motus corporum
simplicium est simplex necessario; dicimus etiam quod motus simplices
sunt duo scilicet motus circularis et motus rectus, et quod motus equalis *270b30*
istorum duorum motuum est ad medium aut a medio. *270b31*

Idest declaratum est iam ex omnibus istis sermonibus predictis quod corpora simplicia sunt quinque, quatuor mobilia motu recto et unum mobile circulariter. Et cum dixit conclusionem, dixit propositiones radicales ex quibus concludebatur, et dixit *Et iam diximus etiam quod motus corporum simplicium* etc., idest et hoc quod diximus declaratum est ex propositionibus propositis scilicet quod motus corporum simplicium oportet quod sit simplex, et econverso scilicet

172 derivaverunt] *om. J* : dimina- *Y* | nomen a] nomina *Z* 173 scilicet] *om. BX* 174 materiam] naturam *BJX* | quintam] veram *YZ* 177 simplicibus quatuor] *inv. B* | contrarium] + eum *W* | habet naturam] *inv. B* : habet materiam *JYZ* 180 eius] suum *JZ* 182 dicitur] erit *YZ* | simpliciter] -plex *Z* 183 credidit] -diderunt *Z* 184 est] *om. BW* | dicitur] erit *JYZ* 185 est] sit *JZ* | motus] + et *WZ*

23. 3 dicimus] dix- *BX* 4 quod] quia *Z* 5 motuum] + aut *XY* 6 declaratum est] *post* iam *Z* | istis sermonibus] *post* predictis *W* 11 propositis] positis *XYZ* 12 sit simplex] *inv. Z*

quod motus simplices oportet quod sint corporum simplicium; et posuimus cum hoc quod motus dividitur primo in duo scilicet circularem et rectum, et quod rectus dividitur in duo scilicet a medio aut ad medium, et uterque istorum aut simpliciter aut in respectu; quapropter motus simplices debent esse quinque, quatuor recti et unus 18vb circu|laris. Corpora ergo simplicia sunt quinque, cum numerus eorum est secundum numerum motuum simplicium.

24. *Et motus circularis non habet contrarium omnino, et potest considerans in hoc suffici ex multis; qorum unum est quoniam si posuerimus motum circularem habere contrarium, erit motus rectus dignior ut sit contrarium.* 270b32 270b34

Cum primum posuit sine demonstratione quod corpus rotundum non habet contrarium, ex quo declaravit quod non habet subiectum, et quod non est generabile neque corruptibile et quod nullum corpus est sextum, vult modo declarare illud quod posuit *Et potest consideranti hoc sufficere ex multis*, et intelligit secundum sufficientiam fidem perfectam non abundantiam estimationis. Et hec demonstratio est accepta a loco maiori sic: si motus circularis habet contrarium, erit dignius ut motus rectus sit eius contrarium quam circularis; et cum motus rectus non est ei contrarium, ergo per locum a maiori neque circularis erit ei contrarius. Et hoc fuit necessarium quoniam motus rectus differt a motu circulari secundum speciem, circularis autem a circulari non differt secundum speciem, et contrarium differt a contrario secundum speciem, ergo illud quod differt secundum speciem dignius est ut sit contrarium quam illud quod convenit secundum speciem. Et manifestum est quod motus rectus non est contra-

---

14 dividitur] -duntur *BX*: dicitur *Y* | scilicet] + in *JZ* | circularem] -cumferentiam *Z* 15 scilicet] + aut *JZ* | aut] et *BX* 18 ergo] enim *B* | numerus] numquam *B* 19 est] sit *JZ* | motuum simplicium] *inv. B*

24. 1 circularis—habet] –res non -bent *JW* | considerans] -rari *Z* 2 suffici] -cere *J*: -cientia *WZ* 3 habere] + motum *JZ* 7 et[1]] *om. BX* | quod[2]] cum *WXY* 8 declarare illud] *inv. JZ* : declarare *W* | consideranti] -rari *BX* 8-9 consideranti hoc] *inv. JZ* 10 abundantiam] -tia *B* 11 loco] + a *BJ* | maiori] -ris *YZ* 13 rectus] eius *B* | est] *om. B*: habet *W* | ei] *om. WX* | contrarium] -rius *XY* 16 differt[2]] *post* contrario *(17) Z* 19 est[1]] + ergo *Z*: + etiam *X*

19ra rius motui circulari, | quoniam duo contraria debent esse in eodem ordine secundum genus quod dividitur per ipsa, et ut sit remotio inter ea ultima in fine remotionis; et remotio inter motus non est nisi secundum locum; et quia non est contrarietas inter locum motus circularis et recti, ergo nec est inter motus: illud enim quod revolvitur circa medium non contrariatur ei quod movetur a medio vel ad medium, sicut contrariat illud quod movetur a medio ei quod ad medium; et hoc notum est per se.

Et etiam si inter corpus celeste et elementa esset contrarietas, oporteret quod corrumperentur adinvicem sicut est in elementis; sed apparet contrarium, corpora enim celestia conservant elementa et continent ipsa et universaliter sunt in eis quasi forma. Et ex hoc omnino scitur quod inter ea et ea que sunt sub eis non est contrarietas omnino; et cum ita sit, dignius est ut inter corpora celestia non sit contrarietas.

25. *Motus autem contrarius concavus qui est in ventre arcus, et motus* *270b34* *gibbosus qui est super dorsum arcus, sunt contrarii, et sunt contrarii motui equali in suis formis et suis compositionibus. Si ergo motus contrarius inveniatur motui equali preter ista, necessario habebit motum circularem contrarium.* *271a3*

19rb Cum posuit ante quod si motus circularis habeat contrarium, dignius est ut sit motus rectus, vult declarare modum secundum quem est dignius, et est quod circularis differt a recto duobus modis oppositis scilicet gibbositate et concavitate arcus: circularis autem a circulari non differt in aliquo modo modorum oppositorum. Dixit *Motus autem concavus* etc., intelligit per contraria opposita secundum relationem; hec enim est dispositio gibbosi cum concavo; et cum dixit quod isti duo modi sunt oppositi, incepit declarare quod differunt a motu equali in diffinitione et forma, et hoc intellexit cum

---

21 per] propter *B* 24 nec est] neque *WX* | motus] -tum *YZ* 25 contrariatur] -riat *XY* 26 quod[1]] + contrarie *W* 27 notum est] *inv. X*: notum *W* 29 quod] ut *WXY* | est] *om. B* 31 sunt] *om. Z* | in] *om. W*: cum *YZ*

25. 3 suis[2]] *om. JZ* | motus contrarius] *om. B* 4 circularem] -laris *BX*: -lari *J* 6 habeat] -bet *WXY* 8 modis] motus *W* 9 autem] *om. B* 10 modo] *om. Z* 11 Motus autem] *inv. W* | etc.] + et *BX* 14 in] *om. Z*: et a *Y* | intellexit] -ligit *JZ*

dixit *In formis eorum et compositione* idest quod cum circulare non est contrarium recto quamvis differat ab eo hiis duabus oppositionibus, dignius est ut non sit contrarium ei circulare.

Deinde dixit *Si igitur inveniatur aliquis motus contrarius motui circulari* etc., idest et cum motus rectus sit dignior ut sit contrarius circulari quam alius, manifestum est quod, si inveniatur aliquis motus contrarius motui circulari preter rectum quasi dicat alius circularis, dignius est ut rectus sit eius contrarius. Et hoc voluit dicere cum dixit *Necessario erit etiam motui circulari contrarium*, idest necessario erit motus rectus contrarius circulari; deinde incepit declarare quod motus rectus non est contrarius, ad concludendum oppositum precedentis scilicet quod motus circularis non habet contrarium omnino.

19va **26.** *Et iam diximus multotiens quod motus equalis est contrarius sibi* *271a3*
adinvicem; superius autem et inferius sunt nomina cadentia super diversitatem locorum et contrarietatem eorum. *271a5*

Idest et significatio quod motus equalis non est contrarius motui circulari est quoniam multotiens diximus quod motus recti sunt contrarii sibi adinvicem, quoniam loca eorum scilicet superius et inferius sunt contraria; et cum motus recti contrarii erunt sibi adinvicem, impossibile est ut sint contrarii circulari, quoniam tunc duo essent contraria uni. Et hoc quod dixit est manifestum visui, quoniam si corpora celestia essent contraria elementis, corrumperetur mundus statim propter augmentum corporum celestium in magnitudine ad elementa; quoniam sicut dicit Aristoteles in primo Meteororum, quoniam si celum esset ignis, destrueret omnia in tempore brevi.

---

15 formis] diffinitionibus *BXY* | compositione] -tionibus *JWZ* | cum] *om.* *WZ* : est *X* 16 oppositionibus] compo- *B* 17 contrarium] -rius *XY* | contrarium ei] *inv.* *JZ* : ei *B* 18 aliquis] *om.* *BX* | circulari] *om.* *JZ* 19 et] *om.* *B* 20 si] *om.* *B* 22 rectus sit] *inv.* *Z* 23 Necessario[1]] -cesse *XY* | erit etiam] *inv.* *Z* 25 contrarius] + motui circulari *JZ* | oppositum] compo- *B*

26. 2 adinvicem] + propter loca *JZG* 6 loca] locorum *B* : loco *W* : loci *X* 7 recti] + sint *JZ* | contrarii] -rient *XY* | erunt] *om.* *JWYZ* 8-9 duo essent] *inv.* *JZ* 9 hoc] + est *Z* | quoniam] quod *B* 12 Meteororum] Metato *XY* 13 quoniam] *om.* *JZ*

27. *Qui ergo dixit quod sermo de motu equali et motu circulari est idem peccavit: motus enim ab A ad B contrarius est motui a B ad A. Et noluit dicere* 19vb | *quod motus equalis habet ultimum: circulus autem non habet finem super puncta; et non intendebat eos per sermonem eius, et sit hec figura.* 271a6 … 271a10

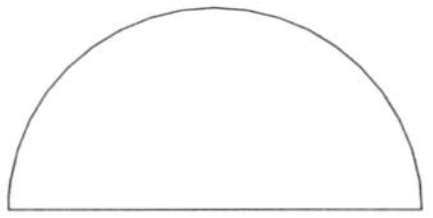

Hec est alia demonstratio preter predictam ad destruendum estimationem dicentis quod motus circularis est equalis in hoc recto, scilicet quoniam motus rectus est contrarius motui recto; et est qui incepit a loco ad quem alius intendit, scilicet quod sic sunt circulari, scilicet quod motus ad aliquam partem esset contrarius motui ab illa parte. Et dixit *Qui ergo dixit quod sermo*, idest quod idem quod contingit in eis scilicet quod contrarietas invenitur in eis similiter. Et cum narravit quod dicens hoc peccat, incepit declarare modum ex quo apparet quod motus rectus differt a circulari in hoc, dicendo *Motus enim ab A ad B* etc., verbi gratia si aliquis ponat quod motus ex A usque ad B super arcum AB est contrarius motui ex B ad A, sicut aliquis potest estimare. Deinde dixit *Et noluit dicere* etc., idest dicens hoc non estimavit hoc quod motus qui est ex A ad B super arcum AB est contrarius mo|tui ex B ad A super illum arcum etc., 20ra nisi quia motus qui est ex A ad B super cordam arcus est vere motus contrarius motui qui est super illam eandem cordam ex B ad A. Et propter hoc estimavit quod sicut invenitur hoc in linea, ita invenitur hoc in arcu, propter coniunctionem linee recte cum circulari.

---

27. 1 equali] + motui circulari non est motus circularis contrarius *Z* | est] *om. Z* 2 noluit] vol- *BXYZ* 3 dicere] + per sermonem suum hoc nisi propter coniunctionem que est linee recte et circularis super eadem puncta et non dixit per sermonem suum *JW*$_2$*Z* | ultimum] + circulum *B* 4 eos] *om. BWXY sed suppl. W*$_2$ 5 figura] + ista *W*: + ita *X* : + BA *BJ* : + l'g a. a. *Y* 6 est alia] *inv. W* | alia demonstratio] *inv. BX*: + iterata *JZ* | preter] post *JZ* | predictam] -tum *Z* 7 scilicet] *om. B* 8 quoniam] + sicut *BJ* | rectus] + sicut *X* 9 incepit] -cipit *WY* 10 esset] est *BX* 11 quod[2]] *om. BX* 12 similiter] + peccavit *JW*$_2$*Z* 14 in hoc] *om. B* 15 ab] *om. BX*: ad *W* 15-6 ad B—ex A] *om. homoiotel. Y* 16 ex A—ad B] *om. Z* 17 noluit] vol- *BXYZ* 18 hoc[2]] *om. BXY* 18-9 super—AB] *om. Z* 19 est—B] *om. B* | est contrarius] *inv. JZ* 20 est[1]] *om. BW* 22 in] *om. Z* 23 hoc] *om. WXZ*

Deinde dedit rationem ex qua apparet quod motus ex A ad B super lineam rectam, et ex B ad A, sunt contrarii motus tantum, et quod motus qui sunt inter ista super lineam arcualem non sunt contrarii. Et fundavit ipsam super duas radices, quarum una est quod contrarium non habet nisi unum contrarium, et secunda quod una contrarietas non invenitur nisi super lineam rectam; et causa illius est quod inter duo puncta non potest cadere nisi una linea recta. Contrarietas ergo non invenitur nisi in longitudine recta: longitudo autem recta est motus recti, contrarietas ergo non invenitur nisi in motu recto; longitudines autem non recte inter duo puncta sunt infinite, quapropter in motu super illas non invenitur contrarietas, quoniam si ita esset, haberet unum infinita contraria. Et omne hoc intendebat cum dixit *Quod motus equalis habet finem*, idest quod opposita earum sint super unam longitudinem tantum; *Circuli autem sunt infiniti super puncta*, idest circuli cadentes inter duo puncta posita; possibile est enim ut ab altero puncto ad alterum exeant linee circulares infinite, et una tantum recta.

20rb3 **28.** *Similiter etiam erit in medio circuli in motu qui est ex C ad D et ex* *271a10*
D ad C; et iste motus existit in diametro super puncta diversa; et hec est
forma. *271a13*

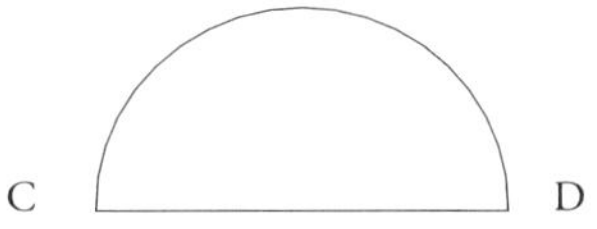

Cum aliquis posset dicere quod non est possibile ut inter duo puncta cadant circulares infinite nisi quando linea circularis non fuerit semicirculi, cum autem erit semicirculi non potest cadere inter ea nisi unus arcus tantum, qui est arcus semicirculi, dixit respondendo

24 motus] *om.* *WXY* 28 nisi] *om.* *WY* 30 potest] possunt *WXY* 31-2 in— nisi] *om. homoiotel.* *B* 31 longitudine] *om.* *W* 32 ergo] *om.* *Z* 34 illas] *om.* *B* 35 esset] *om.* *B* 37 earum] eo- *XYZ* 38 infiniti] + et *B* 39 enim] *om.* *B* 40 recta] + dixit *B* : + dicit *XZ*

28. 1 circuli] -lari *Y* 3 forma] + idest *Z* 4 possibile] im- *WYZ* 5 linea] arcus *W* | fuerit] fuit *W*: sit *JZ* 6 semicirculi[1]] -laris *JYZ* | semicirculi[2]] -laris *J* 7 dixit respondendo] *inv.* *JZ*

Similiter etiam erit in medio circuli etc., idest et sic contingit in duobus motibus oppositis qui sunt ex duobus punctis oppositis qui sunt super semicirculum ut non sint contrarii quamvis non cadat inter illa duo puncta nisi illa medietas tantum. Deinde dixit *Et iste motus existit in diametro super puncta diversa,* et non estimavit dicens quod isti duo motus, qui sunt super eundem arcum qui est semicirculus, sunt contrarii, nisi quia sunt super diametrum, idest quia sunt inter extrema linee recte; sunt ergo oppositi inquantum extrema sunt extrema linee recte, non inquantum sunt duo puncta in circumferentia circuli, quoniam non opponuntur inquantum sunt in circumferentia circuli sed inquantum sunt extrema diametri.

20va Et hoc quod dixit invenitur in omnibus punctis qui sunt in | circumferentia circuli, quoniam cum contraria sunt ea inter que est maxima remotio in loco, et maxima remotio non mensuratur nisi secundum lineam rectam; unde dicitur in descriptione linee recte quod est linea brevissima inter duo extrema copulata. Cum ergo remotio in fine non invenitur nisi in recto inquantum recti, et contraria sunt ea que in fine sunt remotionis, ergo contrarietas non invenitur nisi in linea recta; et cum adhuc fuerit aggregatum quod ea que moventur super extrema diametri sunt contraria, continget necessario quod contrarietas eorum sit propter diametrum non propter circumferentiam.

29. *Et similiter descripsit circulum et posuit motum semicirculi contra-* *271a13*
rium motui medietatis residue; et hec est figura eius. Ponamus ergo circulum integrum, et sit semicirculus B, et moveatur ex H ad C contrario modo motui semicirculi A, qui quidem movetur ex C ad H. Si ergo isti

8 circuli] -lari *W* | sic] si *Z* 10 ut] quod *Z* : et *JX* | sint] sunt *XZ* | cadat] -dant *WYZ* 12 super] + duo *Y* | quod] *om. B* 13 arcum] + d'ntu *B* 14 sunt[1]] *om. W* 15-6 sunt[1]—recte] *om. homoiotel. Y* 15 oppositi] -sita *WZ* 16 sunt] + in *W* 17-8 quoniam—circuli] *om. homoiotel. B* 17 quoniam] quia *Y* | opponuntur] -nitur *WZ* : -netur *X* 20 est] *om. B* 21 remotio[1]] + est *BWXY* | mensuratur] -rata *JZ* : -ra *Y* 23 est linea] *inv. Z* 24 in[2]] *om. Z* | recto] recta *B* 25 in fine] *post* sunt[2] *JZ*

29. 1 similiter] + qui *JG* 2 et] *om. BX* | hec] *om. BY* | est] ex *BWXY* | eius] *om. BJY* | circulum] + et *Z* 4 motui] -tu *B* | quidem] *om. BJX* | ad] *om. BY* | ergo] igitur *Z* *motus fuerint contrarii, non erunt omnes motus circuli motus totius circuli contrarii per illos motus.* 271a19

20vb

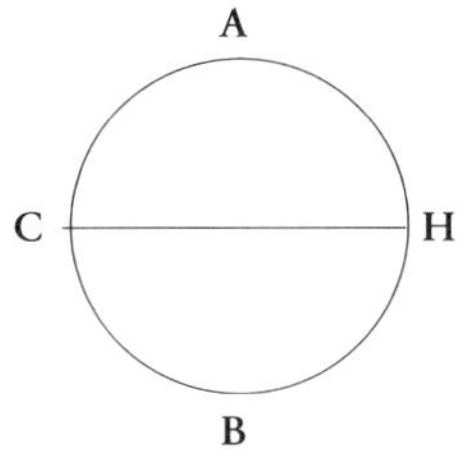

Cum illud quod movetur naturaliter super lineam circularem necesse est ut sit circulare, oportet intelligere ex suo sermone quod illud quod movetur circulariter non est contrarium illi quod movetur circulariter, corpus scilicet rotundarum dimensionum non corpus rectarum; sed quia ipse posuit quod duo mota opposita a duobus punctis super arcum non sunt contraria secundum estimationem nisi inquantum quod mota super lineam rectam sunt contraria, apparet ex sermone suo quod illa duo mota sunt rectarum dimensionum non rotundarum. Rotundas ergo dimensiones impossibile est moveri 21ra super lineam rectam; quapropter modo posuit motum par|tem rotundi, quoniam posset ei aliquis dicere quoniam non est estimandum in duobus motibus ex duobus punctis diversis esse contraria inquantum motus eorum super cordam arcus est contrarius, nisi quando duo mota fuerint rectarum dimensionum. Si autem fuerint arcualia, non erit contrarietas in eis propter contrarietatem motus super lineam rectam, cum sit impossibile in motu arcualis dimensionis moveri per rectam; quapropter dixit *Similiter qui descripsit circulum* etc., idest similiter contingit ei qui descripsit circulum, et posuit mota super ipsum ad partes oppositas duos semicirculos aut duo arcus circuli, quasi ergo dicit primo quod mota in duas oppositas partes super arcum circuli impossibile est ut sint contraria. Aut enim

5 fuerint] + circuli *YZ* | circuli[1]] *om. BJ* | totius] *om. B* 6 contrarii] *om. B* | per] super *B* | motus] + et ista est forma circuli completi sicut vides *J*W_2*G* 8 necesse] -ario *Y*: -arium *J* | suo sermone] *inv. BX* 10 scilicet] *om. BXY* 13 quod] *om. BJ* 17 ei] *om. BX*: eis *WY* 18 contraria] -rietatem *J*W_2*Z* 24 etc.—circulum] *om. homoiotel. B* 25 mota] *om. J*: *post* ipsum *BX*: + vel moveri W_2*Z* | ipsum] eum *Z*: + moveri *J* | oppositas] + scilicet *WZ* | duos semicirculos] *inv. B* 27 contraria] -rii *Z*

erunt dimensionis recte aut arcualis: si recte, non erunt contraria nisi propter motum super diametrum aut cordam, si arcuales (verbi gratia semicirculus ita quod utrumque motorum moveatur contra reliquum donec compleatur circulus) non continget ex concessione contrarietatis in motu semicirculorum ut circuli sint contrarii. Semicirculi enim et universaliter sectiones circulorum non sunt circuli sicut partes linee recte; quapropter motus oppositi super lineam rectam sunt contrarii secundum partes et totum; et si fuerint contrarii secundum partem, continget ut sint contrarii secundum totum; et non ita est in circulis et partibus eorum. Et hoc intendebat cum dixit *Si ergo isti motus sunt contrarii adinvicem, non erunt motus omnes circuli contrarii*, idest si aliquis posuerit motus semicirculi contrarios, non erunt circuli facti a motibus illarum medietatum contrarii; non enim semicirculus est circulus.

21rb7 **30.** *Et etiam motus totius circuli qui est ex A ad B non est contrarius* *271a19*
motui qui est ex A ad C, quoniam motus qui est in circulo est super unum punctum quod non mutat illum locum; et hec est figura circuli. Et iam diximus superius quod motus contrarii sunt super puncta diversa in locis diversis. *271a22*

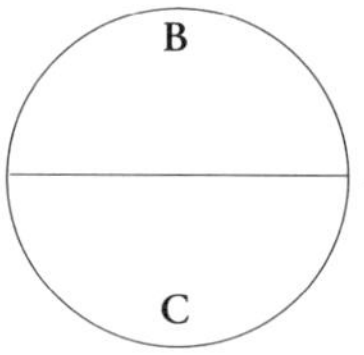

Hec est demonstratio tertia fundata super duas propositiones; quarum altera est quod motus oppositi super unum circulum sunt, quorum quilibet facit umam revolutionem completam et movetur ab

28 erunt[1]] erit *B* | si] *om. Z* 29 arcuales] -lis *B* 30 moveatur] -ventur *Z* 36 partem—secundum] *om. homoiotel. B* 37 ita est] *inv. XZ* 38 motus omnes] *inv. Z* 39 posuerit motus] *inv. B* 40 erunt] *om. W*: erit *BX* 41 enim] + est *BW*

30. 1 Et] *om. XY* 2 est[2]] *om. BWXY* 4 puncta diversa] *inv. Y* 7 sunt] scilicet *BXY* 8 movetur] vide- *Y*: + ab alia est *Z*

uno loco locorum ex quo movetur quilibet eorum ad eundem punctum; motus autem contrarii sunt qui moventur super puncta diversa, ergo motus super circulum non est contrarius motui super circulum. *21va* Et incepit dicere | *Etiam motus totius circuli* etc., idest motus diversi qui sunt in toto circulo non sunt contrarii, verbi gratia sit circulus super quem movetur aliquid a puncto A in partem B donec revertatur ad A, et aliud movetur a puncto A in partem C opposito B donec revertatur etiam ad A: dico quod motus ex A ad B ad C non est contrarius motui ex A ad C ad B ad A: duo enim motus sunt ab eodem puncto A scilicet et ad idem scilicet A; et hoc intendebat cum dixit *Motus enim qui est in circulo est super punctum unum* etc., idest quoniam motus qui est super circulum incepit a puncto et pervenit ad ipsum, et numquam mutat illum locum sed semper a quo movetur ad ipsum revertitur.

Deinde dixit *Et iam diximus superius* etc., scilicet quod ex quo est motus et ad quod sunt contraria; et cum posuerimus has duas propositiones, concludetur in secunda figura quod motus circulares sunt non contrarii, scilicet propositionem dicentem quod motus circularis super rotundum non sunt in loco ad suum oppositum, et dicentem quod motus rectus qui est super dimensionem rectam est ex loco ad suum oppositum.

21vb **31.** *Etiam si motus circularis esset contrarius motui circulari, esset al|ter* *271a22*
eorum otiosus. Et etiam corpus quod circulariter movetur equaliter
movetur in omnibus locis diversis contrariis a quocumque loco incipit. Et *271a25*
iam diximus quod diversitas locorum est superius et inferius, et anterius
et posterius, et dextrum et sinistrum, et quod diversitas motuum est
secundum diversitatem locorum. *271a28*

---

9 movetur] vide- *XY* | movetur quilibet] *inv. Z* | punctum] + alia est W_2Z : + non sunt *J* 11 ergo motus] *om. Z* 12 Etiam] omnia *Z* : a scientia *Y* | motus[1]] mota *Z* 13 sit] si *B* 14 in] ad *BX* 16 ad[2]] *om. W* : per JW_2Z 17 ad[1]] per JW_2Z | ab] ex *B* 18 A scilicet] *inv. YZ* : scilicet AZ *B* | scilicet A] *om. B* 19 unum] *om. B* : quod *X* : + quod *J* : oportet *Y* : unde oportet *Z* 20 incepit] -cipit *WXY* 21 ad] + punctum *B* | illum] ipsum *W* 22 ad ipsum] et ipse *B* 23 quod] *om. B* 25 circulares] -laris *B* 25-6 sunt non] *inv. BX* 26 dicentem] + scilicet *B* | circularis] -lares *JXZ* 27 in] ex *JXY*

31. 1 contrarius—esset] *om. homoiotel. B* 3 movetur] *om. Y* : -veretur *Z*

Hoc quod primo dixit, post declaravit ipsum; et cum incepit dicere hanc propositionem, transtulit se ad aliam demonstrationem, dicendo *Et etiam corpus quod circulariter movetur* etc., idest quod quando corpus rotundum movetur per suas partes in locis diversis, scilicet sursum et deorsum et ante et retro et dextrorsum et sinistrorsum, non continget a motu suarum partium in istis locis ut motus eius sit contrarius, cum non moveatur in illis totaliter sed partialiter; et dixit hoc quia ipse videt et plures Antiquorum quod celum habet dextrum et sinistrum et ante et retro et sursum et deorsum, ut post declarabitur. Et causa propter quam non diversatur motus partium celi propter diversitatem esse eorum in istis locis est quia celum non mutat totaliter suum locum numero quamvis mutet eum secundum formam; et hoc apparuit in Phisicis.

Si ergo partes eius essent contrarie propter contrarietatem loci, tunc natura fecisset otiose quiddam et res esset contraria sibi; et reputo quod non intendat ut hoc sit demonstratio per se, sed intendit declarare per hanc propositionem per quam incepit suum sermonem *22ra* scilicet declarare quomodo contingit illud impossibile si po|suerimus motus circulares esse contrarios per hanc propositionem, et sic erit intelligendus sermo eius ita, et quod diversitas motuum est secundum diversitatem locorum, idest *Et iam diximus* quod motus contrarii sunt qui contrariantur per sex loca: motus autem qui non propter loca contrariantur scilicet circulares non sunt contrarii, quoniam si haberent contraristatem propter loca, contingeret ut aliquid esset contrarium sibi. Et potest esse demonstratio per se, sed manifestum est; et erit intelligere quod corpus rotundum non habet contrarietatem propter cursum in locis contrariis, et quod contraria sunt ea que habent contarietatem propter cursum in locis contrariis.

32. *Et etiam si motuum circuli diversitas esset equaliter omnino, non* *271a29*
moveretur circulus; et si quidam motuum esset fortior quodam omnino, *271a30*

---

7 post] *om.* *VX* 10 locis diversis] suis locis *WYZ* 11 scilicet] idest *JWY* | et[1]] ante *W* | dextrorsum] -trum *YZ* | sinistrorsum] -trum *WYZ* 13 illis] istis *Z* | partialiter] -culariter *BX* 14 ipse] *om. Z* | celum] *om. Z* 15 dextrum] -tro *Y*: -trorsum *X* | sinistrum] -tro *Y*: -trorsum *BX* : + celum *Z* | ut] et *Z* 16 Et] *om. Z* 24 contingit] -venit *YZ* | impossibile] + et *WY* 28 per] propter *XWZ* 31 sed] + et sic *JZ* | manifestum] -tius *JX* 33 propter] per suum *WZ* | locis contrariis] *inv. Z*

non esset motus victus; et si ita esset, esset aliqua pars circuli otiosa ubi motus eius non est per se, ut dicimus quod sotular est otiosus cum non fuerit eo quis calciatus; natura autem nihil facit otiosum. 271a33

Quia motus circularis non differt a motu circulari secundum spe- 22rb ciem, necesse est si habet contrarietatem ut sit in eodem individuo, scilicet quod necesse est ut in eodem corpore rotundo sint due partes contrarie. Et dixit *Et etiam si esset diversitas motuum* etc., idest et cum contingit ex istis locis ut in eodem circulo sint due partes contrarie, contingit ex hoc alterum duorum: aut cum potentia utriusque partium circuli contrariorum sit equalis, et tunc prohibebit altera alteram quapropter non movebitur circulus cum opposite moveantur eius partes, et hoc intendebat cum dixit quod *Si diversitas motuum circuli esset equaliter omnino, non moveretur circulus*: aut altera duarum partium esset aut ut altera duarum partium erit fortior altera, et tunc prohibebitur pars debilior a motu semper; natura igitur tunc fecisset otiosum cum fecit ens sine actione omnino; et hoc intendebat cum dixit quod *Si esset quidam motuum fortior quodam omnino, non esset motus victus*, idest quod si altera duarum partium contrariarum que sunt in circulo fortior altera, non posset moveri pars victa omnino.

Deinde dixit *Si ita esset, et tunc esset aliqua pars circuli otiosa* scilicet victa: motus enim eius semper esset violentus non naturalis, et hoc intendebat cum dixit *Ubi motus eius non est per se*, idest non est naturalis, et esset corpus rotundum otiosum cum non haberet actionem, sicut sotular cum non calciatur; et secundum hanc explanationem intendebat occulte in hac demonstratione propositionem dicentem quod si in corpore rotundo esset contrarietas, oporteret ut in eodem individuo essent due partes contrarie: omnia enim rotunda sunt convenientia secundum speciem.

---

32. 3 non] si *WYZ sed corr.* W_2 | victus] vinctus *B* : motus *Y* 4 dicimus] -ximus *WXY* 5 quis] ali- *JW* 9 esset—motuum] motuum diversitas esset *BJ* | cum] *om. B* 11 cum] quod *JZ* 15 moveretur] -ventur *XY* 16 esset—partium] *om. homoiotel. Z* | aut] debilior *J* | ut] et *J* : non erit *X* : + erit *WY* 17 debilior] *om. BWXY* 18 cum fecit] *om. YZ* 20 si] *om. XY* | duarum partium] *inv. W* 21 que] *dittogr. XY* | circulo] + esset *JZ* : + non *Y* 23 ita esset] *inv. BX* | et] *om. JY* | tunc] *om. Y* : iter *B* | circuli] *om. WXY sed suppl.* W_2 27 explanationem] -positionem *Z* 28 hac demonstratione] hanc -nem *WY* | propositionem] -situm *Z* 30 individuo] + ut *Z*

[22va] Et potest etiam | apud me intelligi quod fundamentum demonstrationis est secundum aliam intentionem propinquiorem ista, et est quod si corpus rotundum esset contrarium corpori rotundo, verbi gratia illud quod movetur ab oriente ad occidentem est contrarium illi quod movetur in orientem ab occidente, necesse est ut essent super unos polos et unum centrum; et si ita esset, contrarient sibi in motu et prohiberent se adinvicem, aut unum eorum esset semper victum et tunc esset otiosum. Et quamvis hec intentio appareat, monstravit ipsam dicendo *Et cum etiam corpus quod movetur circulariter equaliter movetur in locis contrariis a quocumque loco incepit*, idest quod corpora celestia moventur in locis contrariis eodem motu; et si esset in eis contrarietas, esset in partibus diversis et tunc prohiberent se adinvicem. Sed nos videmus ea moveri in locis contrariis motibus non contrariis; quapropter necesse est ut non sint contrarii adinvicem, quoniam si essent contrarii, oporteret ut partes diverse in quibus moventur moverentur super eosdem polos et idem centrum, et essent se prohibentes adinvidem si esset fortitudo in illis equalis; aut si esset vincens in altera, pars victa numquam suo motu moveretur, et si ita esset, tunc natura fecisset otiose. Hoc est ergo quod mihi videtur de explanatione istius loci, et est verum per se et valde est manifestum.

Themistius autem explanavit hunc locum et non addidit super intentiones quas dixit Aristoteles aliquid, neque ostendebat modum ex quo verificabatur demonstratio Aristotelis; et inquit quod Alexander explanavit ipsum alia explanatione, sed non dixit illam; et concessit [22vb] quod est locus valde difficilis. Et fundamentum eius | demonstrationis est super duas propositiones: quarum altera est quod si motus circulares essent contrarii, contingeret quod essent super idem centrum et eosdem polos; et hec propositio est manifesta: spera enim que move-

33 est[1]] *om.* *W* 36 in] ab *BJXY* | orientem] -te *XY*: occidente *BJ* | ab] in *BJX*: ad *Y* | occidente] -tem *Y*: orientem *BJX* 37 unos] duos *JZ* | esset] -ent *WZ*: actione *X* | contrarient] -rientur *BJ*: contraria essent *Y*: contrariat etiam *X* 38 et] ut *YZ* | unum] nullum *BW* | eorum] illo- *XY* | semper] super *Y*: se *B* | victum] iunctum *B* 39 esset] + semper *B* | monstravit] narra- *W*: innuit *Z* 40 cum] *om.* *JZ* 41-2 a—contrariis] *om. homoiotel.* *B* 41 incepit] -perit *XY* 42 moventur] -verentur *XZ* 47 moventur] *om.* *X* | moverentur] *om.* *WYZ* 48-9 essent—et] *om. homoiotel.* *BWXY sed suppl.* W_2 50 fecisset] + aliquid *JZ* | est ergo] *inv.* *J*: ergo *WYZ* 51 et[1]] *om.* *Z* | est[2]] *om.* *JZ* 54 aliquid] *om.* *Z* 57 quod] + non *Z* 58 altera] una *Z* 59 quod] ut *Z* 60 hec] *om.* *Z*

tur super diversum centrum ab eo super quod movetur alia spera non est contraria illi, similiter et que moventur super idem centrum sed super polos diversos, quoniam tales non prohibent se, et contraria sunt que se prohibent; et ex hoc declaratur quod illud quod Aristoteles intendebat per hanc demonstrationem est valde manifestum.

Ego autem longo tempore feci moram in quo non intellexi explanationem istius loci ultimam, sed Deus induxit me ad veritatem; et secundum hanc explanationem erit intentio sermonis Aristotelis quod motus circularis non habet contrarium, non in eo quod est motus abstractus a corporibus naturalibus, sed inquantum est in corpore moto secundum naturam; quoniam si acceptus fuerit motus in imaginatione abstractus a moto, non continget impossibile ponendo eos contrarios et eosdem secundum speciem, verbi gratia cum imaginaverimus duos motus super eundem polum et idem centrum ad duas partes diversas, continget necessario ut prohibeant se et contrarient se; et secundum hunc modum possibile est estimare quod in motu circulari sit contrarietas. Et cum isti duo motus fuerint positi in corporibus naturalibus, revertetur possibile ad impossibile scilicet esse contrarietatem in eadem specie, quia tunc contingeret ut una res destrueret se, et sic natura faceret otiosum; verbi gratia quod secundum considerationem Geometre in superficie et spera possunt contingere se super punctum, et cum fuerint in corpore naturali non possunt se tangere nisi super partem divisibilem.

23ra36 33. *Et cum declarata sunt ista et bene determinata, volumus perscrutari* *271b1*
23rb *de hoc quod remanet. Primo igitur incipiemus utrum aliquod | corpus possit esse sine fine, ut plures Antiqui dicunt; deinde perscrutabimur de mundo utrum sit unus aut plures, quoniam si hoc fuerit aut contrarium, non est parva differentia immo valde magna. Differentia enim cadens in* *271b5* *principiis mundi et in toto mundo facit cadere diversitatem in compre-*

61 diversum centrum] -sa -tra *B* 62 similiter] + illi *Z* 64 quod[1]] *om. BX* 65 demonstrationem] + et *B* 67 ultimam] *ante* istius *Z* 69 non[2]] *om. B* 72 imaginatione] -tionibus *J* : magnitudine *B* | impossibile] : poss- *BWXY sed corr.* W_2 74 eundem polum] *inv. Z* 76 secundum] + diversas continget *B* 78 ad] *om. X* : et *Y* 79 quia] quoniam *B* 80 secundum] si *B* 82 contingere se] se contangere *J* : se tangere *B*

33. 3 possit] posset *YZ* 4 aut[1]] vel *Z* | aut[2]] + fuerit *Z* 5 valde magna] *inv. Z* 6 facit cadere] facere *B* | comprehendendo] appre- *YZ*

hendendo veritatem, hoc igitur est principium omnium diversitatum scilicet propter quas diversificati sunt in narratione totius nature in preterito, et illi qui diversificabuntur in futuro; et non habent tantum diversitatem nisi quia qui parum pertransit de veritate mille millies persaltabit veritatem post, sicut ille qui dixit corpus minimum esse quoniam 271b10 *quando in suo sermone posuerit minimum, movebit et excitabit maximam disputationem inter homines. Et causa illius est quod principium est magnum aut in potentia aut in actu aut in magnitudine; manifestum est igitur quod minimus error in principiis rei multiplicatur in fine. Infinitum autem valde magnum est et habens magnam potentiam et magnam quantitatem maximam in magnitudine: est enim principium rerum. Non est ergo mirum neque extra rationem ut diversitas in hoc sit* 271b15 *valde mirabilis.* 271b16

Cum declaravit quod corpus rotundum est natura quinta, et quod *23va* est neque grave neque leve neque generabile neque | corruptibile, vult modo perscrutari de accidentibus quesitis in eo utrum sit finitum aut infinitum aut unum aut plura. Et hec perscrutatio universalis de esse infiniti in actu firmior: in hoc loco enim sic vertebat perscrutationem suam in corporibus tantum, et non divisit ipsum in potentiam et actum sicut fecit ibi, sed posuit perscrutationem tantum in corporibus quodcumque corpus sit; et dixit *Et primo incipiemus perscrutari* etc., et cum perscrutatio de hoc est propior, quoniam quando declarabitur quot sunt corpora simplicia, fiet perscrutatio de hoc in singulis eorum, deinde perscrutabitur post hoc de composito, quoniam post declarationem quot sunt species corporum simplicium erit demonstratio perfectior et proprior et firmior; cum autem numerus specierum corporum simplicium est ignotus, perscrutatio est dimi-

---

7 igitur est] *inv. X*: ergo est *J*: est *B* | omnium] -nis *BX* | diversitatum] -tatis *BXY* 8 totius nature] *inv. XYZ* 9 diversificabuntur] -versabuntur *XY* 10 qui] aut *YZ* 11 dixit] -cit *YZ* 12 et] *om. BX* 14 in[1]] *om. WYZ* | in[2]] *om. WXYZ* | in[3]] *om. YZ* 15 igitur] ergo *W* 16 autem] enim *BJX* | habens] -bent *B* | et[2]] *om. Z* 18 ut] aut *W*: si *JZ* | in hoc] *post* sit *WY* 22 in] de *YZ* 23 hec] cum *BX*: est *Y* | perscrutatio] + de esse *B* 24 enim] *om. BX* 25 suam] secundo *BY* | potentiam] -tia *X* 26 actum] -tu *XY* 27 Et[2]] *om. X*: + in *BX* | incipiemus] -piamus *Z* | perscrutari] + actionem tantum in corporibus quod *B* 28 cum] *om. JZ* | hoc] homine *Y* | est] *om. Y*: esset *B* | quoniam quando] hic quoniam non *Z* 29 fiet] fiat *BW* 32 cum] dum *BX* 33 perscrutatio] -scruptio *B*

nuta, quapropter revertebatur perscrutari hic de eo. Deinde dixit *Perscrutabimur de mundo utrum sit unum aut multa*, idest et cum perscrutati fuerimus hic de mundo utrum fuerit finitus aut infinitus in magnitudine, perscrutabimur de eo an sit unus aut plures.

Deinde dixit *Quoniam si sit, aut contrario* etc., idest quod diversitas cadens in istis questionibus provenit ad magnam diversitatem: est enim aut in principiis mundi, aut in eius toto; et diversitas in principiis mundi, idest in elementis, aut universaliter in totalitate eius, scilicet utrum sit unus aut plures aut finitus aut infinitus, est causa diversitatis hominum in comprehensione veritatis que est primum principium, et in comprehensione veritatis in omnibus rebus. Et hoc *23vb* quod dixit est manifestum per se, quoniam qui dixit mun|dum esse infinitum in quantitate, non videt ipsum habere principium neque eternum neque non eternum, et qui dixit quod mundi sunt plures, contingit ei credere multa principia. Similiter etiam non est parva differentia inter credentes ipsum esse novum et antiquum; dicentes enim ipsum esse novum non possunt ponere generans nisi corpus: generans autem est alterans, alterans autem est corpus; et sic contingit querere de isto corpore utrum sit generatum aut eternum: si generatum, ergo a corpore generante, et sic in infinitum, si eternum est questio quare generavit ipsum in hac hora et non in alia, quapropter cogetur dicere ipsum motum esse motu eterno, quod est impossibile nisi in corpore quod circulariter movetur ita quod cum fuerit propinquum isti mundo, erit causa generationis, et cum sit remotum in alia hora erit causa corruptionis; contingit ergo ex hoc quod mundus sit pars mundi existentis extrinsecus continentis ipsum eternum; et omne hoc est impossibile.

Deinde dixit *Hoc igitur est principium omnium diversitatum*, idest quod diversitas cadens inter homines naturaliter in principiis mundi et in narratione sue nature est causa diversitatis in veritate in prete-

---

35 Perscrutabimur] + hic *VW*: + h' *XY* | multa] plura *B*: plures *J* | et] *om. JZ* 36-7 perscrutati—magnitudine] *om. homoiarch. X* 36 hic] *om. BJY* | fuerit—infinitus] finitum fuerit aut infinitum *B* 41 aut universaliter] *om. B* 43 hominum] huius *WYZ* | in] *om. B*: ex *Z* 43-4 que—veritatis] *om. homoiotel. B* 45 esse] *om. BX* 46 ipsum] ipse *BX* 48 etiam] autem *Z* 50 novum] + et *B* 51 et] *om. B* | sic] sicut *X*: si *B* 53 in] *om. B* | si] sed *Z* | est] *om. Z* 55 ipsum] *om. XY* | motu] + contrario JYZ_3 58 ergo] igitur *W* | quod] ut *BX* 59 continentis] -tibus *Z*

rito tempore et in futuro et in preterito. Et bene potest sciri hoc non tantum de philosophia sed ex legibus: leges enim non ponuntur nisi secundum estimationem ponentis legem in principiis mundi, aut secundum illud quod videt esse melius hominibus in fide; sed tamen debet ponere fidem propalatam vulgo in principiis similem fidei eius, si videt quod illud, quod videt illud propalandum est homini-
24ra bus, est aliud a vero: si autem videt quod non est pro|palandum vulgo nisi veritas, tunc illud quod ponet in lege est fides eius. Et propter hoc dixit quod ille qui persaltat veritatem parum, erit remotio eius millesies, idest quod qui persaltat veritatem parum in principiis, post removebitur a veritate magna remotione, sicut dixit post Quod minimus error cadens in principio vie itineris inducit errantem post in errorem maiorem; et sicut etiam dicit quod minimus motus accidens malo, similiter minimus error cadens in principiis est causa maioris erroris in eis que sunt post principia.

Deinde dedit exemplum dicendo *Sicut dicentes corpus esse minimum* etc., idest sicut dicentes quod corpora componuntur ex partibus indivisibilibus, quoniam ponendo partem minimam esse movet magnam disputationem et excutit maximam diversitatem in reddendo causas dispositionum entium; et potest intelligi sic, quod hoc movet magnam disputationem: dicere enim partem indivisibilem valde est sufficiens et difficile destruitur: sunt enim plures rationes ex quibus estimatur esse, et alie ex quibus non esse. Hec igitur questio, quamvis sit in radice in qua diversitas est causa diversitatis in pluribus rebus, est in se tamen disputabilis naturaliter, quapropter duobus modis est causa diversitatis, scilicet propter suam naturam et quia est principium; et propter hoc induxit exemplum ex eo; et cum pars minima faciat tam magnam diversitatem, quid ergo estimas de diver-

64 et in preterito] *om. JZ* 67 videt] -detur *Z* 67-8 tamen debet] *inv. JW* 68 propalatam] preparatem *Y*: palam *B* 69 videt[1]] *om. JZ*: -detur *B* | illud[2]] *om. JZ*: + quod *XY* | propalandum] -bendum *Y*: palam *B* 70 propalandum] palam *B* 72 erit] *dittogr. B*: etiam cum *X* 73 qui] *om. WYZ* | persaltat] saltat *Z* 75 itineris] + ille minus W_2Z | errantem] + quod *Z* 76-7 et—malo] *om. J* 76 sicut] sic *Z* | etiam] et *Y*: in *X* | etiam dicit] *inv. Z* 77 cadens] -dit *B* | maioris] magni *X*: magni aeris *B* 79 dedit] dixit *WY* 82 magnam disputationem] *inv. Z* 82-4 et—disputationem] *om. homoiotel. Y* 82 maximam] magnam *BZ* | reddendo] -ducendo *Z* 83 hoc] *om. Z* 84 movet] -vit *X*: continet *Z* 85 sunt] *post* plures *B* 86 alie] *om. Y*: alia *B* | quibus[2]] + estimatur *B* 90 et[2]] quod *Z* 91 estimas] -mat *Z*: est in eis *J*

sitate esse magnitudinis infinite? Deinde dixit *Manifestum est ergo quod minimus error in principie rei est magnus in ultimo*, et non induxit hoc nisi ut omnino esset considerandum de principiis et 24rb habere magnam solicitudinem circa ea; | et ideo sermo de principiis est sermo de omnibus scilicet in potentia.

Deinde dixit *Infinitum autem est valde magnum* etc., idest quod ille qui ponit corpus esse minimum, non solum ponit diversum ab eo qui ponit infinitum, immo oppositum: potentia enim infiniti maxima omnium potentiarum corporis cum sit maioris quantitatis omnium corporum, potentie autem corporum sequuntur magnitudines eorum. Deinde dixit *Est enim principium rerum* etc., idest si enim est, necesse est ut sit principium rerum; et si ita sit, non est impossibile neque extra rationem ut in ipsum cadat diversitas, quoniam perscrutatio dignior est ut sit in rebus huiusmodi, et ut diversitas hominum in eo non sit mirabilis sed naturalis.

34. ⌜*Et cum quidam Antiquorum dixerunt corpus infinitum esse possibile,* [271b17] *volumus modo perscrutari de sermone suo et dicere in eo perfecte. Sed primo revertamur ad illud de quo loquebamur, quousque sit completum antequam incipiamus contradicere eis; deinde incipiemus contra* 24va *sermonem illorum et complebimus con|tra ipsum*⌝. *Dicamus ergo quod* 271b17 *omne corpus aut est simplex aut compositum, corpus ergo infinitum aut est simplex aut compositum. Et etiam si corpora simplicia sunt finita,* 271b20 *ergo compositum ex eis, et hoc manifestum est; corpus ergo compositum ex partibus finitis multitudine et magnitudine est etiam finitum multitudine et magnitudine, quoniam debet esse in multitudine et magnitudine, secundum multitudinem corporum ex quibus componitur et magnitudinem eorum.* 271b23

---

92 infinite] finite *Z* | ergo] igitur *BX* 93 principie] -piis *JW* 94 induxit] -ducit *WYZ* 96 est] et *Z* 97 est valde] *om. B* 98 esse minimum] *om. BX* 98-9 non—infinitum] *om. Y*: infinitum maius ponit quam ille qui ponit corpus (corpus *X*: -pora *B*) esse minimum *BX* 103 est[1]] *om. BJX* | sit[1]] sint *Z* 104 diversitas] + et *B* 106 naturalis] -raliter *B*

34. 1-5 Et—ipsum] *paraphrasis textus Aristotelis translationem Arabicam secuta est* 2 modo] *om. Z*: ergo *Y* | suo] *om. B* 3 loquebamur] -batur *B*: -bor *X* 5 illorum] eo- *W* | complebimus] -vimus *B* 6-7 corpus[2]—compositum] *om. homoiotel. BY* 6 ergo] autem *Z* 7 aut] + est *Z* 8 compositum[1]] oppo- *B* | ergo] enim *J*: est *Z*: + ex eis *X* 9 finitis] in- *WXY* | finitum] in- W_2XZ 10 debet esse] erit *BJ* 11 multitudinem] similitu- *W*

Cum dixit causam formandi perscrutationem de infinito, et cum quidam antiquorum credebant ipsum esse, immo magnificabant ipsum et faciebant ipsum esse principium sicut narravit de eis in Phisicis, dixit *Et cum quidam Antiquorum* etc., idest et cum de causis cogentibus nos ad perscrutandum de infinito est quod quidam Antiquorum dicunt ipsum esse, quapropter oportet nos perscrutari de sermone suo, idest de modo secundum quem dicunt infinitum esse. Antiqui enim dicentes ipsum esse differunt inter se: hii quidem substantiam consistentem dicunt ipsum esse, hii autem corpus; et quia perscrutatio de sermone Antiquorum in eo est alia a perscrutatione de infinito in se, et est utrum corpus infinitum est impossibile aut non, dixit *Sed primo revertamur* etc., idest sed antequam perscrutemur de sermonibus antiquis et de rationibus eorum in hoc, revertamur ad nostrum, quod est querere utrum corpus infinitum sit possibile aut non, deinde perscrutabimur de hoc quod | *24vb* dicunt Antiqui in hoc. Et rectum est secundum sermonem Antiquorum post terminari: contradictio enim contra Antiquos in hoc erit complementum sermonis sed non necessario; quapropter premittendum est necessarium, quamvis in hoc libro non videmus ipsum perscrutari de sermonibus antiquorum in hoc, sed in tertio Phisicorum, et determinatio metaphysice; et in illo loco Phisice est hoc ei principium, et maxime ponentibus ipsum esse substantiam per se.

Deinde dixit *Dicamus modo quod omne corpus* etc., id est dicamus modo in ista questione sermonem continuum, et non curemus modo de contradictione contra Antiquos; et hec propositio a qua incepit in consideratione est manifesta per se: visum est enim quod omnia corpora que sunt inter nos aut sunt composita ab uno corpore aut a pluribus, aut sunt unum scilicet non compositum ex corpore; quoniam quid hii duo modi existunt in hoc mundo manifestum est per se, divisio autem hec est manifesta per intellectum; et non potest aliquis dicere quod nullum corpus est simplex: plura enim corpora dissol-

13 cum[2]] *om. BX* 14 ipsum esse] *inv. B* 16 Et[1]] + etiam *WYZ* | de] *om. WXY* 17 quod] et *Z* 19 dicunt] -citur *Z* | infinitum esse] *inv. Z* 22 de] + secundo *B* 24 non] *om. BXY* | perscrutemur] -tamur *Z* 28 secundum] *om. WXY*: post *Z* 29 enim] autem *Z* 30 premittendum] pretermit- *BZ* 32 determinatio] decimo *B*: undecimo *J* 33 metaphysice] methaforice *Z*: methace *W*: methaphisice *BJ*: metha *Y*: fisice *X*: mathematice *v* 35 dixit] *om. BWXY* 36 in ista] mixta *WYZ* 39 pluribus] + corporibus *JZ* 41 quid] -dam *BYZ*: -dem *X* 42 aliquis] quis *W*

vuntur in alia, sed non procedit hec dissolutio in infinitum, quapropter necesse est ut sint corpora simplicia, et etiam iam declaravit prius corpora simplicia esse propter motus simplices, et quod sunt quinque: hec igitur propositio aut est nota per se aut iam declarata. Deinde dixit *Corpus ergo infinitum aut est simplex aut compositum*, et hoc etiam est manifestum per se, quoniam si omne corpus est aut simplex aut compositum, et posuit aliquis corpus simplex esse, necessario erit aut simplex aut compositum.

Deinde dixit *Si corpora simplicia* etc., idest et manifestum est etc.: *25ra* si omnia corpora simplicia sunt finita in ma|gnitudine et numero, quod necessarium est quod sint composita ex eis sint etiam finita: qui ergo declaraverit nobis quod omnia corpora simplicia sunt finita in magnitudine et numero, declaravit nobis quod omnia corpora sunt finita. Deinde inducit nos ad modum ex quo certificatur hec propositio dicens quod compositum a finito est finitum, et dixit *Corpus enim compositum ex corporibus finitis* etc., et hoc etiam est manifestum per se: compositum enim non intelligitur esse infinitum nisi aut quia unum simplicium ex quibus componitur aut plura uno est infinitum magnitudine, aut quia est compositum ex simplicibus infinitis numero, quamvis sint finita magnitudine. Et hoc intendebat cum dixit *Quoniam erit in multitudine et magnitudine* etc., idest quod non fuit necessarium ut compositum sit infinitum propter simplicia ex quibus componitur cum fuerint infinita, aut magnitudine aut numero aut utroque, nisi quia multitudo et magnitudo compositi sequuntur multitudinem et magnitudinem corporum simplicium; et hec propositio manifesta est in composito ex elementis in actu ut

---

44 hec] ista *Z*: huius *X* 45 est] *om. BJ* | iam] *om. BX* | declaravit] -vimus *BX* 46 prius] + quod *B* 48-50 et—compositum] *om. homoiotel. B* 49 etiam est] *inv. Z* | est manifestum] *inv. XY* | est aut] *inv. JZ* 50-1 et—compositum] *om. homoiotel. X* 50 posuit] -suerit *Y* | simplex[2]] finitum *JW*$_2$*Z*: infinitum *J*$_2$ | necessario] *om. W* 52 Si corpora] *om. YZ* | simplicia] celestia *B* | est] *om. W*: esse *JZ* 54-5 quod[1]—ergo] *om. WYZ* 54 necessarium est] *inv. X* | sint] *om. J* | etiam] *om. J* 55 declaraverit] -ravit *X*: -rat *WY* | simplicia] *om. WY* 56-7 in—finita] *om. homoiotel. WYZ* 56 declaravit] -rat *X* 57 inducit] -xit *JZ* 59 etiam] *om. BJZ* 60 se] + est *Z* 61 aut quia] aut quod *W*: *inv. BJ* 62 aut] *om. X* 62-3 aut—magnitudine] *om. homoiotel. Y*: aut magnitudine *W*: aut omnia multitudine *Z* 63-4 Et—magnitudine] *om. homoiotel. X* 64 in] *om. B* | multitudine … magnitudine] *inv. W* | et magnitudine] *om. Y* 65 compositum sit] *inv. W* | infinitum] + nisi *X* 67 compositi] -tione *Z* 68 sequuntur] *om. X*: -quitur *B* 69 ut] scilicet *B*

compositio dictionis ex literis. Et similiter videtur esse in composito secundum mixtionem: magnitudo enim admixti non est nisi propter magnitudinem rerum mixtarum; si ergo mixtum fuerit infinitum, ergo aut unum ex quibus componitur aut est infinitum magnitudine aut plura uno, aut omnia finita in magnitudine sed infinita numero.

25rb26 **35.** *Oportet nos ergo querere utrum sit possibile quod aliquod corpus* 271b23
simplex sit infinitum magnitudine, aut sit impossibile, et ponere perscru-
tationem nostram primo de primo corporum simplicium, deinde de aliis. 271b25
Et dicamus quod necessarium est ut totum corpus rotundum sit finitum
per hanc demonstrationem quoniam si corpus rotundum sit infinitum,
ergo linee exeuntes a medio erunt infinite etiam. Ergo et diversitas linea- 271b30
25va *rum infinitarum infinita est, et est intelligere per di|versitatem distan-*
tiam linearum abinvicem: ergo non inveniretur magnitudo ex extrinseco
inter lineas infinitas: hec ergo distantia non habet finem necessario, cor-
pus enim habens finem est finitum. Et etiam si distantia linearum sit
distantia qua alia maior non possit reperiri, ut diximus quod numerus 272a1
infinitus est ille qui caret termino in multitudine, similiter ergo infinita
distantia est carens termino in distantia. Et etiam si impossibile est ut
corpus infinitum abscindatur pertransitum et corpus infinitum fuerit
ens, necessario erit diversitas et distantia infinita; et cum ita sit, mani-
festum est quod impossibile est quod sit motus circularis. Visus autem tes- 272a5
tatur contrarium: videmus enim celum moveri motu circulari; et iam
declaravimus hoc ratione cum diximus quod motus circularis est quorun-
dam corporum simplicium. 272a7

Cum declaravit quod si corpus infinitum est, aut igitur est compositum aut simplex, incepit prius perscrutari hoc in simplicibus, et primo de nobiliori eorum quod corpus est quintum; et dixit quod *Totum corpus rotundum est finitum necessario* etc., idest quod corpus quod circulariter movetur, cuius esse est declaratum, necesse est ut sit

---

70 esse] *post* composito *Z* 73 est] *om. B*

35. 3 corporum] *om. Z*: -poris *X* 4 ut] *om. W*: quod *Z* 7 per] *om. BWXY* 8 abinvicem] adin- *WYZ* | inveniretur] -nitur *BJ* | ex] *om. WYZ* | extrinseco] -seca *XYZ* 9 distantia] diversitas *BXY* 11 diximus] -citur *BJ* 12 termino] *om. Y*: fine W_2Z | infinita] -ti *BX* 16 quod sit] *om. XY* | testatur] attes- *WY* 20 igitur] ergo *Z* 22 corpus est] *inv. BJX* 23 Totum corpus] *inv. Z*

finitum magnitudine. Deinde dixit *Per hanc demonstrationem* etc., et hoc manifestum est, quoniam si ponatur rotundum infinitum, contingeret ut linee exeuntes a centro sint infinite cum infinitum *25vb* | careat circumferentia et linee exeuntes a centro non terminantur nisi per circumferentiam.

Deinde dixit *Diversitas autem linearum* etc., idest quod si linee exeuntes a centro sunt infinite, necessarium est ut distantie que sunt inter illas sint infinite; et hoc necessarium est quoniam quando a centro exierunt due linee, necesse est ut quanto magis exierunt due linee, tanto magis magnificetur distantia inter eorum extrema. Si autem iste due linee sint infinite exeundo, necesse est ut distantia inter ea sit infinita; cum enim distantia terminatur, aut ponatur distantia finita inter extrema istarum duarum linearum, necesse est ut due linee terminentur. Deinde explanavit diversitatem, et dixit *Et intelligere est per diversitatem* etc., et cum dixit quod distantia que est inter lineas infinitas necesse est ut sit infinita, dixit *Non invenitur ergo magnitudo ex extrinseco inter lineas infinitas* etc., idest quod si distantia inter lineas infinitas est infinita, continget ut non inveniatur magnitudo finita ex extrinseco corporis rotundi inter lineas infinitas secundum quod sunt infinite, quoniam si inter eas esset magnitudo finita, oporteret ut quod est inter eas sit finitum; et si quod est inter eas esset finitum, essent linee finite.

Deinde reversus fuit ad declarandum quod ista distantia que est inter lineas infinitas est infinita, et dixit *Erit ergo hec distantia* etc., idest quod necessarium est quod hec distantia sit infinita; quapropter et magnitudo que est illic sit infinita: magnitudo enim habens finem est finita idest terminata; et innuit per hoc ad hoc quod dicitur de diffinitione esse infiniti, scilicet quod est maximum omnium quantitatum, quasi igitur dicit quod non est necesse ut ista distantia sit infi-

---

27 contingeret] -get *XY* | sint] sunt *Y* 27-8 sint—centro] *om. homoiotel. X* 28-9 et—circumferentiam] *om. homoiotel. B* 31 sunt[1]] sint *WX* | necessarium est] necessario *Y* 31-2 necessarium—infinite] *om. homoiotel. BX* 33 exierunt[1]] -rint *WXY* 34 eorum] ea- *BJX* 35 sint] sunt *Z* 37 extrema] *post* necesse *Z* | duarum linearum] *inv. JY* 38-9 et[1]—diversitatem] *om. homoiotel. X* 38 Et[2]] *om. YZ* 39 intelligere est] *inv. J* | per] *om. BWY* 41-2 etc.—infinitas] *om. homoiotel. BX* 42 est] sit *W*: esse *B* 43 ex] et *B* 44 esset] est *W* 46 esset] *om. Z*: sit *Y* 49 sit infinita] sit finita *W* : finita sit *YZ* 50 sit] est *JZ* 51 idest terminata] *om. B* | innuit] intue *B*: intendit *WZ* | ad hoc] *om. JZ* 52 esse] enim *X*: + enim *WZ*

26ra nita | nisi quia ista distantia sit maior omni distantia posita, et quod ita est necessario est infinitum. Quapropter dixit post *Et etiam si distantia linearum est distantia* etc., idest manifestum est quod, si distantia istarum linearum infinitarum est talis, quod nulla alia distantia inveniatur maior, quod est infinita; et si est infinita, quod nulla alia distantia inveniatur maior, sicut numerus infinitus est in quo nullus finis invenitur in multitudine, et econverso. Et non voluit ostendere conversum nisi ut ostenderet nobis diffinitionem converti cum diffinito, idest quoniam sicut numerus est infinitus qui caret fine et econverso, ita etiam distantia est infinita illa que caret fine et econverso. Et non induxit hoc exemplum nisi ad demonstrandum conversionem esse in diffinitione distantie infinite sicut in numero infinito; et ita format suum sillogismum: distantia que est inter lineas infinitas est tanta quod nulla alia invenitur maior, et quod est huiusmodi est infinitum, ergo distantia que est inter lineas infinitas est infinita.

Et cum declaravit quod necessarium est quod ultima distantia que est inter istas lineas est infinita, et quod propter hoc continget ut magnitudines que sunt inter eas sunt infinite, coniungit ad has aliam propositionem ex quibus contingit impossibile, dicendo *Etiam si impossibile est quod corpus infinitum abscindatur transitu* etc., idest si manifestum est quod impossibile est quod corpus infinitum abscindatur idest pertranseatur complete, et posuerimus corpus rotundum infinitum esse, ex cuius positione contingit ut diversitas inter lineas exeuntes a centro sint infinite, manifestum est ut sit impossibile quod sit motus circularis; contingeret enim, quando corpus rotundum moveretur ab aliquo puncto | quousque reverteretur ad idem, 26rb quod linee pertransirent distantias infinitas, quoniam unaqueque

53-4 infinita—sit] *om. homoiotel. B* 54 sit] *om. X*: est *J* 55 dixit post] *inv. Z* 56 est[2]] *om. YZ* 57 est talis] *inv. Z* 58 quod[1]] quia *JZ* 58-9 quod[1]—maior] *om. homoiotel. B* 58 et—infinita] *om. homoiotel. X* 60 voluit] + nobis *BX* 61 conversum nisi] *inv. WYZ* | ut] *dittogr. Y*: et ut *W* 63-4 ita—econverso] *om. homoiotel. B* 63 etiam] et *X* | distantia est] *inv. Y* | est infinita] *inv. X* 67 invenitur] -niatur *WY* | quod est] *inv. Z* 68 inter] + istas *BX* | infinitas] *om. BXY* 70-1 cum—et] *om. homoiotel. X* 71 istas] *om. WYZ* | quod] *om. YZ* 74 impossibile] poss- *WY* | transitu] per- *JZ* | si] *om. B* 76 posuerimus] -suimus *JX* 77 positione] pro- *Z*: com- *B* : ratione *W* 79 quando] quod *Z* 80 moveretur] -vetur *BX* | reverteretur] -tetur *W*: -tantur *X* earum revertetur ad suum locum, et etiam contingeret ut permutarentur, scilicet ut unaqueque earum moveatur quousque veniat ad locum sui comparis. Et quia distantia inter eas est infinita, continget necessario ut pertranseat magnitudinem infinitam in tempore finito, sicut demonstratur in sexto Phisicorum; cum ergo corpus rotundum infinitum fuerit positum, impossibile est ut moveatur motu circulari completo.

Deinde dixit *Sed visus testatur contrarium, videmus enim quod celum movetur motu circulari*, scilicet completo; et quia celum non est nisi per rationem, et fuit etiam declaratum per rationem esse huiusmodi motum, dixit *Et iam declaravimus hac ratione superius cum diximus* etc., idest et iam declaravimus corpus esse quod naturaliter movetur motu circulari. Inducamus ergo huiusmodi rationem secundum doctrinam aspectibilem, et sit corpus rotundum infinitum cuius centrum est H et cuius concavum circulus AB; et extrahemus lineam HB in infinitum, et HA; distantia ergo maxima est infinita inter eas. Cum ergo linea AH movebitur donec superponatur super lineam HB, necesse est ut pertranseat spatium infinitum in tempore finito, quanto magis cum movebitur donec revertatur ad suum locum, scilicet quot spatia infinita debet pertransire in tempore finito: pertransit enim spatia infinita in numero; et quodlibet illorum est infinitum in magnitudine in numero, quoniam linee exeuntes a centro sunt infinite. Sit ergo inter quaslibet duas lineas aliqua distantia: si ergo linee
26va sint infinite in numero, continget ut hec | spatia sint infinita in numero, et quia omnes due linee istarum linearum sunt infinite, continget ut distantia que est inter ea sit infinita quantitate, quod est valde impossibile et inopinabile.

---

82 contingeret] -get *XY* | permutarentur] -tentur *BX*: -taretur *Y* 83 ut] *om. BX*: quod *W* | veniat] per- *Z* 84 comparis] operis *BX* 86 rotundum] *om. WYZ* 87 positum] *om. Z* 89 Sed] si *Y*: sicut *Z* 91 et—rationem] *om. homoiotel. X* | etiam] *om. JYZ*: et *W* 91-2 rationem[2]—motum] huismodi motum esse rationem et *Z* 92-3 hac—declaravimus] *om. homoiotel. XY* 92 hac] hoc *JW* 93 idest] *om. WZ* 94 circulari] -lariter *Z* 95 et] ut *B* 96 concavum] + est *W* 97 in] *om. BY*: etiam *X* | est infinita] *post* eas *BX* 100 cum movebitur] commove- *B* 101 pertransit] -eat *W* 102-3 et—numero] *om. homoiotel. Y* 102 illorum] eo- *W* 104 Sit] et W_2*Z* | ergo[1]] *om. BZ* | quaslibet duas] *inv. X* : queslibet *B* 105-6 continget—numero] *om. homoiotel. B* 106 quia] quoniam *YZ* | due] *om. Y*: differente *Z* | linee] *om. Z* 107 ut] + omnis *J*W_2*Z* | ea] eas *Z*

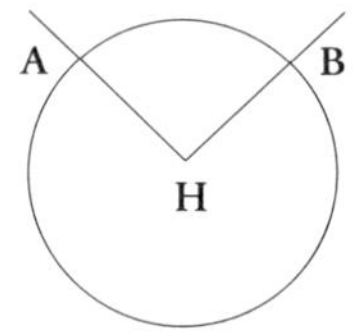

26va18 **36.** *Et etiam si ex tempore finito accipiatur tempus finitum, remanet* 272a7 *tempus finitum, et habet principium; et cum tempus rei habet principium ergo et motus habet etiam principium, et propter hoc magnitudo habet etiam principium, et etiam in quo est res; et secundum hoc erit in* 272a10 *omnibus aliis etiam. Sit ergo linea infinita ACH, sed extremum H solum sit infinitum, et sit alia linea TB infinita in duobus extremis, et sint due linee in circulo quasi corde, sicut in ista figura. Dicamus ergo quod linea ACH abscindens circulum ex C abscindit lineam BT donec moveatur* 272a15 *motu circulari: tempus autem abscisionis ACH iam fuit finitum. Motus enim celi et omnium que sunt in eo est motus circularis finitus, qua-* 27ra *|propter et tempus acceptum ex eo in quo est motus circularis abscindentis circulum, et lineam infinitam in utrisque extremis est finitum; et cum ita est, est ergo principium abscisionis linee ACH per lineam TB, quod est impossibile. Manifestum est igitur quod corpus infinitum impossibile est ut moveatur motu circulari; et cum ita est, manifestum est quod impossibile est ut mundus sit infinitus.* 272a20

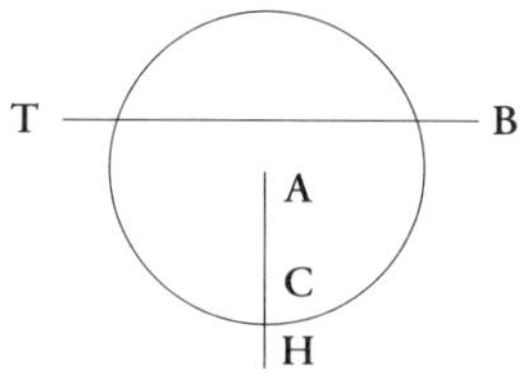

Hec est alia demonstratio preter predictam, et ipse proponit huic demonstrationi propositiones, quarum una est quod *Si accipiatur ex tempore finito tempus finitum, remanet tempus finitum et habet princi-*

36. 1 tempus finitum] *om. W* 3 hoc] + etiam *Z* 4 habet etiam] *inv. WXY* | quo] qua *B* 5 omnibus aliis] *inv. BJ* | infinita] + linea *W* | extremum] -traneum *B* 6 sint] sunt *WZ* 7 sicut] + est *JZ* | in[2]] *om. BXY* | ergo—linea] *om. BXY*: + ab *Z* 8 abscindit] -dat *W*: -det *J* 9 finitum] -tus *B* 10 finitus] in- *B* 11 circularis] *om. XYZ* 17 ipse] *om. B* 19 tempus[2]] *om. XY*

pium, idest quod hoc tempus remanens habet principium et finem; hec enim est dispositio cuiuslibet finiti; et hoc intendebat cum dixit *Et etiam si ex tempore finito accipiatur* etc. Deinde posuit aliam propositionem et dixit *Et si tempus rei habet principium* etc., idest mani-
27rb festum est ex Phisicis quod | si tempus in quo est motus sit finitum, habens principium et finem, et motus qui est in eo erit finitus, et habebit principium et finem, similiter etiam res mota debet esse finita, ut declaratum est in Phisicis. Et hoc intendebat cum dixit *Et magnitudo propter hoc habet principium*, et etiam locus et spatium, et hoc intendebat cum dixit *Et etiam in quo est res*, scilicet idest similiter contingit ut spatium in quo est motus finitus sit finitum, et habens principium et finem; et propter hoc explanavimus magnitudinem idest rem motam. Deinde dixit *Et secundum hoc est in omnibus aliis*, idest quod iste propositiones sunt in omnibus mobilibus, scilicet quod si moveantur in tempore finito, sunt finita in magnitudine finita et in motu.

Et cum posuit maiores propositiones istius sillogismi, dedit propositiones quibus coniunctis concludatur impossibile, et utebatur in hoc exemplo scilicet argumento exemplari loco sermonis universalis ut diceret brevius, quasi igitur diceret Et cum hec propositiones sint manifeste cum posuerimus corpus rotundum esse infinitum, possibile est ut in circulo cadente in eius concavo ponamus cordam et extrahamus extrema eius in infinitum, et imaginemur hanc cordam esse extra centrum: dividet ergo circulum in duo equalia. Cum ergo imaginati fuerimus lineam exeuntem a centro in infinitum in partem circuli in qua non est linea infinita mota ad partem corde scilicet per motum circuli, illa autem linea sit quiescens aut mota ad contrarium illius partis, necessario pertransibit lineam infinitam in utriusque extremitatibus in tempore finito. Tempus enim revolutionis finitum

20 remanens] *ante* hoc *B* | habet] -bens *B* 22 etiam] *om. Z* 23 et[1]] + etiam *B* | Et[2]] quod *YZ* 26 etiam] *om. B* | res mota] *dittogr. B* 27-9 Et[2]—etiam] *om. BY* 28 magnitudo] *post* hoc *Z* | spatium] + similiter debet esse finitum W_2Z 29 Et etiam] *om. WX* 33 mobilibus] motibus *Z* 34 quod] + et *Z* 35 motu] + finita in spatio W_2Z 37 concludatur] -ditur *BXYZ* 38 hoc exemplo] *inv. Z* | argumento] a ratione *XZ* | loco] *om. B* 39 diceret[1]] deberet *YZ* | igitur] *om. WYZ* | diceret[2]] -cat *B* 43 esse] *om. BX* | extra] *om. Y*: per *W* | ergo[1]] igitur *BX* 45 infinita mota] -am -am *B* 48 enim] + omnium W_2Z

est, et tempus in quo pertransivit linea infinita (in altera duarum extremitatum suarum partem circuli, cui parti | 27va non occurrit linea infinita in duobus extremis) est finitum necessario; ergo illud quod remanet est finitum necessario, et est tempus in quo pertransivit linea finita in altera extremitatum lineam infinitam in duabus extremitatibus, ergo linea pertransiens in illo tempore est finita; omne autem finitum habet principium et finem, linea ergo posita infinita est finita; sed posita erat infinita, deductum ergo est ad impossibile, et hoc impossibile accidit ex predicto falso posito, ergo falsum est ponere corpus rotundum esse infinitum.

Sed Aristoteles induxit demonstrationem in exemplo et dixit *Sit ergo linea ACH* etc., idest verbi gratia si corpus rotundum sit infinitum, possibile est ponere in suo concavo circulum in quo sit linea infinita exiens a centro: sit ergo centrum huiusmodi circuli A, et sit linea exiens ab A ad C ad H, et sit infinita in parte H. Deinde dixit *Et sit alia linea TB infinita in duabus extremis*, idest et ponamus aliam lineam infinitam in duabus extremis extra centrum et in parte opposita parti in qua est linea infinita exiens a centro: sit ergo linea TB. Deinde dixit *Dicamus ergo quod linea ACH* etc., idest dicamus ergo quod linea exiens a centro in infinitum scilicet ACH, cum movebitur quousque compleat revolutionem, necessarium est ut pertranseat lineam infinitam in duabus extremis scilicet T B, cum extremum eius fuerit quiescens scilicet centrum; et fuerimus imaginati ipsam motam super circumferentiam circuli, et illam lineam fuerimus imaginati quiescentem scilicet infinitam in duabus extremis. Deinde dixit *Et iam fuit tempus linee ACH infinitum*, idest et tempus in quo movet linea ACH quousque perficiat suam revolutionem, necessario est finitum. Deinde dedit causam in esse hoc secundum tempus finitum dicendo *Tempus enim motus celi*, idest hoc tempus, scilicet tempus in quo revolvitur | 27vb linea ACH que est infinita in duabus extremis non contingit ut sit finitum nisi quia posuimus ipsam motam circulariter,

49 pertransivit] -sibit *J* : -sit *B*   51 duobus] dua- *BX*   56 erat] erit *Y* : erant *Z* | deductum] et educi *W*   57 falsum] si non *B* : finem non *X*   58 esse infinitum] *inv. BX*   59 induxit] -cit *BX*   63 ad[1]] in *W* : per *Z*   64-5 idest—extremis] *om. homoiotel. JX*   64 et[2]] *om. B*   67 ergo[1]] *om. BJ*   71 fuerit] sit *Z*   71-2 ipsam—imaginati] *om. homoiotel. X*   72 super] circa *Z*   73 scilicet] *om. Z* | duabus] duo- *BJX*   74 movet] -vetur *WY*   76 secundum] *om. BJXY*   77 idest] + et *BX*   78 duabus] duo- *BXZ*   79 circulariter] -culi *Z*

lineam scilicet per motum celi, sed quia celum et omne corpus rotundum contingit ut tempus sue revolutionis perfecte sit finitum, similiter contingit in omnibus mobilibus in ipso, ut ista linea quam posuimus mobilem in eo.

Deinde dixit *Quapropter tempus acceptum ex eo* etc., idest quia tempus revolutionis est finitum, et tempus in quo linea infinita in altera duarum extremitatum pertransit lineam infinitam in duabus extremis est pars eius, necesse est ut hoc tempus sit finitum: pars enim finiti est finita necessario. Et hec propositio secundum quod apparet est alia a propositione quam primo posuit: illa enim propositio fuit quod cum a tempore finito abscinditur tempus finitum remanet finitum, hec autem est quod pars finiti est finita; et est propinqua illi, et ambe sunt vere et habent eandem potentiam. Deinde dixit *Et cum ita sit, principium igitur pertransitus linee ACH per lineam BT* etc., idest et si hoc tempus, in quo pertransivit linea ACH per lineam TB, sit finitum, ergo utraque linea est finita. Et hoc intendebat cum dixit *Invenitur ergo principium pertransitus linee ACH per lineam BT*, quoniam cum tempus fuerit finitum, necesse est ut pertransiens et pertransitum sint finita, ergo necesse est ut inveniatur principium pertransitus unius earum ab altera; infinitum enim non habet principium, scilicet quoniam si posuimus ipsas esse infinitas, non inveniretur punctus qui sit principium occursus unius earum alteri. Si autem posuerimus eas finitas, necesse est ut iste *28ra* punctus inveniatur, et est punctus qui est in extremo linee; | linee enim que sunt huiusmodi non occurrunt sibi primo nisi super puncta extremorum; et visum est, ex quo impossibile est invenire hunc punctum in duabus lineis infinitis, quod est impossibile pertransire unam illarum per alteram neque in tempore finito neque in tempore infinito; et hec est alia demonstratio a predicta, et Aristoteles utetur ea post.

80 quia] *om. BWXY* 82 contingit] -venit *YZ* | in ipso] *om. BX* 86 duabus] duo- *BXYZ* 89 primo posuit] *inv. Z*: proposuit *BY* 93 sit] + invenitur *BX* | igitur] sibi *Z*: ergo *B* 94 tempus] -pore *Z* | linea] + infinita *B* 96 ergo] igitur *JX* | principium pertransitus] *inv. B* 97 quoniam] + ut *B* 99 ab] ad *Z* 100 ipsas esse] *inv. WY* 102 earum] eo- *Z* | eas] ipsas *BJX* 103 linee] *haplogr. BX* 104 sibi] nisi *BX* 105 impossibile est] *inv. BX* 106 est impossibile] *inv. JW* | pertransire] -sit *B*

37. *Et etiam corpus infinitum impossibile est moveri, et hoc manifestum* 272a21 *est ex hoc quod dicemus. Sint ergo due linee quarum altera sit A et altera B, et sint finite; et eat linea A finita ad lineam B finitam, sicut vides in figura; et eat linea B finita ad lineam A finitam sicut vides. Quoniam, quando fuerit huiusmodi, erit separatio utriusque earum a sua compari* 272a25 *insimul, quoniam sicut alter duorum motuum occurrit alii, sic alter primo. Si ergo due linee moventur insimul una contra alteram, separabi-* 28rb *tur una earum ab alte|ra in illa parte separatione veloci; et si altera earum moveatur et altera non, et mobilis eat ad quiescentem, erit separatio earum tarda, quamvis linea mobilis iens sit unius dispositionis in velocitate.* 272a28

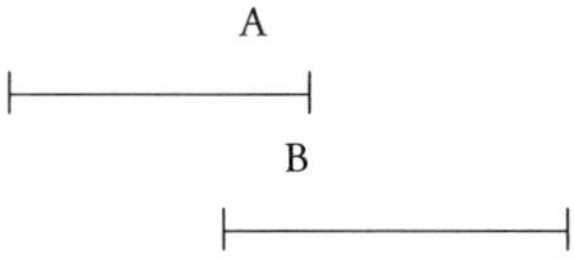

Apparet ex hoc sermone quod est universalior quam quod intendebat declarare, et est quod corpus rotundum impossibile est ut sit infinitum quoniam motus eius esset infinitus in tempore infinito, et motus corporis celestis esset finitus. Et ideo videtur nobis quod ipse intendebat inducere demonstrationem generalem, scilicet quod omne infinitum non potest moveri omnino: *Sint ergo due linee* etc., idest et sint ergo due linee mote per duas dimensiones equidistantes in duabus partibus diversis. Deinde dixit *Et eat linea A finita ad lineam B finitam sicut vides*, idest et moveatur linea A finita ad partem B quousque equidistet ei: similiter imaginemur etiam lineam B finitam moveri ad partem oppositam ad quam movetur linea A. Deinde dixit *Quoniam cum fuerint huiusmodi* etc., idest quod utraque illarum exibit ad equidistantiam sue comparis in eodem instanti. Deinde dedit causam in hoc et dixit *Quoniam si occurret alteri* etc., et separatio

37. 2 dicemus] -ximus *Z* | Sint] sunt *XZ* | altera[2]] + sit *BZ* 3 et[1]—A] *om. Z* | sint] sunt *W* 3-4 in—vides] *om. homoiotel. X* 4 figura] -ram *Y* | eat] erat *B* 7 due linee] *inv. B* 9 eat] erat *XY* 10 linea] in ea *B* | mobilis iens] *inv. Z* 13 impossibile] poss- *BXY* 14 esset] *om. BX* | in] ex *WY* 15 esset] est *XY* 17 Sint ergo] *inv. BX* 20 sicut] *om. Z* | A] *om. XY* 21 etiam lineam] *inv. X*: lineam et *B* 23 quod] de *B* 24 comparis] operis *B* | instanti] *om. WYZ* 25 si] sic *Y*: sicut *X* | occurret] -ritur *WY*: -ratur *X*

eorum non sit insimul nisi quia sicut altera earum separatur ab altera, sic ista a prima.

Et cum dixit quod separatio utriusque illarum ab altera fuit insimul, dixit *Quoniam sicut alter duorum motuum occurritur* etc., idest et hoc fuit quia sicut alter duorum motuum venit ad suum comparem et secat illum, sic alterum venit ad istum et secat eum. Et causa illius
28va est | quod illud quod secat aliquid secatur ab eo, et quod mensurat aliquid mensuratur ab eo: hoc enim est ex revolutione, sive alterum sit motum sive alterum sit quiescens. Quapropter dictum fuit post hoc quod *Si due linee moventur insimul una contra alteram* etc., idest quod separatio utriusque earum a sua compari, quando insimul ambe moventur, est velocior quam quando altera est mota, altera vero quiescens cum fuerit motus mote illarum cum quiescente equalis motui eius cum altera mota. Quapropter dixit *Si autem altera earum moveretur, altera autem non*, idest quod pertransitus non erit tardior nisi quando mota vadens ad partem quiescentem fuerit in eadem dispositione velocitatis quam habebat cum movebatur ad partem moti, et cum proponit hanc propositionem.

28vb19 **38.** *Manifestum est igitur quod impossibile est ut motus finitus eat et pertranseat motum infinitum in tempore finito; et cum impossibile est,* 272a28
ergo pertransit eam in tempore infinito; et iam declaravimus hoc in libro 272a30
quem facimus de Motibus, et determinavimus ipsum bene. Modo autem dicamus quod sive motus finitus iverit ad motum infinitum sive motus infinitus ad motum finitum, non est ibi differentia, quoniam quando alter duorum motuum separabitur ab altero, et tunc alter ab isto, sive sit 272b1
quiescens sive non; sed tamen separatio earum est velocior si ambe

26 eorum] illo- *JZ* | sit] fit *BJX* 27 sic ista] sicut illa *Z* 29 sicut] *BXY* | occurritur] -rit *JXZ* : -runt *Y* 30 motuum] -torum *JXYZ* 31 Et[3]] *om. B* 33 eo] illo *BJX* 34 sit[1]] *om. B* | fuit] est *W* 35 moventur] -verentur *XY* 36 utriusque] *om. W* | earum] illa- *Z* | insimul ambe] insimul ambo *W*: ambo insimul *Z* 37 mota] + altera *Z* 38 fuerit] fuit *Z* | motus] + alterius *J*W_2*Z* 39 altera[2]] -ri *Z* 40 moveretur] -vetur *Z* : -veatur *J* 41 nisi] ut *B* : via *Y* 43 proponit] prepo- *Z*

38. 1 igitur] ergo *WZ* 2 cum] *om. Z* : si *J* | impossibile] poss- *BWZ* 3 ergo] + ut *WZ* | eam] eum *JZ* | iam—hoc] hoc declaravimus iam *B* 5-6 sive[2]—finitum] *om. homoiotel. WY* 5 motus[2]] *om. BJX* 6 motum] *om. BJX* | finitum] in- W_2*Z* | ibi] idem *Z* | differentia] circulariter *Y* : dicere *Z* 7 sit] *om. JZ* 8 earum] *om. WXY* | est] eius *B* | ambe] -bo *WZ* *moveantur. Et secundum hoc forte non est impossibile ut separatio moti* 29ra *super quiescentem sit velocior | separatione moti super motum motu opposito diverso; et hoc est quando duo mota motu opposito eunt tarde,* 272b5 *et motus moti super quiescentem fuerit velocior. Revertamur modo et dicamus quod sermo noster non est impossibilis, cum diximus etiam quod motum separabitur a quiescente velocius a separatione eius que est cum movetur ad motum, quoniam possibile est ut linea A moveatur ad lineam B, et ipsa moveatur etiam motu tardo tardior quam motu eius ad lineam quiescentem. Revertamur etiam et dicamus quod si illud tempus* 272b10 *est infinitum in quo pertransit motus finitus motum infinitum et separabitur ab eo, et in quo ivit motus infinitus et pertransivit motum finitum, tunc partes eius sunt infinite necessario. Manifestum est igitur ut corpus infinitum impossibile est ut totum moveatur aut pars eius, quoniam si totum moveatur aut pars eius, etiam erit hoc necessario in tempore finito, quod est impossibile, quoniam totum celum revolutum est et ivit in tempore finito, et ivit pertransitum eius in omni quod est intra* 272b15 *ipsum, scilicet linea finita que est super quam est A. Et certe manifestum est modo ut corpus rotundum fit infinitum.* 272b17

29rb

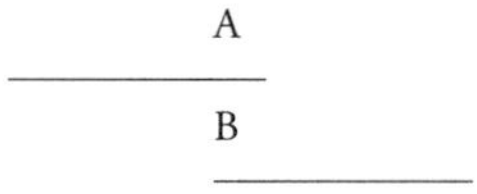

Cum posuit quod cum linea secaverit lineam secundum equidistantiam per motum unius, et quietem alterius (aut motum eius ad oppositum motus alterius linee), et fuerit utraque finita, necessario secantur adinvicem in tempore finito, et separatio illarum abinvicem ab equidistantia aut contactu est in tempore finito, dixit *Manifestum*

10 quiescentem—super] *om. homoiotel. B* 11 eunt] exeunt *Z* : erunt *BJ* 12 fuerit] fuit *Z* : fiunt *Y* | modo] ergo *XZ* 13 diximus] -xerimus *YZ* | etiam] *om. Z* 14 motum] + etiam non *Z* 14-5 velocius—movetur] *om. BJWXY sed suppl.* W_2 15 A] *om. BX* | moveatur] -vetur *Z* 16 moveatur etiam] *inv. WY* : moveatur *Z* | tardo] tarde qui est *Z* | motu] -tus *Z* 17 etiam] ergo *BZ* 20 eius] *om. B* | ut] quod *BJX* 21 ut—pars] ut moveatur neque totum pars *BJ* 22 etiam erit] *inv. BX* 23 finito] in- *JWY* 24 pertransitum] -tus *BJ* 25 scilicet] + quod *W* | est^1] *om. JXY* | super] sunt *B* 26 rotundum] + non *Z* 27 cum^2] *om. BJ* | secaverit] -vit *Z* 28 quietem] -escentem *WY* | motum^2] -tus *JZ* 31 ab] ad *JYZ* | equidistantia] -tiam *JWY* | dixit] + igitur *WZ* : + ergo *Y*

est igitur quod impossibile est etc., idest cum ista proposito sit manifesta, manifestum est quod cum posuimus duas magnitudines secundum figuram predictam, quarum altera sit finita, altera autem infinita, impossibile est ut pertranseat magnitudo finita per suum motum magnitudinem infinitam in tempore finito. Et hoc manifestum est ex conversione propositionis quam posuit, quoniam conversum eius est quod magnitudo que pertransit aliam magnitudinem per motum eius in tempore finito est finita; et iam posuimus hanc magnitudinem infinitam esse scilicet alteram duarum magnitudinum; ergo contingit impossibile etc.

Et cum declaravit quod due magnitudines huiusmodi, scilicet quarum altera est infinita et altera finita, non possunt per|transire se in (*29va*) tempore finito, dixit *Et si hoc igitur est impossibile* etc., idest et cum illud quod pertransit motu aliquam magnitudinem aut movetur super aliquam magnitudinem, necesse est ut pertranseat ipsam aut in tempore finito aut in tempore infinito; sed iam declaratum est quod impossibile est esse motum infinitum super spatium infinitum, aut in directo linee infinite aut motum infinitum super spatium finitum, aut in directo linee finite in tempore finito; remanet igitur ut pertranseat ipsum in tempore infinito. Deinde dixit *Et iam declaravimus hoc* etc., idest iam declaratum est in sexto Phisicorum quod motus infinitus non est nisi in tempore infinito; et hic liber vocatur de Motu, aut quia finis eius est perscrutari de motu, aut quia plura que continentur in eo sunt de motu, aut quia finis eius sunt ista duo.

Deinde dixit Volumus declarare quod impossibile est in talibus istis duobus motis quos posuimus pertransire se adinvicem in tempore infinito nedum in tempore finito; et cum dixit *Quod sive motus finitus est iens* etc., idest manifestum est per se quod nulla differentia est in hac positione, scilicet quod alterum duorum motorum sit finitum, alterum autem infinitum, et etiam alterum eorum motum

---

32 igitur] *om. WYZ* | idest] *om. WY* 33 posuimus] -suerimus *BX* 36 magnitudinem infinitam] –ne –ta *BX* 37 conversione] + per contrapositionem *Z* | quoniam] propter *B* 39 motum] -tus *B* 41-2 ergo—cum] *om. W* 41 contingit] *om. Z* | impossibile] + est *Z* | etc.] *om. BX* 44 impossibile] poss- *BX* 48 esse] *om. B* | infinitum] fin- *JZ* | super] aut *B* 49-50 aut—finite] *om. homoiotel. W* 49 spatium finitum] *inv. Z*: spatium *B* 51 in] *om. W* 52 idest] ut *W*: et *Y* | iam] *om. Z*: nam *X* | sexto] quinto *Y*: *post* Phisicorum *B* 54 est] *om. WY* | quia[2]] *om. B* 55 sunt[2]] est *W* 57 motis] motibus *Z*: modis *B* 59 est[1]] sit *W*

ad alterum et alterum quiescens, aut quod sit motum finitum illud quod movetur ad magnitudinem infinitam, et quod sit magnitudo infinita illud quod movetur ad finitum. Et pronuntiavit motus loco rei mote quoniam sequitur rem motam in privatione finiti in eo quod dixit, sive sit motus finitus idest sit motus qui est in re mota finita; similiter intellexit cum dixit *Sive sit motus infinitus*, idest qui est in re mota infinita.

Et cum narravit quod nulla differentia est inter has propositiones, dedit causam et dixit *Quoniam quando alter duorum* | 29vb *motuum separabitur ab altero* etc., idest cum nulla differentia sit inter has propositiones in eo quod intendimus declarare quod, cum altera illarum duarum magnitudinum separabitur a suo compari, continget necessario ut altera separetur ab illa, sive sit finitum illud quod movetur sive infinitum. Deinde dixit *Sive moveatur sive non*, idest et nulla differentia est in hac propositione, vel si ponatur altera istarum quiescens et altera mota aut ambe mote.

Deinde cum narravit quod similes sunt in eo quod intellexit declarare, incepit dicere differentiam inter istas, ut appareat ex hoc quod in hoc non sit aliqua vis, et dixit *Sed tamen separatio earum est velocior quando ambe moventur*, idest quod differentia, que est quod si unum duorum fuerit motum et alterum quiescens aut ambo mota, nullam facit vim in eo quod intendimus declarare, quia cum altera istarum magnitudinum separabitur a suo compari, suum compar separabitur ab eo. Deinde dixit *Et secundum hoc forte non est impossibile* etc., idest quod cum ambe magnitudines moveantur una ad oppositionem sui comparis, et hoc erit cum motus rei mote duarum magnitudinum fuerit velocior motibus earum adiunctis insimul, scilicet quando utraque illarum movetur ad oppositum sui comparis. Et

---

62 alterum[1]] + motum aut *VWZ (sed* aut *del.* W_2*)* | et] ad *Z*   63 et] aut *JZ*   64 movetur] *om. B*   65 finiti] -ta *Z*   66 sit motus[1]] *inv. Z* | est] *om. W*   67-8 similiter—infinita] *om. homoiotel. BX* | qui est] *om. Y*   69 est] *om. WY*   71 has propositiones] *inv. WY*   72 quod[2]] quia *WY* | illarum] ista- *XY*   73 separabitur] -ratur *JWY* | necessario] -arium *BWY*   74 altera] -erum *JWY*   75 et] *om. BJZ*   76 est] *post* hac *W* | vel] *om. X*: et *B*   77 aut] ut *YZ*   79 dicere] *om. WY* | differentiam] causam *Z* | istas] eas *BX*   81 quod[1]] + ista *XY* | est] + velocior *B*   82 aut] ut *YZ*   83 declarare] + et *JXY* | quia] quod *JY*: est quod *X*   83-4 quia—magnitudinum] *om. B*   83 altera] -ram *X*   84 istarum] *om. X*: + duarum *WY* | magnitudinum] *om. X*   85 eo] ea *JW*   86 quod] quia *Z*   87 oppositionem] compo- *YZ*: contrario positum *W* | cum] quod *B* | motus] motu *WXYZ*   89 oppositum] -tionem *X*: operationem *Y*

hoc intendebat cum dixit *Et hoc erit quando due res mote motu opposito erunt tarde* etc., scilicet quando due res mote secundum oppositionem moventur tarde, et res mota que movetur super quiescentem movetur velociter, scilicet velocius quam motus coniunctus duarum rerum motarum oppositum.

Et cum dixit hoc universaliter, incepit uti in hoc exemplo, dicendo *Revertamur modo et dicamus quod sermo noster impossibilis non est* etc., *30ra* idest dicamus quod iste sermo non est impossibilis in duabus | istis magnitudinibus quas ponimus unam finitam et alteram infinitam, quando confessi fuerimus quod altera illarum pertranseat alteram, scilicet quod accidat illi quod movetur ex eis ut separetur a quiescente super quod movetur velocius, quam separetur illud quod movetur ex eis ab altero moto scilicet quod movetur in oppositione sui comparis. Et cum dixit quod hoc non est impossibile, dixit *Possibile est enim ut moveatur linea A ad lineam B* etc., idest et hoc est possibile, quoniam possibile est imaginari rem motam scilicet A, que movetur ad aliam rem motam scilicet B, et illa res mota movetur ad oppositum prime et secat utraque illarum suum comparem in longiori tempore quam in quo secat res mota rem quiescentem; et cum hoc est possibile in duobus motis finitis, possibile est etiam quando alterum fuerit finitum et alterum infinitum, cum concesserimus quod altera illarum pertransit alteram.

Et cum declaravit hoc, revertebatur ad illud in quo stabat, idest quod si res mota infinita secaverit in pertransita quiescentem infinitam, aut res mota infinita quiescentem finitam, omnino secat ipsum in tempore infinito, et dixit *Revertamur etiam et dicamus quod si hoc tempus sit infinitum* etc., idest et cum posuerimus quod tempus in

---

90 res mote] *inv.* *B* 91 oppositionem] -situm *JZ* 92 moventur] -vetur *WX* 94 oppositum] -tionem *BX* 95 in] *om.* *JZ* 96 modo] *om.* *X* : ergo *BYZ* | noster] + est *B* | non est] *om.* *B* : *ante* impossibilis *X* 97 duabus istis] *inv.* *BX* 98 ponimus] -suimus *XYZ* 99 quando] quoniam *WY* 100 accidat] -dit *YZ* | quiescente] descendente *B* 103 quod] *Z* | est] + etiam *JW* 104 est enim] *inv.* *XZ* | hoc] + non *JZ* | possibile] im- *JWYZ* 105 quoniam] + non *JZ* | possibile] im- *JWYZ* 105-6 A—scilicet] *om. homoiotel.* *B* 105-7 A—utraque] *om.* *W* 107 oppositum] -tionem *BXY* | illarum] ha- *Z* : ea- *Y* 108 et] + hoc *B* 109 hoc] + non JW_2Z | possibile[1]] im- *Y* 111 illarum] ea- *Z* 112 declaravit hoc] *inv.* *Z* | revertebatur] *ante* hoc *B* : *post* illud *Z* | illud] aliud *B* 113 infinita] finita *XY* | pertransita] -tum *BJ* : -tu J_2 114 ipsum] ipsam *JX* 115 infinito] finito *XY* 116 sit] fuerit *B*

quo secat res mota finita magnitudinem non finitam sit infinitum, aut tempus in quo secat res mota infinita magnitudinem finitam aut spatium finitum, continget universaliter ut res mota infinita secet magnitudinem finitam in tempore infinito: secat enim partem ante totum. Et similiter contingit in re mota finita que ponitur secans, scilicet quod secat rem motam infinitam in tempore infinito, ideo quod 30rb pars eius necessario secat rem motam infinitam in tempore in|finito; et cum pars secatur in tempore infinito, et quod secatur in tempore infinito est infinitum, necesse est ut partes motus per quas secat suum totum sint infinite. Et hoc intendebat cum dixit quod partes eius necessario erunt infinite, scilicet partes motus universaliter. Et potest intelligi quod cum posuerimus partem vel corpus infinitum secantem vel consequentem rotundum, necessario erit finitum ex concavo et infinitum ex convexo, et cum ita sit, contingit ut partes illius corporis sint infinite; cum enim portio rotundi sit infinita ex convexo et finita ex concavo, necessarium est ut omnis pars eius sit infinita ex convexo. Si ergo posuerimus aliquam magnitudinem ex illa secantem magnitudinem infinitam, continget ut secet totum eius in tempore infinito, et similiter pars partem illius magnitudinis; et cum ita sit, erit totus motus infinitus et similiter quelibet pars eius; et quasi ista explanatio sit convenientior ei quod post dicet, et prima convenientior verbis suis. Quod autem dicit post est hoc: *Manifestum est igitur quod corpus infinitum impossibile est ut moveatur, neque totum neque pars eius, quoniam si totum moveretur aut pars eius, erit necessario in tempore infinito*, idest quod corpus rotundum, si ponatur infinitum ex parte convexi, continget ut totum et partes sint infinite, ut declaratum est in prima demonstratione cum imaginate fuerint partes distincte exeuntes a centro in infinitum. Et potest

---

117 finita] in- *XY* | non] *om. WX* 117-8 non—magnitudinem] *om. homoiotel. BYZ* 117 finitam] in- *WX* | sit] *dittogr. X* 119 continget] -geret *B* 121-2 que—infinito] *om. homoiotel. Y* | scilicet quod] *inv. Z*: scilicet *BWX* 122 secat] -catur *B* | rem—infinitam] re mota infinita *BX* | in—infinito] *om. BWXZ sed suppl.* W_2 123 eius] ipsius *WXY* 126 sint] sicut *Z*: sive *X* 127 motus] eius *YZ* | universaliter] -lis *XY* | Et] *om. Z* 128 quod] *om. JX* 129 secantem] societatem *B* 129-30 secantem—infinitum] *om. homoiotel. Y* 129 consequentem] consecantam *J*: secantem sectam *X*: seccam *B* | rotundum] -dam *X* | finitum] in- *X* 134 secantem] sequentem *Y* | magnitudinem] *om. Z* 135 similiter pars] *inv. Z*: + huis JW_2Z 136 totus] + mundus *JZ* 137 quasi] quod *Z*: quia *BX* 138 est] *om. BJX* 139 impossibile est] *inv. Z*: impossibile *Y* 140 erit] *om. W* 141 idest] scilicet *Z*

intelligi ex hoc sermone secundum primam explanationem quod quando posuerimus quod corpus rectum sive rotundum secet corpus infinitum in tempore infinito, aut secat corpus infinitum corpus finitum, continget ut secet partem eius in tempore infinito; propor- 30va tio enim infiniti ad totam magnitudinem quam secat, et | ad minimam partem eius, est eadem proportio; et cum ita sit, impossibile est ut infinitum secet aliquid in tempore finito, neque partem neque totum, quoniam si ponatur quod secet ea, erit in tempore infinito, scilicet totum et partem: sed celum secat omnia super que revolvitur in tempore finito, et similiter secat partem eius super quam revolvitur; ergo est infinitum.

Et secundum hanc explanationem est intentio sermonis eius, qui est *Manifestum est igitur quod corpus infinitum impossibile est ut moveatur totum aut pars eius*, idest quod corpus infinitum impossibile est ut moveatur super aliquam magnitudinem, neque super totam neque super partem; et secundum hoc erit intentio sermonis eius *Quoniam celum totum revolvitur et ivit circulariter* etc., idest cum celum sit infinitum, continget ut secet magnitudinem positam super terram, verbi gratia A in tempore infinito; similiter contingit ut secet totam terram cuius hec magnitudo est pars scilicet tempore infinito. Et hoc est contrarium sensui: videmus enim totum celum secare lineam positam super faciem terre in tempore finito, et similiter totam terram. Et secundum explanationem postremam erit intentio quod nos invenimus partem celi secare partem terre in tempore finito, et totum terram totam. Et si esset celum infinitum, contingeret ut pars eius secaret partem terre in tempore infinito: pars enim eius est infinita ex convexo; et similiter contingeret ut secaret totum eius totam terram in tempore infinito; et totum hoc est contrarium sensui. Et quia hic sermo continet in se duos modos demonstrationis scilicet universalem et particularem, secundum quod declaratur ex eo

---

145 quod] quoniam *Y*: et *B* 146 posuerimus] -suimus *JZ* | rectum … rotundum] *inv. B* | secet] *om. B*: secat *X* 147 secat] -cet *BJX* 150 impossibile est] *inv. B* 152 ea] eam *Z* 155 ergo] + non *JZ* | infinitum] finitum *WY* 156 intentio] *post* eius *W* | sermonis eius] *inv. Z* 163 similiter] -pliciter *B* | contingit] -get *Z* 164 scilicet] + in *BZ* | infinito] *om. B* 166 in—finito] *om. WYZ* | et] *om. WZ* 167 secundum] *om. W* 168 nos] non *B* 169 terram] *om. B*: *post* totam *J* | contingeret] -get *Z* 171 est] + si *B* 172 totam terram] *inv. Z* 174 eo] + scilicet *BJ*

30vb quod impossibile est corpus infinitum simplici|ter moveri neque infinitum, rotundum enim impossibile est ut per motum eius secet aliquid, quoniam cum res mota non possit moveri nisi secet ante totum: infinitum autem neque secet totum neque partem; manifestum est igitur quod est immobile, idest et cum ita sit, dixit *Et certum manifestum est modo quod corpus infinitum est nullo modo*, idest quod hec intentio est manifesta ex hoc sermone valde et bene.

39. *Et etiam linea finita non potest esse infinita: si autem possibile sit,* 272b17 *ergo in altera duarum extremitatum scilicet extremitate procedente secundum longitudinem infinitam; et similiter superficies etiam impossibile est ut sit infinita in omnibus partibus nisi in dimensione infinita;* 31ra *superficies autem est prior figura, tunc enim | impossibile est ut sit quadratum infinitum in aliqua dimensione, neque circulus neque orbis* 272b20 *neque circulus sit infinitus, sicut est impossibile ut sit figura pedis infinita. Si ergo neque orbis neque circulus sit infinitus neque motus circularis omnino infinitus.* 272b22

Et hec est demonstratio quarta preter tres predictas, et vult dicere quod linea non potest imaginari infinita propter suas duas partes ex quibus constituitur, scilicet quia est longitudo sine latitudine; sed non imaginatur esse infinita nisi propter unam tantum scilicet longitudinem que est in ea quasi materia et genus. Illud autem quod est in ea quasi differentia, scilicet alia pars descriptionis, facit ipsam esse finitam scilicet esse longitudinem sine latitudine; et similiter superficies non est infinita nisi ex parte materie que est longitudo et latitudo: ex parte ergo forme, que est sine profunditate, est finita. Et similiter corpus non potest imaginari esse infinitum nisi propter

176 enim] cum *BJX* | motum] -tus *B* 177 possit] posset *WYZ* 178 autem] aut *JY* | secet] -cat *XY* | partem] -tes *Z* 179 igitur] *om. XY* | certum] -te *BX* 180 est[1]] *om. WY* | modo] *om. JY* 181 est manifesta] *inv. BX* | valde] *om. YZ*

39. 1 possibile sit] *inv. BX* 3 superficies] -ciem *B* 4 dimensione] dimissio- BW_2 | infinita[2]] fin- *Z* 5 est prior] *inv. XZ* 6 aliqua] *om. Z*: qua *W* | dimensione] dimissio- *B* 7 neque—infinitus] *om. BJX* | circulus sit] *inv. W* 8 orbis] *om. J*: circulus *Z* | neque[2]—infinitus] *om. homoiarch. W* 10 Et[1]] *om. Z* | est demonstratio] *inv. B* | predictas] + causas *B* 11 suas duas] *inv. B*: duas *X* 13 imaginatur] -netur *Z* 18 ergo] autem *BJX*

magnitudinem que est in eo quasi genus: propter autem illud quod est in eo quasi forma, et est numerus dimensionum, scilicet quod descriptio corporis quia est magnitudo divisibilis in tres dimensiones, non potest imaginari habere quartam dimensionem neque habere dimensiones infinitas, et non est possibile ut imaginetur in eo infinitum nisi ex parte generis, quod est magnitudo. Et cum totum hoc sit sicut narravimus, et genus figure est superficies, manifestum est quod non potest imaginari infinitum nisi propter genus quod est superficies, non propter differentiam que est ab una linea vel a pluribus. Et
31rb similiter figura cor|porea non potest imaginari esse infinita nisi ex parte generis scilicet acceptio in sua descriptione idest corpus simplex: ex parte autem differentie per quam est figura, impossibile est imaginari esse infinitam nisi esset possibile imaginari finitum esse infinitum secundum quod est finitum, scilicet quod dictum est in diffinitione eius quod est corpus contentum superficie aut superficiebus; et cum corpus rotundum infinitum non sit, impossibile est ut corpus celeste sit infinitum. Et ideo dixit Aristoteles *Dicamus etiam quod linea finita impossibile est ut sit infinita*, idest quod linea impossibile est ut sit infinita simpliciter, quoniam aliquo modo est finita.

Et cum dixit quod impossibile est invenire lineam infinitam omnibus modis, dixit modum ex quo possibile est imaginari ipsam esse infinitam, et dixit *Si igitur possibile sit, erit in altera duarum extremitatum* etc., idest quod linea, inquantum privatur a latitudine, impossibile est ut imaginetur infinitum, et est modus quod est in ea quasi differentia: inquantum autem longitudo, possibile est ut imaginetur infinita, et est pars que est in eo quasi genus. Et cum dixit quod linea non imaginatur infinita nisi uno modo, dixit quod *Superficies etiam impossibile est ut sit infinita in omnibus dimensionibus nisi in dimensione infinita*, idest quod superficies etiam non potest imaginari infinita in omnibus dimensionibus: et non est possibile hoc in ea scilicet secundum estimationem, nisi in duabus

---

22 quia] que *YZ* | divisibilis] -viditur *YZ* 23 dimensionem] dimissio- *B* | neque] nec *Y*: nedum *BX* | habere[2]] *om. Z* 24 imaginetur] -nemur *JZ* 30 acceptio] -cepta *Z*: -cepti *J*W_2 32 esset] esse *B* | esset possibile] *inv. Z* 33 est[2]] + superius *B* 34 aut] vel *Z* 36 Et—Aristoteles] *om. BY* 37-8 idest—infinita] *om. homoiotel. B* 41 modum] -dus *B* | ipsam] ipsum *XY* 42 infinitam] -tum *XY* | igitur] ergo *BX* 44-6 modus—est[1]] *om. homoiotel. BZ* 46 eo] ea *BJ*

dimensionibus tantum scilicet longitudo et latitudo, que sunt in ea quasi genus: ex profunditate autem non imaginatur cum careat ea; et hec privatio est in ea quasi differentia. Deinde dixit *Superficies est*
31va *prior figura*, idest secundum quod differentia | est prior: corpus est differentia, corpus enim figuratum est quod continetur ab una superficie aut a pluribus, superficies igitur sunt in diffinitione corporis quasi differentie. Deinde dixit *Tunc enim impossibile est ut quadratum sit infinitum* etc., idest et cum corpus figuratum habuerit superficies terminatas, impossibile est corpus figuratum infinitum esse: impossibile est enim quadratum esse infinitum in aliqua parte nec aliqua figura quoniam hoc esse non remaneret quadratum. Deinde dixit *Quoniam neque circulus neque orbis* etc., idest impossibile est ut imaginetur corpus rotundum infinitum, sicut est impossibile ut imaginetur pes infinitus: non enim est pes nisi per eius figuram, et si imaginetur esse infinitus in aliqua parte, tunc non erit figuratus.

Deinde dixit *Si ergo orbis non est neque circulus infinitus, et etiam neque motus circularis infinitus*, idest et si aliquis posuerit ipsum in re mota infinita in tempore infinito: potest enim aliquis dicere quamvis ponatur quod celum impossibile est ut sit infinitum, quia motus eius est in tempore finito, tamen non est remotum ut extra ipsum sit corpus rotundum continens ipsum cuius motus est infinitus et in tempore infinito? Quapropter dixit respondendo quod cum omne rotundum sit finitum, necesse est ut motus eius sit finitus et in tempore finito.

31vb **40.** *Et dicamus etiam quod aliquid quod impossibile est ut corpus quod* 272b23
movetur circulariter impossibile est esse infinitum. Sit igitur centrum C, 272b25
et sit linea AB infinita, et linea HZ infinita etiam, et moveatur linea

---

55 corpus] + autem *Y*: -poris autem *JZ* 56 differentia] + est superficies *JZ* 57 a] *om. B* 60-1 infinitum esse] *inv. Z* 61 est enim] *inv. Z* | esse infinitum] *inv. BX* | in] *om. B* 63 orbis] cibus *B* 64 est impossibile] *inv. Z* 68 Si—orbis] *om. WYZ* 69 infinitus] *om. B* 72 tamen non] *inv. BX* | est[2]] erit *JZ* 75 eius] *om. Z*: finitus *Y*

40. 1 aliquid] *om. Z*: aliud *BX*: illud *Y* | quod[2]] ut *BWZ* | est] *om. BW* 2 impossibile est] *om. JZ* | esse] erit *Y*: sit *Z* 3 infinita[1]] *om. B* | linea[2]] *om. Z* | infinita etiam] *inv. Z* | moveatur] + etiam *BJ*

CD: quoniam, cum movebitur, impossibile est separari a linea H omnino, sed erit simile linee DZ, et secat lineam H in puncto Z. Et cum ita sit, impossibile est ut motus infinitus sit in corpore rotundo omnino. *272b28*

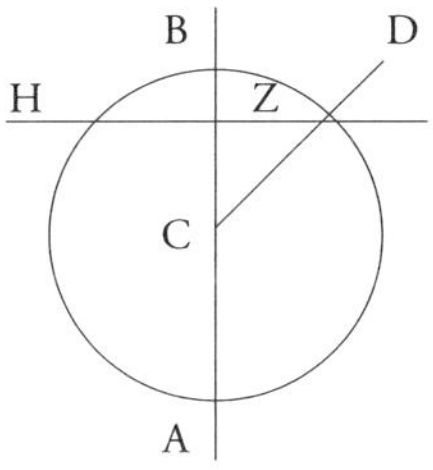

Hec est demonstratio quinta quod corpus infinitum impossibile est moveri circulariter: sit ergo C centrum, idest manifestum est cum posuerimus quod corpus non movetur circulariter, necesse est ut habeat centrum sive sit finitum sive infinitum; ponatur ergo infinitum, et centrum eius sit C. Deinde dixit *Et sit linea AB infinita* etc., idest et cum ita sit, possibile est ut ponatur linea infinita transiens
32ra per eius centrum et infinita in dua|bus extremitatibus; sit ergo illa linea AB. Deinde dixit *Et sit linea equalis* etc., idest extrahamus super hanc lineam non super centrum perpendiculare, et sit etiam infinita in duobus extremis; sit ergo linea HZ. Deinde dixit *Et moveatur linea CD*, idest et extrahamus a centro C lineam infinitam in extremo, sit ergo CD secantem lineam HZ; et imaginemur hanc lineam CD moveri per motum corporis quod circulariter movetur, et duas lineas secantes se quiescentes. Deinde dixit *Quoniam cum movebitur, impossibile est separari a linea H omnino*, idest et cum imaginati fuerimus lineam exeuntem a centro (que est infinita in uno extremo scilicet CD, que secat lineam infinitam in duobus extremis scilicet HZ, motam et secantem eandem lineam HZ), manifestum est quod

4 a linea] lineam *BXY* 6 sit] + sicut narravimus similiter figura est superficies manifestum est etiam quod imaginari infinitum nisi propter grave quod est et etiam est superficies non propter differentiam que est ab una linea et est universaliter in partibus et similiter etiam figura corporea non *B* 7 omnino] + et forma quidam exemplificamus sicut vides *W_2ZG* 8 demonstratio quinta] *inv. B* 9 ergo] igitur *Z* | idest] + et *JZ* 11 centrum] + suum *B* 14 et] *om. JZ* 17 duobus] dua- *WXY* 18 et] *om. BX* 19 secantem] -cans *JZ* 20 circulariter] + et *B*

numquam separabitur ab illa linea; si enim separabitur ab ea, non separabitur ab ea nisi in ultimo puncto eius, sed secundum quod est infinita non habet ultimum punctum scilicet in parte D. Deinde dixit *Sed erit simile linee DZ*, idest sed oportet secare arcum similem arcui DZ. Deinde dixit *Et secat lineam in puncto Z*, idest quod linea CD secat lineam H in puncto Z et secat totum circulum; ergo necesse est ut tota linea infinita secet lineam HZ infinitam, quod est impossibile.

Deinde dixit *Et cum ita sit, impossibile est* etc., idest quia motus infinitus non perficitur et motus corporis rotundi perficitur, quia ergo motus corporis infiniti est infinitus, et motus corporis rotundi est finitus, ergo corpus rotundum impossibile est esse infinitum.

32rb7 41. *Et etiam si celum sit infinitum, et motus eius circularis est, necessa-* 272b28
rio secat in tempore finito corpus infinitum. Sit ergo celum fixum, et sit 272b30
aliud intra ipsum simile in magnitudine motum: dico quod si corpus
quod est intra celum est infinitum, et totum celum moveatur circulariter,
tunc corpus infinitum pertransibit suum simile in tempore finito. 273a1

Hec est demonstratio sexta idest quod si celum sit infinitum et motus eius sit circularis, necessario corpus infinitum pertransibit spatium infinitum in tempore finito. Et hoc intendebat cum dixit *Necessario secat in tempore finito corpus infinitum*, et intendebat per corpus vacuum in quo imaginatur esse celum.

Deinde incepit declarare et dixit hanc consequentiam *Sit ergo celum fixum* etc., idest imaginemur celum fixum in vacuo equali ei, et hoc intendebat cum dixit *Et sit aliud intra ipsum sibi simile in magnitudine*. Deinde dixit *Dico quod si corpus intra celum* etc., idest quod si celum sit infinitum, vacuum etiam et dimensio in quo est celum erit infinitum; cum ergo celum moveatur in hoc vacuo infinito quousque compleat revolutionem, tunc pertransibit corpus

---

26-7 separabitur[2]—ea] separabatur ab eo *B* 26 ea] eo *X* 27 ab ea] *om. J* 29 secare] *om. B* 35 et—perficitur] *om. homoiotel. BW* | motus] + infinitus *Z* 36 infinitus] *om. Z* 37 corpus rotundum] *inv. Z*

41. 1 sit] esset *JX* 5 suum] *om. J* : sive *B* 10 esse] omne *Y* 11 et dixit] *post* consequentiam *JZ* 12 imaginemur] -netur *B* | ei] *om. B* 13 ipsum] *dittogr. B* 15 quo] qua *XZ* 17 compleat revolutionem] -pleare volunt *B*

infinitum vacuum infinitum. Et hoc intendebat cum dixit *Tunc pertransibit corpus infinitum simile ei in tempore finito*, idest tunc | *32va* pertransibit corpus infinitum aliud corpus infinitum in tempore finito ut dictum est in Phisicis. Et quidam dicunt quod hec demonstratio non est vera nisi secundum sermonem quod locus est dimensio et inane, et cum nulla dimensio et inane sit, in quo celum sit, ergo imaginari hoc est falsum; impossibile ergo non sequitur nisi ab hac positione falsa, non ab alio quod vult destruere. Dicamus ergo quod cum ponetur corpus infinitum et ponetur quod omne corpus habet locum, continget necessario ut locus sit inane, cum inane diminuatur secundum diminutionem corporis et crescat equaliter: in continenti autem non sequitur ut corpora sint equalia secundum equalitatem eius, quasi ergo non sequitur hoc ab ista positione nisi quia concessa est et concluditur impossibile; et non posuit ipsam quoniam est possibilis in se, sed quia sequitur a sermone eorum scilicet ponentium infinitum.

42. *Et possumus convertere sermonem nostrum: si tempus in quo per-* *273a1*
transivit illud corpus celum fuerit finitum, erit necessario magnitudo etiam in qua pertransivit finita; pertransibit necessario corpus sibi simile
in magnitudine. *273a4*

Hoc non est conversum sermonis predicti, immo conversum sue *32vb* potentie. Sermo enim quod Si tempus in | quo pertransivit illud corpus inane fuerit finitum, necessario erit illud inane; et corpus quod est in eo finitum est conversum sermonis, quod si corpus et inane in quo est fuerit infinitum, necessario erit tempus in quo pertransit ipsum infinitum. Hoc quidem non fuit positum in demonstratione predicta cum induxit ipsam in modo ducendi ad impossibile non modo recto; quoniam ista demonstratio, cum inducatur hoc modo,

---

18-9 vacuum—infinitum] *om. homoiotel. YZ* 19-21 idest—finito] *om. homoiotel. YZ* 21 quidam dicunt] *inv. Z* 23 cum nulla] illa *Z* | et^{3}] *om. BX* | celum sit] *inv. J* : celum *W* 26 ponetur corpus] *inv. Z* : quo est *W* 27 continget] -git *Z* 30 ista] illa *Z* | quia] quod *YZ* 31 est possibilis] *inv. Z*

42. 1 possumus] posuimus *BWX* | si] secundum *W* 2 finitum] in- *WXYZ* 3 pertransibit] + autem *JZ* 5 Hoc] + autem *Z* | est conversum] *inv. Z* 6-8 in—eo] *om. Z* 9 est] cum *B* | infinitum] -tus *B* 10 ipsum] ipse *B* 11 ducendi] dic- *BXY* | non] in *B* 12 cum inducatur] *om. B*

erit in sillogismo hipothetico; et cum destruxerimus consequens, sequitur oppositum antecedentis, scilicet quod cum destruxerimus ipsum pertransire in tempore infinito, ut visum est, continget necessario ut corpus inane non sit infinita; et ideo posuit quod tempus finitum sit, et coniungit ei positionem quam intendebat destruere; quapropter contingit impossibile quod est contradictorium illius quod sequitur in hipothetico dicente quod si corpus infinitum pertransit spatium infinitum, necessario erit in tempore infinito. Impossibile ergo consequens, quo utebatur Aristoteles in hac demonstratione, est contradictorium consequentis hipothetici, quod est in potentia illius sermonis de quo diximus quod posuimus convertere ipsum.

43. *Manifestum est modo quod celum est finitum, et quod corpus rotun-* *273a4*
dum impossibile est ut sit sine ultimo et sine fine. Modo ergo volumus
dicere de aliis corporibus simplicibus. *273a6*

33ra Manifestum est ex sermone nostro quod celum est finitum, et non tantum celum sed declaratum est quod omne corpus rotundum impossibile est ut sit infinitum, quodcumque corpus sit. Oportet nos igitur declarare hoc in aliis corporibus simplicibus que recte moventur, cum hoc enim erit verificatum quod omne corpus simplex est finitum.

44. *Volumus modo dicere de aliis corporibus quod impossibile est quod* *273a7*
sint infinita, neque corpus vadens ad medium neque a medio; motus
enim diversus est superius et inferius, et motus diversi sunt in diversis
locis. Dicamus ergo quod si alterum contrariorum fuerit terminatum,
alterum etiam erit terminatum, medium autem terminatum est; quod *273a10*
enim vadit ad inferius ab omni parte impossibile est ut pertranseat

16 corpus] + et *WXY* | sit] sint *WX* 19 in] et *JYZ* 20 infinito] finito *BWX* | Impossibile] + est *Z* 22 est[1]] + enim *Z* 23 posuimus] possumus *JZ*

43. 2 ut] quod W_2*XYZ* 3 simplicibus] + dicit *BZ* 5 celum] finitum *W* 6-7 nos igitur] *inv.* *Z* 8 cum] *om.* *BJ* | hoc] tunc *J* : causa *B* | verificatum] -tus *B*

44. 1 Volumus] + autem Z | dicere] *om.* *WY* | quod[2]] ut *JYZ* 2 sint] sit *B* 3-4 diversis locis] *inv.* *BJX* 5 alterum] *om.* *B* | autem] quod *B* 6 est] *om.* *BWY*

33rb *medium neque prolongetur ab eo; | si ergo medium est terminatum, et superius est terminatum necessario; et si loca sunt terminata, corpora erunt finita omnino.* 273a15

Vult declarare quod corpora simplicia recti motus sunt finita, et primo declaravit quod corpus inferius scilicet terra est terminatum, et similiter corpus superius scilicet ignis aut aliud; deinde declarat quod ea que sunt inter ista sunt terminata. Cum ergo dixit *Neque corpus motum ad medium* intendit terram, et cum dixit *Neque corpus tendens a medio* intendit ignem, cum ergo dixit *Motus enim diversus secundum contrarietatem et est superius et inferius*, idest quod loca ad que erunt hec corpora mota diversantur secundum contrarietatem et est superius et inferius. Deinde dixit *Et itiones diverse sunt in diversis locis*, idest quod hoc fuit quia motus contrarii sunt ad loca contraria.

Et cum posuit quod motus horum corporum simplicium contrarii sunt ad loca contraria, dixit *Quod si alterum contrariorum fuerit terminatum, alterum erit terminatum*, idest cum radix predicta est posita, scilicet quod loca istorum corporum sunt terminata, ponamus secundam radicem dicentem quod si alterum duorum contrariorum fuerit terminatum, necessario erit secundum terminatum. Et hoc manifestum est ex descriptione contrariorum que sunt in fine remotionis; et cum utrumque eorum maxime distet, necesse est ut sit equaliter in contrarietate, scilicet ut neutrum sit fortius reliquo. Sed cum posuimus alterum eorum finitum et alterum infinitum, tunc non contrariabunt sibi equaliter, quoniam forma contrarietatis que est in altero eorum non erit equalis forme contrarietatis alterius
33va secundum quod sunt con|traria. Et manifestum est per se ut sint in eodem gradu contrarietatis, et si non, non in fine distabunt; possibile enim esset addere super minorem ut esset magis contrarium, et tunc idem haberet plura contraria; et illud quod erat positum in maxima

---

7 prolongetur] -tegetur *Y*: -tegatur *W* | si ergo] signo *B* 8 superius] + etiam *WY* 10 motus] + non *Z* | sunt] sint *B* | finita] in- *YZ* 11 primo] post *Y*: prius *BW* | declaravit] -rat *JX* | terminatum] -tam *B* 12 aut] vel *B* 15 ergo] *om. JZ* 16 et[1]] *om. JZ* 16-8 idest—inferius] *om. homoiotel. B* 18 diverse] -si *Z* | diversis locis] *inv. BX* 19 quod] est *BX* | quia] quod *WX* 22 alterum erit] *inv. Z* 25 erit secundum] *inv. X*: erit *BW* 27 eorum] isto- *Z* | sit] sint *BX* 29 et] *om. B* 30 sibi] + ad *B* 31 erit] est *JZ* | contrarietatis] *om. B*

distantia non est ita quod est contrarium positionis. Et etiam non essent opposita secundum quod sunt extrema, scilicet quod proportio alterius eorum in contrarietate ad alterum non esset sicut proportio alterius ad primum; sed notum est per se quod proportio ignis ad aquam in contrarietate est sicut aque ad ignem.

Et cum posuit istam radicem, dedit tertiam radicem dicentem medium scilicet esse terminatum; deinde induxit demonstrationem eius et dixit *Medium autem est terminatum* etc., et intendit per medium locum inferiorem: est enim medium, manifestum est quod est terminatum, quasi dicat Et significatio quod locus vocatus inferius est terminatus quia est medium, cum omnes lapides non pertranseant ipsum; cum ergo locus inferius est medius, medium autem est terminatum, ergo locus inferius est terminatus. Et significatio quod corpora gravia non pertranseunt medium est quod corpora gravia descendunt ex omni parte celi descensu consimili, quapropter videtur quod eundem locum intendant; et universaliter si pertransirent medium, moverentur a medio non ad medium. Et hec propositio apparet ex casu lapidis secundum angulos rectos ad superficies ponderosas que sunt super terram ex omnibus partibus celi, ex quo videtur quod omnia intendant punctum unum ut Antiqui dixerunt in spera; hoc enim non accidit nisi in spera. Hoc enim non videtur esse manifestum nisi in spera, sed tamen hec propositio non videtur esse manifesta nisi concedatur celum esse spericum: qui autem
33vb dicunt celum non esse spericum | non concederent quod sit in celo locus inferior loco. Et cum sit ita, continget ut descensus lapidis non sit super lineas equidistantes omnino, quapropter non concurrunt omnino; et cum non concurrunt, pertransibunt medium; ergo locus inferior non erit terminatus; ergo necessarium est ponere partes celi esse diversas; et tunc verificabitur nobis hec propositio, sed cum hoc

---

36 distantia] differentia *YZ* | etiam] ideo *VW* 39 ad primum] prima *Z* | quod] quia *Z* 44 inferiorem] + quia *J*W_2*Z* 45 Et] *om.* *BX*: quod *Z* | significatio] signum secundum *Z* | vocatus] -tur *Z* 48 est terminatum] *inv.* *Z* | terminatus] + est *Z* | significatio] + est *BZ* 49 pertranseunt] -eant *JZ* 53 lapidis] -dum *BY* | ad] *om.* *BWY*: super *X* 55 ut] et *YZ* 56 spera[1]] + sed modo hoc *B* 57 esse] *om.* *W* 59 dicunt] *om.* *XY* | non esse] *inv.* *Y*: esse *X* | concederent] -det *BX*: -dunt *Z* 60 sit ita] *inv.* *Z* 61 concurrunt] -rent *WXY* 62 concurrunt] -rent *WY*: + non *Z* 63 erit] est *W*: erat *Z* 64 diversas] + sphericas *J*: + partes plani sunt sed W_2: + partes plani sunt sed sperice ordinatus *Z* | cum] tamen *JX*

videtur esse manifestum per sensum. Celum enim videtur aut spera perfecta aut pars spere scilicet medietas spere; et cum motum sit per se, manifestum est quod partes illius sunt diverse; et cum videmus quod omnes lapides descendunt secundum eundem situm ex partibus diversis celi, scimus quod descendunt super lineas concurrentes in uno puncto, ille ergo est locus inferior.

Deinde dixit *Si ergo medium est terminatum, ergo superius est terminatum*, idest et cum necessarium est quod locus inferior sit terminatus, necessarium est ut locus superior sit terminatus inquantum est contrarius; iam enim diximus quod cum alterum contrariorum sit terminatum, alterum erit terminatum. Et potest aliquis dicere Cum locus inferior non videtur terminatus, donec concedatur quod celum sit pars spere; sed hoc posito, positum est esse terminatum arcum; iam ergo accepta est res in declaratione sui. Dicamus ergo quod locus superior, quem vult declarare esse finitum, non est quod est inter terram et celum; manifestum est enim quod hoc est finitum, sed vult declarare quod iste locus cum loco celi non est infinitus; celum enim si sit in loco est in superiori, et si iste locus est infinitus, continget ut locus inferior sit infinitus; sed locus inferior est finitus, ergo et locus superior.

34ra14 **45.** *Et etiam superius et inferius sunt terminata, ergo medium terminatum; si ergo non sit terminatum, invenietur motus infinitus; sed iam diximus quod impossibile est motum esse infinitum, et cum ita sit, medium ergo est terminatum, ergo quod est intra ipsum et quod potest esse in eo est terminatum, scilicet quod corpus tendens ad superius et ad inferius impossibile est quod sint in eo; alterum enim eorum movetur a medio et alterum ad medium.* *273a15–273a21*

65 manifestum] verum *Z* | enim] + ut *B* 67 quod] que *YZ* 68 descendunt] -dentes *BW* | situm] sicut *BXY* 69 quod] + non *Z* | descendunt] -derunt *W* 71 ergo[2]—terminatum] *om. homoiotel. BY* 73 locus superior] *inv. Z* 73-4 terminatus—sit] *om. homoiarch. X* 73 est[2]] + superius *Y* 74 quod] et *Z* 76 terminatus] *ante* non *Z* : *om. BWXY sed suppl.* W_2 77 esse] + omnem *JZ* 78 ergo[1]] enim *Z* | res] *om. B* 79 finitum] in- *Z* | est[2]] *om. BY* 81 cum loco] *post* celi *Z* 83 et] *om. XY*

45. 2 ergo] igitur *WXY* 5-6 et ad inferius] *om. BWXY* 6 sint] sit *JXZ* 7 alterum] *om. B*

Cum declaravit quod terra et ignis sunt finiti quia loca eorum sunt finita, vult declarare quod aqua et aer sunt finita, dicendo *Et etiam si superius et inferius* etc., idest cum declaratum est quod locus inferior et superior sunt terminati idest finiti, necesse est ut locus qui est inter ea sit terminatus etiam, et hoc intendebat per medium. Cum enim duo extrema sint terminata, necesse est ut medium sit terminatum, *34rb* quoniam si illud quod est inter locum terre et locum | ignis esset infinitum, et ignis moveretur a terra usque ad suum locum, contingeret quod numquam veniret ad suum locum: esset ergo in perpetuo motu, et sic motus rectus esset infinitus, quod declaratum est esse impossibile in libro predicto.

Deinde dixit *Et cum ita sit, medium est terminatum*, idest et cum impossibile est ut inter locum ignis et terre sit motus infinitus, necessarium est ut iste locus sit finitus scilicet medium, idest quod est inter duo extrema. Medium enim in hoc sermone dicitur secundum tres intentiones, secundum locum inferiorem et secundum medium spere et secundum medium quod est inter duo extrema contraria; et hoc intendebat cum dixit *Et medium terminatum*. Deinde dixit *Quod ergo est intra ipsum et quod potest esse terminatum*, et necessarium est quod existit modo in eo, et que erunt in futuro seculo sint terminata idest finita. Deinde monstravit que sunt illa corpora, et dixit *Scilicet quod corpus tendens ad superius* etc., idest et cum impossibile est ut in hoc loco sit motus, neque corporis quod a medio movetur scilicet ignis neque corporis quod ad medium movetur scilicet terre, quia est finitus et motus existit in eo, manifestum est quia est finitus; ergo corpora que sunt in eo sunt finita.

34va **46.** *Manifestum est igitur ex eis que diximus quod impossibile est ut sit* *273a21*
corpus pondus infinitum; et omnino impossibile est ut aliquod istorum

8 declaravit] -raverit *B* | sunt[1]] sint *B* 9 aqua ... aer] *inv. W* | si] *om. XY* 10-11 inferior ... superior] *inv. JXZ* 12 intendebat] + cum dixit *B* 16 quod] ut *Z* 17 motu] -tus *Z* 19 Et[1]] *om. B* | et[2]] *om. B* 22 sermone] loco *Z* 24 contraria] *dittogr. BJ* 25-6 Quod ergo] *inv. JZ* 26 esse] + in eo *J* : + in eo est *X* | et[2]] *om. B* 27 existit] -stunt *XY* | sint] sit *XYZ* 28 monstravit] de- *BXZ* | sunt] sint *BX* 30 neque] in *W* | quod] et *Z* 32 finitus[1]] in- *BX* | finitus[2]] in- *B*

46. 1 eis] hiis *JZ* | sit] *om. J* 1-2 sit—ut] *om. homoiotel. X* | sit—aliquod] *om. B* 2 pondus] ponderosum sit *J*

corporum gravium sit infinitum; corpus enim infinitum infinitam habet
gravitatem, et similiter de levitate. Si igitur gravitas infinita sit, ergo 273a25
levitas infinita; et hoc erit si inveniatur corpus ascendens in infinitum. 273a26

Hec est demonstratio fundata super tres propositiones: quarum una est quod omne corpus motum motu recto habet gravitatem aut levitatem, secunda est quod omne corpus infinitum grave aut leve habet gravitatem aut levitatem infinitam, tertia est et impossibile est gravitatem aut levitatem esse infinitam. Et dixit *Manifestum est igitur ex hiis que diximus*, et potest intelligi quod iam declaratum est quod motus rectus impossibile est ut sit infinitus, ergo impossibile est ut gravitas sit infinita; quamvis ipse post addidit in declaratione istud. Deinde dixit *Et omnino est impossibile* etc., idest et cum declaratum sit quod impossibile est gravitatem esse infinitam, manifestum est corpus grave impossibile esse infinitum. Deinde dixit *Corpus enim infinitum infinitam habet gravitatem*, idest quod necessarium est quod cum gravitas infinita non sit, ut corpus grave infinitum non sit: corpus enim infinitum habet infinitam gravitatem. Deinde dixit *Et similiter de levitate*, idest et similiter dixit de levitate et corpore levi sicut de gravitate et corpore gravi: si enim gravitas est infinita et levi-
34vb tas, et si impos|sibile in uno et in altero. Deinde dixit *Et hoc erit si inveniatur corpus ascendens in infinitum*, idest quod non invenietur levitas infinita nisi inveniatur corpus ascendens super alia corpora in infinitum, similiter non invenietur corpus leve in infinitum nisi inveniatur levitas infinita.

47. *Et verificatio eius est secundum quod narrabo. Sit ergo gravitas cor-* 273a27
poris infinita, deinde perscrutemur utrum sit possibile aut non. Ponamus ergo modo corpus infinitum lineam AB, et sit sua gravitas C, et extraha-

3 gravium] *om. Z* | enim] est *B* 4 levitate] levi *BWYZ* | igitur] enim *XZ* 5 inveniatur corpus] *inv. JX* 7 aut] seu *B* 8 grave] -vitate *B* 9 et] quod *J*W_2*Z* 11 diximus] + idest dicemus *B* 13 istud] illud *Z* 14 est impossibile] *inv. JZ* 15 est^2] *om. B* 17 infinitam] *om. B* 18 quod] *om. JXY* : ut *W* | cum] *om. B* : sit *W* | corpus grave] *inv. BX* 19 habet infinitam] *inv. XY* : habet *W* 20 dixit] *om. X* : -ximus *JY* : -camus *B* | levitate^2] + diximus *X* | levi] + et *B* 21 est] *om. WXYZ* 23 inveniatur] -nitur *B* | idest] + et *Z* | invenietur] -niatur *YZ* : -nitur *J* 24 ascendens] + et *B* | alia] aliqua *Y* : omnia *JX* 24-5 in infinitum] *om. W*

47. 1 ergo] *om. BW* 2 infinita] -tas *Z* 3 ergo modo] *inv. XZ* : modo *B*

mus a corpore infinito magnitudinem infinitam, et sit BD, et gravitas 273a30
eius sit H. Erit ergo gravitas H minor gravitate C: gravitas ergo minoris corporis minor est gravitate maioris corporis. Et numeremus maiorem gravitatem per minorem quousque sit similis ei in gravitate; et similiter 273b1
35ra *numeremus etiam corpus maius per minus, scilicet per corpus B|D corpus BZ. Possibile est enim extrahere a corpore infinito quodcumque voluerit extrahens extrahere; erit ergo proportio H ad C sicut proportio BD ad BZ; et hoc volumus declarare.* 273b3

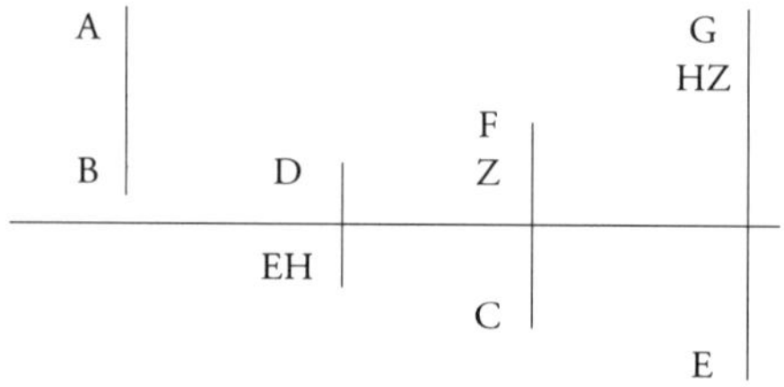

Vult demonstrare quod omne corpus grave aut leve infinitum necessario habet infinitam gravitatem aut levitatem, et potentia huius demonstrationis est in sillogismo hipothetico sic formato: si corpus infinitum habens motum rectum sit, necessario habet gravitatem aut
35rb6 levitatem infinitam; sed neque gravitas neque levitas est infini|ta, ergo neque habens corpus motum rectum est infinitum.

Sed quia continuatio et similiter destructio non sunt manifeste per se, incepit declarare continuationem per viam ducentem ad impossibile, deinde declaravit destructionem; et dixit *Si ergo gravitas corporis infiniti infinita*, idest ponamus contradictorium eius, quod est quod corpus infinitum habet potentiam infinitam, deinde queremus

---

4 infinito] finito *Z* 5 Erit—H] *om. homoiotel. B* | ergo[1]] enim *J* | gravitas[1]] tunc *J* 6 minor est] *inv. WZ*: minorem *B* | gravitate] -tem *B* | numeremus] invenire- *B*: + cum *Z* 6-7 maiorem gravitatem] *inv. J*: gravitate maiore *Z* 7 per] *om. Z* | minorem] + gravitatem insistentas *Z* | ei] *om. BZ* 8 numeremus] invenire- *B* | numeremus etiam] *inv. Z*: numeremus *J* 9 est] *om. WYZ* 11 declarare] + de literis istis sicut vides W_2*ZG* 12 grave] + infinitum *Z* | infinitum] *om. BWXY sed suppl.* W_2 13 et] + est *Z* 14 sillogismo hipothetico] *inv. W* | si] sicut *B* 15 habens] -bet *B* 15-6 gravitatem ... levitatem] *inv. WY* 20 declaravit] -rabit *BX* | ergo] *om. W* 21 infiniti] *om. BJX*: + est *Z* | infinita] finita *JW*

utrum sit impossibile aut non. Ponamus ergo corpus grave esse infinitum et sit AB, et sit gravitas finita existens in hoc corpore linea exiens ab ea scilicet C, et accepit loco gravitatis lineam, quoniam linea est divisibilis. Deinde dixit *Et extrahamus a corpore finito magnitudinem finitam scilicet BD*, idest et secemus ex corpore finito corpus finitum, et hoc possibile est, verbi gratia secemus ex linea AB, que est infinita, lineam BD finitam; et hoc quod dixit possibile est in corpore infinito, sive sit infinitum in una parte sive infinitum ex omnibus partibus. Deinde dixit *Et sit gravitas eius H*, idest ponamus gravi-
35va tatem istius corporis finiti lineam D ad H. | Deinde dixit *Erit tunc H minor C*, idest quod ex hoc sequitur necessario quod gravitas corporis finiti quod est BD, scilicet H, est minor gravitate corporis infiniti quod est AB, scilicet C. Deinde reddidit causam in hoc et dixit *Gravitas enim corporis minoris est minor gravitate maioris corporis*, idest quod cum manifestum est quod gravitas partis est minor gravitate totius, finitum autem est pars infiniti, manifestum est quod gravitas finiti est minor gravitate infiniti.

Deinde dixit *Et similiter numeremus gravitatem maiorem per minorem* etc., idest et mensuremus maiorem gravitatem per minorem multotiens, idest duplicemus minorem gravitatem multotiens quousque sit equalis maiori gravitati aut maior. Deinde dixit *Et similiter numeremus corpus maius per minus* etc., idest et cum imaginati fuerimus mensurationem maioris per minorem gravitatem quousque consumatur, manifestum est quod mensuramus etiam per corpus minus in quo est gravitas minor scilicet corpus BD, magnitudinem ex corpore maiori finitam: possibile est enim secare ex corpore infinito secundum mensuram minoris infinities necessario; ergo cum pervenimus ad gravitatem equalem gravitati corporis infiniti, perveniemus ad corpus finitum. Et hoc intendebat cum dixit *Possibile est*

---

23 esse] *om.* *JZ* 26 Et] *om.* *BZ* | finito] in- $B_2JW_2Z_2$ 29 corpore] aliquo *WY* : quolibet *Z* 30 infinitum[1]] finitum *JXY* 31 Et] *om.* *B* 33 sequitur necessario] *inv.* *B* 34 finiti] in- *B* 35 reddidit] reddit *JZ* : dedit *BX* 36 corporis minoris] *inv.* *Z* | minoris—corporis] *om. homoiotel.* *B* 38 est pars] *post* infiniti *Z* 40 maiorem ... minorem] *inv.* *B* 41 mensuremus maiorem] *inv.* *J* : mensuremus *B* 43 aut] et *JW* | maior] + minori *JZ* 44 numeremus] + etiam *Z* 45 gravitatem] *om.* *B* 46 manifestum] + etiam *W* | est] + etiam *Z* 48 finitam] in- *BWXY* : + que quidem est BZ JW_2Z | infinito] finito *W sed corr.* W_2 49 minoris] -rem *B* | infinities] -niti *B* | ergo cum] *inv.* *Z* 50 pervenimus] -nerimus *YZ* | infiniti] *om.* *B*

extrahere a corpore infinito quodcumque voluerit extrahens extrahere ex eo, idest possibile est enim dividere ex eo equale minori corpori, et non cessare hoc quousque perveniamus per proventum ad gravitatem C, que est infinita, et ad corpus finitum quod sit pars corporis infiniti. Cum enim additio procedit super corpus minus in infinitum, et potentia augmentatur per augmentationem eius, necessario perveniemus ad corpus finitum, scilicet cum pervenerimus ad potentiam equalem potentie C, | *35vb* quam posuimus infiniti.

Deinde dixit *Erit ergo proportio H* etc., idest et ponamus corpus ad quod perveniemus per duplicationem corporis BD minoris corporis BZ: gravitas ergo istius corporis necessario est gravitas C, que erat infiniti; erit enim proportio gravitatis H, que est gravitas corporis minoris quod est BD, ad gravitatem C que est corporis infiniti, eadem cum proportione corporis minoris quod est BD, ad corpus finitum maius quod est BZ. Et secundum permutationem proportionis erit proportio gravitatis H ad corpus BD sicut proportio gravitatis C ad corpus BZ; sed gravitas H est totius corporis BD minoris, ergo gravitas C est totius corporis BZ maioris; sed iam posuimus istam gravitatem esse corporis infiniti, erit ergo eadem gravitas corporis infiniti et finiti insimul, quod est impossibile. Et hoc intendebamus declarare, sed inducit hoc impossibile post hoc, et ipse quasi ponit quod cum duplicata est minor gravitas quousque perveniat ad equale gravitati maiori, ncessarium est ut perveniat apud duplicationem ad corpus finitum. Deinde declaravit secundum viam proportionis quod necessarium est ut gravitas maior sit omnium partium istius corporis finiti, scilicet quod gravitas C sit corporis BZ.

---

53 possibile] im- *B* | enim] tantum *Z* | equale] -li *Z* 54 non—hoc] non hic cessare *Z* | perveniamus] -niemus *W* | proventum] punctum *BW* 55 C] + scilicet *Z* | infinita] -ti *J*: -to *WY* | sit] si *B* 56 additio] addeo *BX* 57 eius] *om. B* 58 pervenerimus] -niemus *BJ*: -nimus *Y* 61 corporis] *om. X*: -pus *BJ* 62 istius] illius *YZ* 63-4 corporis minoris] *inv. W* 64 minoris—corporis] *om. homoiotel. X* 64-5 ad—BD] *om. homoiotel. Y* 67 ad—BD] *om. WY* | sicut] sic *BJ*: sed cum *Z* 69 minoris] + gravitatis *Z* | maioris] + corporis *Z* | iam] + declaratum *B* 71-2 Et^{2}—impossibile] *om. homoiotel. Y* 72 sed] + ipsum *B* 75 declaravit] -rat *WXY* 77 quod] *om. BJ*

36ra27 **48.** *Revertamur etiam et dicamus quod si magnitudo corporum sit* 273b3 *secundum gravitatem eorum, et gravitas minoris corporis sit minor et gravitas maioris sit maior, erit gravitas corporis infiniti et corporis finiti* 273b5 *equaliter, quod est impossibile. Et etiam si gravitas maioris corporis est maior gravitate minoris corporis, erit gravitas BZ maior gravitate BZ, erit ergo gravitas corporis finiti maior gravitate corporis infiniti, et erit* 36rb *etiam gravitas | corporum inequalium in magnitudine eadem, quod est impossibile.* 273b10

Vult demonstrare quot impossibilia existentia in hac conclusione sunt et numerare etiam impossibilia contingentia huic positioni, ex quibus principiis contradicit hoc impossibile, et dixit *Revertamur modo et dicamus*, idest et cum declaratum est quod sequitur ab ista positione inveniri gravitatem infinitam in corpore infinito et finito, revertamur ergo et dicamus impossibilia consequentia ab ista conclusione. Cum enim notum est per se quod magnitudo corporum est secundum eorum gravitatem, scilicet quod corpora superant se in gravitate secundum superationem eorum in magnitudine maxime eiusdem speciei, et quod gravitas minor est minoris corporis et maior maioris et econverso etiam scilicet quod corpus minus habet minorem gravitatem et maius maiorem, manifestum est quod sequitur ab illa predicta positione contradictorium horum principiorum scilicet quod potentia maioris corporis et minoris sit eadem idest equalis, immo impossibilius quod potentia finiti et infiniti sit equalis, que est maior duplicationibus infinitis.

Deinde incepit dicere aliud impossibile consequens ab hac positione inopinabilius, et dixit *Et etiam si gravitas maioris corporis sit maior gravitate* etc., idest quia manifestum est per se sicut diximus quod corporis maioris est gravitas maior gravitate minoris corporis, manifestum est quod possibile est duplicare corpus minus quousque aggregetur corpus maius corpore finito cuius gravitatem invenimus

---

48. 3 sit] est *B* | infiniti … finiti] *inv. Z*   5 BZ[1]] HC *(obsc.) B*: BZH J_2 | BZ[2]] + quod *B*   6 ergo] *om. W*   9 demonstrare] declara- *Z*   12 ista] illa *Z*: hac *J*   13 infinito … finito] *inv. JXYZ*   14 ergo] *post* dicamus *B* | ista] illa *Z*   17 superationem] exsupera- *W*: exupera- *X*: extiperatione *B*: experientiam *Y*   21 illa] ista *Y*   23 finiti] finita *B*   25 aliud impossibile] possibile *Y*   27 gravitate] *om. Z*   27-8 etc.—gravitate] *om. homoiotel. X*   27 quia] et *W* | est] + etiam *WY*   28 est gravitas] *inv. Y*   30 aggregetur] -gatur *Z*

equalem gravitati corporis infiniti; et cum ita sit, manifestum est 36va | quod corporis huiusmodi gravitas est maior gravitate corporis cuius gravitas est equalis gravitati infiniti. Sit ergo hoc corpus BZH; et BZ, quod est minus BH, est illud in quo secutum est in sermone predicto ut gravitas eius sit equalis gravitati corporis infiniti; et cum gravitas magnitudinis BZH sit maior gravitate magnitudinis BZ, gravitas autem magnitudinis BZ est equalis gravitati corporis infiniti; ergo gravitas BZH finita est maior gravitate infiniti non tantum equalis, quod est impossibile. Et hoc reducitur in hoc scilicet quod gravitas corporum diversorum sit eadem, quod est impossibile; quapropter dixit post *Et etiam erit gravitas corporum inequalium eadem.*

49. *Et etiam quamvis corpus sit non consimilium partium, tamen nulla* 273b10 *differentia est in eo quod diximus; sermo enim de ipso similis est sermoni de corpore consimilium partium. Possumus enim accipere de corpore infi-* 273b24 *nito multas partes equales in gravitate, sicut fecimus in BD; si ergo volumus addemus, et si voluerimus diminuemus, cum acceperimus de illo corpore, sicut vides in ista figura.* 273b26

Cum non sequatur quod gravitas maioris corporis sit maior nisi in eodem corpore consimilium partium, in corporibus autem diversis 36vb non sequitur verbi gratia quod | corpus plumbi est gravius maiori se ferreo et maximo ligneo, quapropter aliquis potest dicere Forte corpus infinitum cuius potentia est finita erit non consimilium partium; quapropter non sequitur ut proportio gravitatis ad gravitatem sit sicut partis ad partem verbi gratia quod sit idem corpus continuum cuius alia pars sit ferrea et alia plumbea et alia cuprea. Aristoteles igitur dicit respondendo quod nulla differentia est in sequendo impossibile vel si ponatur hoc corpus esse consimilium partium vel non;

31 gravitati] -tatem *B* 32 corporis[1] … gravitas] *inv. BX* 34 quod] et *Z* 35 gravitati corporis] *inv. W* 37 est equalis] *inv. B* 41 dixit] -ximus *XYZ* | etiam erit] *inv. B*

49. 2 similis est] *inv. W* 3 Possumus enim] posuimus *B* 5 voluerimus] -lumus *WYZ* | acceperimus] -pimus *BY* | de] ab *Z* : ex *BJY* 7 sequatur] -itur *BX* | corporis sit] *inv. B* : sit *Y* | nisi] ut *Z* 11 cuius] cum *WY* | non] *om. XY* 12 quapropter] propter quod *BX* | ut] *om. XY* 13 quod] qui *W* | sit] si *B* 14 alia[1]] aliqua *YZ* : altera *J* : autem *X* 15 differentia est] *inv. BJ* notum est enim quod corpus dividitur secundum gravitatem, et quod possibile est nobis dividere ipsum in partes equalis gravitatis, quamvis non sint equalis magnitudinis, et quod si duplicaremus tales partes in infinitum, perveniremus ad corpus cuius potentia sit equalis cuicumque potentie finite posite; et cum ita sit, deductum est ad impossibile predictum.

Et cum dixit *Quamvis corpus non sit consimilium partium, nulla differentia est in eo quod diximus*, potest intelligi quod si partes istius corporis infiniti, cuius gravitatem posuimus finitam, sint dissimiles adinvicem idest non eiusdem speciei, quoniam cum fuerint dissimiles non augmentabitur in eo scilicet corpore gravitas secundum proportionem augmentationis corporis. Sed in primo sermone accepit ipsum augmentari secundum eandem proportionem, quoniam quamvis non augmentaretur secundum eandem proportionem, tamen possumus pervenire in eo duplicando ad corpus cuius potentia erit equalis cuicumque potentie posite per se absque usu illius proportionis predicte. Cum enim corpus dissimilium partium non contingit quod proportio sui minoris partis ad maiorem sit sicut gravitas ad gravitatem; et ideo dixit *Possumus enim accipere ex corpore infinito multas partes equales in gravitate*, diminuendo et augmen-
37ra tando | in partibus diversis quousque aggregetur ex hoc habens gravitatem equalem gravitati alterius. Et hoc intendebat cum dixit *Si ergo voluerimus addemus, et si voluerimus diminuemus*, quasi igitur diceret quod bene possumus accipere ex corpore cuius partes sunt inequales partes equales in actu, quoniam in nobis est diminuere aut addere in eo quod accepimus. Et hec demonstratio fundata est in eo quod declaratum est in Prioribus Analeticis, quod falsum impossibile non sequitur ex falso possibili: quicumque enim ponitur hic sine diminutione est falsum possibile.

---

17 enim] igitur *B* 18 equalis] -les *BXY* 19 quod] *om. Z* 20 in] *om. W* 21 posite] *om. Z* : potestate *Y* 23 non sit] *inv. BX* 26 cum] non *B* 28 primo] ipso *B* |ʾ accepit] -ceptum *BJ* 30 augmentaretur] -tetur *WZ* | eandem] *om. B* 31 possumus] posuimus *BX* 32 erit] erat *WXY* 35 Possumus] posuimus *B* | enim] igitur *Z* 36 augmentando] augendo *JWZ* 37 ex] in *Z* 38 ergo] igitur *BX* 39 quasi igitur] ergo *Z* | diceret] -cit *Z* 40 inequales] equales *YZ* 41 in actu] *om. Y*: gravitatis *J* 42 fundata est] finita est *Y*: est finita *Z*: facta est *X* 43 Prioribus] Primis *BXZ* | Analeticis] + scilicet *WY* | quod] quoniam *Z* | falsum] + est scilicet *B* 44 ponitur] -nuntur *Z*: -nit *J* 45 est falsum] *inv. B*

50. *Manifestum est igitur ex hoc quod diximus quod impossibile est* 273b26
quod gravitas corporis infiniti sit finita. Si ergo impossibile est ut sit finita, necesse est ut sit infinita; et si hoc est impossibile, scilicet gravitatem esse infinitam, impossibile est corpus esse infinitum. 273b29

Et cum declaratum est quod impossibile est corpus esse infinitum habens gravitatem finitam, et necessarium est ut omne grave sit aut finitum aut infinitum, necesse est si corpus sit infinitum ut sua gravitas sit infinita; et si post hoc declaratum fuerit impossibile est esse gravitatem infinitam, erit manifestum corpus esse infinitum.

37rb7 **51.** *Et etiam dicemus quod impossibile est gravitatem esse infinitam; et* 273b29
demonstratio eius est ex hoc quod dicemus. Et est quod si quantitas huius gravitatis movetur per hoc spatium in hoc tempore, movebitur alia quantitas gravitas in minori tempore et etiam in minimo tempore. Dicamus modo quod coniunctio temporum adinvicem est secundum coniunctionem gravitatis adinvicem; sed coniunctio temporum differt a coniunctione gravitatis; si enim medietas gravitatis movetur per aliquod spatium 274a1
in aliquo tempore, movebitur duplum illius gravitatis finiti per illud spatium per medietatem illius temporis. 274a2

In hoc sermone narrat unam propositionem huius demonstrationis. Ponit enim quod cum aliqua gravitas movetur per aliquod spatium in aliquo tempore, et posuerimus aliam gravitatem maiorem, continget quod movebitur per illud spatium in minori tempore et quod proportio maioris gravitatis ad minorem est sicut proportio minoris temporis ad maius; et hoc intendebat cum dixit *Sed coniunc-*

---

50. 2 infiniti] *om. BWXY sed suppl.* W_2 | sit[1]] *om. BXY* | finita *JZ*] infin- *cod. cett.* 3 necesse] -ario *YZ* | ut sit] *om. BXY*: *post* infinita *Z* | infinita] finita *W* | et—impossibile] si ergo impossibile est ut sit et hoc est *BJG* 4 est] et *Y*: + et *W* 5 Et] *om. BJ* | esse] *om. Z* 6 finitam | *BX*] infin- *cod. cett.* 8 fuerit] + quod *J*W_2 | est] *om. B* 9 infinitum] finitum *JZ*

51. 1 gravitatem esse] *inv. BJX* 2 dicemus] -cimus *B* 3 quantitas] -tatis *BWX* 4 gravitas] -tatis *JY* | etiam] *om. BW* | minimo] minori *BX* 5 modo] etiam *Z* 6 sed] si *Z* 6-7 sed—gravitatis[1]] *om. X* 7 medietas] -tatis *B* 8 finiti] in- *X*: finite *JZ* 10 propositionem] -portionem *XY* 12 aliam] illam *W sed corr.* W_2 | maiorem] + illa W_2*Z* 13 continget] -git *JZ* 14 quod] quia *WY* | maioris gravitatis] *inv. JW*

tio tamen temporum differt a coniunctione gravitatis, et dixit *Movebitur autem quantitas gravitatis in minori tempore*, idest movebitur aliqua quantitas maior prima per illud spatium primum in minori tempore.
37va Deinde dixit *Et etiam in mi|nimo tempore*, idest et cum acceperimus tertiam quantitatem maiorem secunda, inveniemus eam moveri per illud idem spatium in tempore minori secundo, et similiter de quarto et sic in infinitum.

Et cum posuit quod hoc accidit, scilicet gravitas et levitas secundum parvitatem et magnitudinem in corporibus diversis quando movebuntur in eodem medio, incepit dicere quod apparet ex hoc sensu quod necesse est in hiis gravitatibus quod proportio gravitatis ad gravitatem sit sicut temporis ad tempus, et dixit *Quod coniunctio temporum est secundum* etc., idest quod cum coniunxerimus tempora adinvicem, inveniemus proportionem inter ea sicut proportio gravitatum adinvicem; sed tamen proportio maioris gravitatis ad minorem non est sicut proportio maioris temporis ad minus, sed econverso ut diximus.

52. *Et etiam gravitas finita movetur per omne spatium finitum in tem-* 274a2
pore finito. Et cum ita sit, si gravitas infinita inveniatur, necesse est ut aut moveatur aut non; si igitur moveatur per aliquod spatium, secun-
37vb *dum motum quantitatis | gravitatis finite, sed tamen movebitur per illud* 274a5
spatium in minori tempore; gravitas enim movetur in velocitate et tarditate secundum magnitudinem corporis. Cum enim corpus fuerit magnum, tunc gravitas eius movebitur in minori tempore tempore motus gravitatis minoris corporis; tempus autem minus attribuitur gravitati maiori. Et iam diximus superius quod gravitas infinita omnino non movetur; nulla enim est proportio inter gravitatem finitam et infinitam,
nam gravitas finita movetur per spatium finitum in tempore finito. 274a9

---

16 tamen] in *Y*: *om. BZ* | temporum] -re *XY* 17 autem—movebitur] *om. homoiotel. JW* | autem] aliqua *Y* | movebitur] -bunt *Y* | aliqua] alia *BJY* 18 maior—primum] gravitatis *W* | maior] minor *X* | prima] puta *J* 19 Deinde—tempore] *om. homoiotel. Y* | minimo] minori *X* 21 secundo] tertio *B* 22 in] *om. Z* 23 hoc] hiis *JZ* 25 movebuntur] -bitur *WZ* 28 cum] *om. B* 29-30 inveniemus—adinvicem] *om. homoiotel. X* 29 gravitatum] -tatis *BJ*

52. 3 moveatur[1]] -vetur *XY* 4 gravitatis] *om. Z* 5 tarditate] gravitate *BWY sed corr.* W_2 7 tempore[2]] *om. X*: quam *ZG* 10 proportio] comparatio *BX*

Idest dicamus etiam quod principiorum positorum ad declarandum hoc est quod omnis gravitas finita pertransit spatium finitum in tempore finito. Deinde dixit *Et cum ita sit, si inveniatur* etc., idest et cum hee propositiones fuerint concesse et coniungatur eis gravitatem esse infinitam, necesse est ut moveatur et ut non moveatur: moveatur propter propositionem ut non moveatur propter impossibile quod sequitur ad ponendum illam mobilem. Deinde incepit declarare quomodo contingit ut non moveatur, et dixit *Si igitur moveatur per aliquod spatium* etc., idest quod si gravitas infinita moveatur per spatium equale spatio per quod moveatur gravitas finita, continget ut moveatur per ipsum in aliquo tempore, quamvis sit valde minimum; omne enim motum movetur in tempore.

Deinde incepit notificare quomodo sequitur ab hac positione ut moveatur, et non in tempore finito omnino, et quod positum est
38ra | mobile, et sequitur ab eo ut moveatur, et non in tempore est positum mobile et immobile, et dixit *Gravitas enim movetur in velocitate et tarditate* etc., idest necesse est ut gravitas infinita moveatur et non moveatur ex hoc; velox enim est quod in tempore minori movetur tempore in quo movetur tardum; et corpus gravius velocius est minori; et proportio gravis ad grave est sicut proportio velocitatis ad velocitatem et sicut proportio temporis ad tempus, scilicet quod temporis minoris est maior gravitas, et quod secundum magnitudinem eius est parvitas temporis. Deinde dixit *Et iam diximus superius* etc., idest cum positum est quod proportio gravitatis ad gravitatem in gravibus mobilibus est sicut proportio temporis ad tempus, si posita fuerit gravitas infinita, manifestum erit ex eo quod diximus scilicet quod si posita fuerit gravitas infinita omnino non movebitur. Omne enim quod movetur movetur in tempore, et omne tempus habet proportionem ad aliud tempus maius que est sicut proportio gravitatis ad

---

12 Idest] *om. JY* | etiam] *om. J* : *post* quod *B* 13 quod] quia *B* | finitum] in- *BWX* 14 inveniatur] -niamus *B* 15 propositiones] -portiones *YZ* 16 et—moveatur[3]] *om. homoiotel. X* 16-7 et—moveatur] *om. homoiotel. B* 16 ut[2]] cum *Z* | moveatur[2]] -veantur *Y* | moveatur[3]] *om. WYZ* 17 propter[1]] per *J* | propositionem] pos- *WY* | ut] et *JXY* 21 moveatur] -vetur *XY* | continget] -git *WZ* 23 motum movetur] *inv. B* 25 et[1]] *om. B* 26 est positum] *inv. JZ* : positum *W* 30 velocius est] *inv. W* 31 grave] -vem *WXY* | velocitatis] gravitatis ad gravitatem et tarditatis *Z* 35 gravibus] –vitatibus *VX* 37 erit] est *Z* | diximus] + superius *Z* 38 gravitas] *om. VWYZ* | gravitas infinita] *inv. J* 39 movetur[1]] *om. B* : + non *Y*

gravitatem; et cum inter gravitatem infinitam et finitam nulla sit proportio, continget ut nulla proportio sit inter duo tempora, quapropter in nullo tempore movebitur infinitum; et hoc intendebat cum dixit *Nulla enim proportio est* etc. Deinde dixit *Nam gravitas finita movetur per spatium* etc., idest gravitas finita movetur per omne spatium finitum in tempore finito, infinita autem gravitas in nullo tempore movetur per spatium, quod est impossibile: declaratum est enim in sexto Phisicorum quod omne quod movetur in tempore movetur, et omne tempus est divisibile.

38rb12 53. *Gravitas autem infinita non movetur omnino per illud spatium* 274a9
neque in minimo tempore; et si movetur, nihil proficeret in ponendo 274a10
quod gravitas infinita movetur, quoniam si ita esset, esset alia gravitas
finita que moveretur per illud quo mota est gravitas infinita; continget
ergo quod gravitas finita et infinita moveantur per idem spatium in
eodem tempore, quod est impossibile. Si enim gravitas infinita movetur
per aliquod spatium in aliquo tempore finito, erit etiam alia gravitas 274a15
finita que movetur per aliquod spatium in illo tempore, quod est impossibile. 274a16

Cum declaratum est quod si sit gravitas infinita, necesse est quod moveatur in instanti, quod est impossibile, vult declarare quod etiam sequitur impossibile quamvis concesserimus quod movetur per spatium in tempore; et hoc intendebat cum dixit *Et si moveretur, nihil proficeret*, idest si concesserimus quod gravitas infinita pertransit spatium finitum in tempore, nihil proficit ei qui voluerit ponere possibile
38va scilicet quod gravitas infinita possibile est moveri, et | quod non sequitur impossibile ex ponendo ipsam esse mobile.

---

41 infinitam … finitam] *inv. JZ* 45-6 finita[2]—gravitas] *om. homoiotel. BWXY sed suppl.* W_2 46 autem] + est *Z* : + et W_2 | gravitas] + autem *XY* : + que *Z* : + infinita *B* 48 in[1]] ex *BX* | sexto] septimo *W*

53. 1 spatium] *dittogr. Z* 2 minimo] minori *BJ* | si] nisi *Y*: suum *B* 4 moveretur] –vetur *Y* 7 in—etiam] *om. B* | aliquo] alio *WZ* | tempore] + sit *Z* | finito] *om. X* | etiam] et *J* 7-8 gravitas finita] *inv. B* 8 que] *om. B* | movetur] *om. B*: perambularet *JG* | per] *om. BJ* | aliquod] idem *BJ* 11 moveatur] -veantur *Z* 11-2 vult—impossibile] *om. homoiotel. B* 11 vult] + enim *X* | quod etiam] *inv. J* : + est *Y* 12 sequitur impossibile] *inv. Y*: impossibile *X* 14 idest] et *XZ* 15 proficit] -ciet *JX* : + et *WY* : + in *Z* 16 possibile] im- *BX*

Deinde dedit impossibile quod sequitur ex positione eius quod est gravitatem esse infinitam et mobilem, concedendo quod motus eius est in tempore, et dixit *Quoniam si ita esset, esset alia gravitas magna* etc., idest quoniam si posuerimus gravitatem infinitam moveri per spatium finitum in tempore finito, erit possibile inveniri gravitatem finitam que moveretur per illud spatium in illo eodem tempore; quapropter continget inveniri gravitatem finitam et infinitam que movetur per idem spatium in eodem tempore; quod est impossibile.

Deinde incepit dicere quomodo invenitur hoc possibile, et dixit *Si enim gravitas infinita* etc., idest et hoc contingit quod si posuerimus quod gravitas infinita pertransit spatium finitum in tempore finito, possibile erit inveniri aliam gravitatem finitam pertranseuntem per illud idem spatium in illo eodem tempore. Et hoc quod dicit apparet secundum viam proportionis, quoniam cum posuerimus gravitatem parvam finitam, necessario pertransibit illud spatium in tempore maiori tempore in quo pertransivit ipsa gravitas infinita; cum ergo acceperimus ex gravitatibus aliquam cuius proportio ad istam gravitatem finitam sit sicut proportio minoris temporis ad maius, scilicet temporis in quo pertransivit spatium gravitas infinita, ad tempus in quo pertransivit minor gravitas illud spatium, necesse est ut ista gravitas maior finita, scilicet cuius proportio ad minorem gravitatem est sicut proportio minoris temporis ad maius, secet illud spatium, pertranseundo in tempore in quo pertransit ipsa gravitas infinita; quod est impossibile.

38vb15 54. *Ostensum est ergo et declaratum quod impossibile est gravitatem esse* *274a16*
infinitam, similiter et levitatem infinitam; et cum ita sit, manifestum est quod neque corpora gravia aut levia possunt esse infinita. Et etiam declaratum est ex hiis sermonibus particularibus corpus nullo modo posse *274a20*

18 impossibile] + est *B* 20 esset] *haplogr. B* 23 finitam] + et infinitam *WX* | illo] *om. JW* 23-5 quapropter—tempore] *om. homoiotel. W* 24 continget] -git *J* | moveretur] -ventur *Y*: -vetur *X* 26 possibile] im- *J* 29 aliam] aliquam *Y* 30 dicit] -xit *BX* 33 maiori] + illo *Z* | ipsa] ipsum *BJ* : ipse *X* 36-7 spatium—pertransivit] *om. homoiotel. Y* 38-9 ad—proportio] *om. homoiotel. B* 39 proportio] *om. W* 40 ipsa] ipsum *BJX*

54. 1 est^{2}] *om. BJXZ* | gravitatem esse] *inv. BJX* 2 similiter] sic *Z* 4 particularibus] partibus *WZ* | nullo] in illo *B* | posse] *om. YZ*

esse infinitum. Et etiam declaratum est universaliter in libro quem fecimus de principiis mundi: illic enim diximus quomodo corpus est finitum et quomodo infinitum, hic autem diximus secundum alium modum et secundum speciem. 274a24

Recapitulat ea que dixerat, scilicet quod nulla est gravitas infinita et quod eandem demonstrationem habet gravitas et levitas; et cum hoc sit, declaratum est, et iam erat declaratum quod si corpus infinitum grave et leve inveniatur, necesse est ut gravitas aut levitas eius sit infinita, manifestum est ex hiis duobus fundamentis quod corpus 39ra grave aut leve infinitum impossi|bile est esse. Et hoc intendebat cum dixit *Manifestum est et impossibile est ut sint corpora gravia aut levia infinita*, idest quod cum hec propositio scilicet gravitas aut levitas infinita impossibile est esse infinita, coniungatur propositioni hipothetice que prius declarata est, scilicet quod si corpus grave aut leve sit infinitum, necesse est ut gravitas eius aut levitas sit infinita, manifestum est quod corpora gravia aut levia impossibile est esse infinita. Et quia ex hoc apparet quod omnia corpora sint finita (corpora enim simplicia sunt tria, et in unoquoque eorum declaratum est quod sit finitum), omne igitur corpus est finitum, et quia mundus constat ex hiis simplicibus finitis quantitate et numero, ergo et mundus necessario finitus est.

Et quia hoc declaratum est in Phisicis, alio modo demonstravit hoc; et ne aliquis estimaret esse repetitionem eiusdem, et dixit *Et etiam declaratum est ex hiis sermonibus particularibus* etc., idest dicamus etiam quod differentia inter hanc demonstrationem, quam diximus in libro in quo diximus causas universales entium, est quod hec demonstratio est particularis, illa autem universalis, scilicet quod questio hic est particularis et illic universalis. Deinde dedit modum

---

7-8 alium—secundum] *om. homoiotel. B* 9 est gravitas] *inv. XY* 10 habet] -bent *WY* 11 est—declaratum] etiam *JZ* | iam] in que *B* 13 fundamentis] finitis *Z* 14 infinitum] *post* esse *Z* 15 et] *om. X*: quod *JZ* 16 propositio] -portio *Z* | scilicet] *om. Z* | aut] et *Z* 17 infinita[1]] *om. BJX* | est] *om. B* | infinita[2]] *om. WYZ* | coniungatur] -gantur *B* 19-20 manifestum—infinita] *om. homoiotel. BWXY* 20 aut] ut *Z* 22 declaratum] determina- *JZ* 24 et[2]] *om. YZ* 26 declaratum] demonstra- *Z* 27 et[1]] *om. WXY* | et[2]] *om. JZ* 30 est] *om. WYZ* 31-2 illa—particularis] *om. homoiotel. Y* | scilicet—universalis] *om. homoiotel. X* 32 dedit] dicit *BXY*

per quem demonstratio illic fuit universalis et hic particularis, et dixit *Illic enim diximus quomodo corpus est infinitum* etc., idest quia in Phisicis consideravimus de infinito et declaravimus quomodo potuit esse et quomodo non, in potentia quidem, in actu autem non, hic autem declaravimus quod iste mundus est finitus in actu. Et etiam propositiones per quas fuit hoc declaratum illic sunt universaliores istis, et etiam subiectum est universalius illic hic subiecto: illic enim declara-
39rb tum est quod im|possibile est inveniri corpus naturale simpliciter infinitum cuiuscumque fuerit nature, hic autem declaratum est quod iste mundus est finitus in actu tantum; et hoc fuit declaratum hic per ea que sunt propria unicuique corpori, et illic per universalia. Et hoc intendebat cum dixit *Hic autem diximus secundum speciem et alium modum*, idest hic autem diximus hanc questionem specialiter et magis proprie quam in priori libro et secundum alium modum demonstrationis.

55. *Volumus igitur etiam perscrutari alio modo de mundo utrum sit unus aut plures, dicendo quod si impossibile est ut corpus universale sit infinitum, possibile est tamen plures esse celos. Forte igitur aliquis perscrutabitur de hoc et queret ipsum finitum, dicendo sicut mundus continens nos est ens permanens, non est prohibitum quin etiam sint alii mundi plures, immo non infiniti in multitudine.* *274a24–274a28*

Cum declaratum sit quod partes istius mundi sint terminate numero et etiam finite, declaratum est necessario quod iste mundus est finitus. Et quia anima non desiderat post hoc magis quam scire
39va utrum sit finitum aut infinitum, scilicet scire | utrum sit unus aut plures, dixit quod vult perscrutari de hac questione et vult studere in ea propter nobilitatem illius. Et quia est principium magnum in positione principiorum mundi (mundi enim si essent plures, princi-

---

33 per] propter *XY* | illic fuit] illic fuerit *XY*: fuerit illic *JZ* 34 est] *om. BZ* 35 potuit] ponit *Y* 39 subiectum] *om. B* | est—illic[1]] illius universalius est *B*: illic universalius est *X* | illic[1]] illi *J sed corr.* J_2 | illic hic] hoc illic *Y* | hic subiecto] *inv. J*: hoc subiecto *V* | illic[2]] illis *B* 42 hic] *om. WYZ* 44 secundum] *om. B*

55. 1 igitur] *om. BJ* 3 est tamen] *inv. W* 5 sint] *om. Z*: *post* mundi *(6) WY* 6 plures] + sunt *Z* | immo] uno *JX* | in] *om. BX*: + multiplicatione vel *Z* 7 sint] sunt *XYZ* 9 non] *om. XY*

pia essent plura), dixit etiam *Si corpus universale impossibile est esse infinitum, possibile est ut celi sint plures*, idest potest aliquis querere et dicere quod quamvis sit declaratum quod corpus quod est totum istius mundi impossibile est ut sit infinitum, quid ergo prohibet ut sint plures mundi? Quoniam celum dicitur tribus modis: dicitur enim de toto mundo, et etiam de corpore rotundo scilicet celesti, et etiam de superiori orbe, et hoc in lingua Grecorum secundum quod Aristoteles dicet post. Deinde dixit *Non infiniti in multitudine*: declaratum est enim universaliter quod hoc impossibile est scilicet corpus infinitum esse in actu. Deinde dixit *Sicut mundus continens* etc., hec est ratiocinatio accepta extrinsecus et non per se, et movet hominem naturaliter ad perscrutandum de hoc, et dat superficialiter hoc quasi esset possibile; videmus enim quod de individuis pluribus invenitur plus quam unum, quapropter aliquis estimat hoc de mundo cum mundus sit unum individuum, et vulgus opinatur esse possibile, et plures Antiquorum estimaverunt esse necessarium.

39vb 56. *Sed tamen oportet nos incipere prius et perscrutari quid est corpus* *274a29*
infinitum. Et dicamus quod omne corpus aut est finitum aut infinitum,
aut totum fuerit consimilium partium aut dissimilium partium; si non
consimilium partium, aut erit ex forma finita aut erit ex forma infinita. *274a33*

Cum declaratum est quod mundus est finitus, deinde voluit declarare utrum sit unus aut plures; et iam declarabat etiam in libro predicto de infinito simpliciter, et illud quod declaratum fuit illic de esse infinitum in potentia fuit perfecte et bene dictum; esse autem corpus naturale in actu et infinitum impossibile declaratum est etiam illic universaliter; sed quia illic non declaratum fuit ab eo quot sunt corpora simplicia neque quot sunt loca numeralia, neque quod quodlibet

---

14 etiam] et *BXY* | corpus universale] *om. JZ* | est] + universale *JZ* 15-6 querere ... dicere] *inv. B* 16 quod[3]] *om. B* 18 Quoniam] quod *Z* | dicitur[1]] *post* modis *JYZ* | tribus] partibus *B* | dicitur[2]] *om. Z* 19 enim] *om. YZ* 21 Aristoteles dicet] *inv. W* | infiniti] -tum *B* 27 hoc] *om. Z*

56. 2 infinitum[2]] + si infinitum *Z* 3 fuerit] erit *BJX* 4 finita] *om. Y*: in- *B* | infinita] finita *B* 6 etiam] *om. JZ* 8 perfecte] -tum *JZ* | autem] aut *XYZ* 9 et] *om. BXY* | est etiam] *inv. Z* | illic] *om. W* 10 declaratum] + enim *BZ* 11 quot] *om. B* | numeralia] natura- *BJX* | quod] *om. B* | quodlibet] qui- *B*

illorum est finitum proprie sicut est declaratum hic; et fundamentum demonstrationis in ea est super hoc quod si inveniatur corpus infinitum, aut erit compositum aut simplex: hic autem declaratum est ab eo quot sunt species simplicium corporum, et etiam declaratum est in unoquoque eorum ipsum esse finitum. Sed quia ita est, vidit quod si iteraverit illam demonstrationem quam induxit in predicto libro, erunt propositiones eius magis confirmate, et erit demonstratio eius perfectior et proprior in hac questione; ideoque dixit *Sed tamen oportet prius perscrutari quid est corpus infinitum*, idest et cum perveniemus ad hoc de cognitione specierum mundi corporum, et quod unumquodque eorum est infinitum, oportet nos antequam perscrute-
40ra mur utrum sit unus, oportet nos | iterare perscrutationem universalem scilicet utrum corpus infinitum sit simplex; declaratio enim istius intentionis est firmior declaratione predicta. Deinde incepit dividere sillogismum hipotheticum disiunctum, et est idem cum eo qui est in Phisicis sed firmior, dixit *Omne corpus aut est finitum aut infinitum*, si infinitum aut simplex aut compositum, si compositum aut ex simplicibus finitis in specie idest forma, aut infinitis in specie; et hoc intendebat cum dixit *Aut erit forma finita aut infinita.*

57. *Dicamus ergo quod impossibile est ut sit ex formis infinitis, et hoc* 274a33
manifestum est ei qui concesserit et conceperit propositiones predictas positas et sillogismos convenientes fundamentis. Dicimus enim si primi motus 274b1
infiniti sunt, erunt etiam forme corporum simplicium finite necessario, |
40rb *nam motus corporum simplicium simplices sunt, motus autem simplices* 274b5
sunt finiti, et omne corpus naturale habet motum necessario. Dicamus ergo quod si corpus infinitum sit ex formis finitis, erit quelibet partium eius infinita necessario, ut aqua et ignis, quod est impossibile omnino; iam enim declaravimus quod neque gravitas neque levitas sunt infinite. 274b8

---

12 illorum] eo- *YZ* 13 ea] eo *WX* 14-5 ab—est] *om. homoiotel. B* 16 vidit] + et *Z* 17 iteraverit] numera- *B* : voluit *X* | predicto] primo *Z* 18 confirmante] -mantes *B* 20 quid est] *om. X*: qui sit *B* 21 cognitione] corruptione *Z* 23 unus] + scilicet *JZ* 24 sit] est *B* | simplex] + aut compositum etc. *J*W_2*Z* | enim] igitur *Z*: autem *B* 27 aut^{1}] *B* | est finitum] *inv. Z*: finitum *J* | aut^{2}] + est *XY* 27-8 si infinitum] *om. BZ* 29 idest] et *JY*

57. 2 ei] eis *B* : illi *J* | conceperit] cep- *Z* | propositiones] *om. W*: duas *Y*: radices W_2*ZG* | predictas] + duas *W* | positas] *om. B* 4 etiam forme] forme etiam *post* simplicium *J* | finite] in- *J* 8 ut] et *W*

Cum dixit omnes modos destruendos sui sillogismi, incepit destruere quemlibet eorum ut concluderet hoc quod intendit, tamen tacuit hoc quod manifestum est esse destructum, similiter tacuit hoc quod est iteratum; et hic non est unus sillogismus sed plures hipothetici, quasi ergo diceret quod omne corpus aut est finitum aut infinitum, manifestum est quod si impossibile fuerit esse infinitum, necessario erit finitum. Deinde dixit Sed si fuerit infinitum, aut erit simplex aut compositum; sed simplex impossibile est esse, omne enim simplex aut est grave aut leve aut neutrum; declaratum est autem quod quodlibet eorum est finitum si non est simplex, necesse est esse compositum, si compositum et infinitum aut ex simplicibus et infinitis in specie: et omnia ea erunt finita in magnitudine aut ex finitis in specie, aut quedam eorum aut omnia huiusmodi scilicet infinita in magnitudine. Sed impossibile est ut sit ex simplicibus infinitis in specie, iam enim declaratum est quod modi simplicium corporum sunt tres; similiter impossibile est etiam ut unum eorum aut plura uno sint infinita, quia declaratum est quod omnia simplicia
40va sunt finita; | corpus igitur infinitum, cum impossibile est esse simplex aut compositum (et omne corpus aut compositum est aut simplex), necessario tale corpus non erit. Deinde dixit *Ex formis infinitis*, idest ex corporibus simplicibus infinitis. Deinde dixit *Et hoc manifestum est illi qui concesserit* etc., idest et esse corporum simplicium finitum in specie et forma, manifestum est ei cuius natura est comprehensiva primas propositiones quas diximus, et cognoverit modos sillogismorum quos posuimus concludentes recte; et hoc intendebat cum dixit *Et sillogismos convenientes in situ fundamenti*, idest positos in figura positione convenienti primis propositionibus in veritate.

Deinde dedit propositiones ex quibus apparet corpora simplicia esse finita, et dixit *Dicimus enim si primi motus* etc., idest et unus ser-

---

11 concluderet] -dentes *WYZ* | tamen] cum *Z*: non *W* 12 quod] *om. B* 13 hipothetici] *om. Z* 18 aut[2]] + est *B* 19 autem] enim *W* 20 esse] *om. W*: est *B* | si] sed *Z* 21 et[1]] *om. BJX* | erunt] essent *B* 21-3 aut—magnitudine] *om. homoiotel. X* 21-2 ex—aut[1]] *om. homoiotel. BWY sed suppl.* W_2 22 finitis] infiniti *Z* | aut[1]] *om. Z*: et JW_2 | omnia] *om. B* 24 corporum] *om. WX sed suppl.* W_2 25 sunt tres] *inv. BJX* 26 sint] sunt *B*: fuit *Y* 28 compositum[2]—simplex] est simplex aut compositum *JZ* 29 dixit] *om. WY* 30 Deinde] *om. XY* | dixit] *om. WZ* 31 esse] omne *BX* | finitum] -ta *BXY* 34 posuimus] com- *BY* 35 sillogismos] si nullus *B* | fundamenti] finiti *WY*: simplici *Z*: fundamentis *J* 37 apparet] -rent *Y*: oportet *BX* 38 primi motus] *inv. BX*

monum quos prius posuimus ad declarandum hoc est quod, si motus simplices sint finiti, necesse est ut forme corporum simplicium sint finite. Deinde dedit causam consequentie, dicendo *Motus enim simplicium corporum sunt simplices*, idest quoniam necesse est ut quilibet modorum simplicium habeat motum simplicem, cum omne corpus naturale habet motum et in corpore simplici non inveniatur motus compositus. Et cum dedit causam istius continuationis, posuit antecedens et conclusit consequens, et dixit *Et motus simplices sunt finiti*, idest iam declaratum est superius quod motus simplices sunt finiti immo tres tantum; et cum ita sit, necesse est ut corpora simplicia sint finita dixit; et hoc quod dixit quod *Omne corpus naturale habet motum necessario*, dixit hoc ne aliquis diceret quod non sequitur ut
40vb numerus corporum simplicium sit secundum numerum | motuum, quia impossibile est corpus naturale simplex esse immobile, sed semper quiescens: ipse vero dicit quasi respondendo quod hoc impossibile est secundum fundamenta posita in hac arte, et si fuerit hoc possibile in toto illius corporis simplicis, non erit possibile in partibus, sicut est de terra, que secundum totum est quiescens tamen secundum partes mobilis.

Deinde dixit *Si corpus infinitum est ex formis finitis* etc., idest et cum declaratum est quod modi simplicium sunt finiti in numero, si aliquod corpus inveniatur compositum ex eis infinitum, necesse est ut hec simplicia ex quibus componitur sint finita in magnitudine cum sint finita in specie, scilicet quod unum eorum aut plura uno sint infinita in magnitudine quoniam si simplicia essent finita in specie et in magnitudine, necessario et compositum esset finitum. Deinde fecit rememorationem de hoc quod iam declaravit, quod corpus simplex infinitum in magnitudine est impossibile esse, et dixit *Quod est impossibile, iam enim declaravimus* etc., idest quoniam si

---

39 declarandum hoc] *inv. Z* | hoc] *om. BWY* | est] *om. WXY* 41 dedit] dat *JZ*: dixit *Y* 42 corporum] *om. Z* 42-3 sunt—simplicium] *om. B* 43 modorum] corporum *JZ* | modorum simplicium] *om. X* 46 conclusit] -cludit *JYZ* 47 simplices] *post* finiti *B* 48 est] *om. Z* 52 impossibile] poss- *BWXY* 53 semper] + esse *Z* | vero] ergo *BX* 54 secundum] *om. B* 54-5 secundum—possibile[1]] *om. X* 56 de] *om. XY* | quiescens] + idest *B* 57 partes] -tem *B* 58 est ex] *inv. B* 59 est] + et *Z* | modi] quatuor *Z*: + corporum *J*W_2*Z* | sunt] sint *Y*: sicut *B* | si] scilicet *BY* 61 finita] in- *J*W_2*XZ* 62 scilicet] sed *Z* 64 compositum] cum possit *B* 66 infinitum] finitum *WY* | esse] *post* infinitum *Z*

corpus compositum infinitum sit, aut erit necessario compositum ex gravi et levi, aut secundum commixtionem aut secundum vicinitatem partium parvarum; corpus enim celeste nulli est elementum sicut ex nullo est elemento. Et videtur quod non dixit quod sequitur in unoquoque eorum, ex quibus componitur corpus infinitum esse infinitum in magnitudine, nisi quia elementa sunt contraria; et cum unum eorum sit infinitum, continget ut alterum sit infinitum, nisi infinitum corrumperet finitum.

Si igitur aliquis dixerit quod impossibile est inveniri corpus compositum infinitum ex elementis finitis in forma et magnitudine, sed *41ra* infinitis in | numero, idest inveniri individua eiusdem speciei infinita, verbi gratia terre infinite, dicemus respondendo quod destructio istius sermonis apparet ex potentia illius sermonis. Si enim terre infinite invenirentur, possibile esset inveniri ex eis una terra contracta et additione adinvicem; et cum ita sit, possibile erit inveniri gravitatem infinitam. Et ipse propalabit hoc post et declarabit ipsum perfecte cum declaraverit quod locus elementorum est unus numero; et post hoc declarabitur quod unus est mundus.

58. *Et etiam si forme sint infinite, et loca earum erunt infinita in* *274b8* *magnitudine necessario; et cum ita sit, et motus eorum erunt infiniti,* *274b10* *quod est impossibile si concesserimus propositiones primas scilicet quod* *41rb* *corpus vadens ad inferius impossibile est ut in infinitum va|dat, et quod corpus vadens ad superius vadat, impossibile est ut in infinitum vadat, quoniam impossibile est ut sit quod impossibile est esse: hic autem sermo susceptus est equaliter in qualitate et quantitate et situ. Et dico si corpus* *274b15* *non sit album aut unius cubiti aut currens, impossibile est transmutari*

---

68 infinitum] finitum *WY* | compositum] -suit *B* 69 et] ex *B* 71 sicut—elemento] *om. homoiotel. B* | elemento] -tum *XZ* | videtur] habet *W* 73 infinitum] *om. J*: finitum *BWY* | et] + quod *W* 74 sit[1]] est *BX* 76 corpus compositum] *inv. B* 80 illius] istius *WX* | sermonis[2]] demonstrationis *BX* | enim terre] *inv. X*: ratio enim *B* 81 contracta] -tractu *B* : -tacta *J*: -tactu *W sed corr.* W_2: -tactum *X* 82 additione] *om. BZ*: a diminutione *Y*: *post* adinvicem *W* | adinvicem] + vel aggregata a diminutionibus scilicet partibus infinitis *Z* | erit] esset *BY* 83 declarabit] -ravit *XZ* 84 declaraverit] -ravit *XZ* | et] + cum W_2Z

58. 1 infinite] finite *WY* | earum] eo- *XY* 4 in] *om. B* 4-5 et—vadat[2]] *om. homoiotel. BWYZ* 6 est[1]—esse] *om. BWY sed suppl.* W_2 | ut sit] *om. X* | esse] *om. X* 7 est] *om. BX* 8 sit] + necessario *Z* | currens] in Egipto *G* *in aliquod ex hiis que diximus, et impossibile est ire ad locum ad quem impossibile est pervenire.* 274b17

Vult declarare in hoc capitulo quod impossibile est quod aliquod corpus simplex sit infinitum in magnitudine per aliam demonstrationem, quamvis sit universalior predicta, et dixit *Et etiam si forme sint infinite*, idest et si forme corporum simplicium sint infinite in magnitudine, necesse est ut loca earum sint infinita, et si hoc est, motus eorum infiniti necessario erunt aut in seipsis aut in suis partibus, necesse est enim quod partes corporum simplicium mobilium moveantur in infinitum, non enim movetur nisi ad locum totius; cum igitur locus totius sit infinitus, continget quod partes movebuntur in infinitum; motus autem illius quod in infinitum movetur necessario est infinitus; pars enim movetur ad totum, non secundum quod est totum sed secundum quod est in suo esse naturali; et locus totius corporis et partis est idem.

Et cum posuerit quod necessarium est, si elementa sint infinita in magnitudine, quod motus partium eorum sint infiniti, dixit *Quod impossibile est* etc., idest et ponere motum infinitum impossibile est, secundum quod apparet apud illum qui concesserit primas proposi-
41va tiones quas naturaliter habemus in hoc; | videtur enim quod naturaliter est nobis notum quod omne motum secundum rectitudinem non movetur nisi ad aliquid ad quod cum pervenerit quiesceret, nisi motus eius esset otiosus. Et cum ita sit, manifestum est quod, cum videmus lapidem motum ad inferius et ignem ad superius, sciemus quod non moventur nisi ad loca extra que sunt, et cum pervenerint ad loca illa quiescunt: videtur igitur quod hoc intendebat cum dixit *Si nos concesserimus et conceperimus propositiones primas*. Et secundum

---

9 aliquod] -quid *WZ* : -quo *Y* 11 impossibile—quod] *dittogr. V* | quod²] *om. Y* 13 Et²] *om. B* 15 earum] eo- *JXYZ* 15-6 et—partibus] *om. BWY sed suppl.* W_2 15 est²] et W_2Z 16 necessario erunt] *inv. Z* : necessario W_2 | seipsis] se *Z* 17 est] *om. WY* | enim] *om. XY* | corporum simplicium] *inv. BX* 18 movetur] -ventur *JXY* 19 igitur] ergo *BX* 22 locus] + partis et locus W_2Z 23 corporis] *om. BX* : temporis *Y* | et partis] *om. WYZ* 24 posuerit] -suit *BW* | est] + quod *BX* 25 sint] sunt *Z* 26 motum] -tus *B* 28 videtur] -demus *BY* 30 quiesceret] -cet *WXZ* : -cat *J* 32 videmus] -dimus *B* 33 moventur] -vetur *BX* | pervenerint] -nerit *BJ* : -niunt *Z* 34 loca illa] *inv. BX* | videtur] + enim *Z* | igitur] enim *Y*

hanc explanationem erit sermo eius scilicet quod corpus vadens ad inferius aut ad superius impossibile est ut vadat in infinitum, et ex hiis propositionibus et potest intelligi, quod qui concesserit propositiones primas in hoc genere, credet quod corpus vadens ad inferius aut ad superius impossibile est ut vadat in infinitum; et hec explanatio est convenientior manifesto sermoni ipsius. Post enim dat medium terminum quo declarat hoc *Et est impossibile, enim impossibile est*, idest quod ponere motum infinitum est ponere mobile ad finem ad quem non potest pervenire: impossibile est igitur ponere motum huiusmodi, impossibile est enim incipere esse quod impossibile est ut sit; et hac eadem demonstratione utebatur in fine sexti Phisicorum.

Et cum dedit propositionem (dicentem quod impossibile est ut sit) impossibile est incipere, incepit verificare hanc propositionem inducendo omnia predicamenta, et dixit *Et hic sermo susceptus est et concessus equaliter in qualitate et quantitate et loco et situ*, idest quod hec propositio vera est in omnibus predicamentis: vera est igitur in *41vb* ubi et motu sicut in qualitate et in aliis predicamentis. | Deinde dedit exemplum et dixit *Si impossibile est ut corpus sit album neque unius cubiti neque currens terminetur, impossibile est transmutari in aliquod ex hiis que diximus*, idest verbi gratia de qualitate, idest si impossibile est quod aliquod corpus fiat album, impossibile est transmutari in album; et similiter si impossibile est quod aliquod corpus sit aliquorum cubitorum, impossibile est transmutari in aliqua hora inesse illorum cubitorum; et similiter si impossibile est quod aliquid perveniat currendo et ambulando ad aliquid, non invenietur incipiens currere ad illum locum.

---

37 ad] *om. B* | ut—infinitum] ut infinitum vadat *B* 37-40 et—infinitum] *om. homoiotel. WXYZ* 38 qui] *om. B* | concesserit] -cessit *B* 39 genere] *om. B* 40 explanatio] -positio *BX* 42 enim] *om. X*: est *praem. BJW*: esse *praem. Z*: + quod *Y* 42-3 impossibile est] *om. JX* 43 motum] + est *B* 44 finem] + et *Z* 49 est] *om. Z* 51 et^{2}] *om. Z*: in *WY* | et^{3}] *om. WY*: + in *JW* 52-3 vera^{2}—predicamentis] *om. homoiotel. B* 52 in^{2}] numero *Y* 57 quod] ut *WZ* | fiat] fuit *Y*: sit *BJ* 59-60 impossibile—cubitorum] *om. homoiotel. Y* 59 est] *om. Z* 60 quod] ut *W* 61 et] per *Z* | invenietur] -nitur *J*: -niretur *X* 62 illum locum] *inv. Z*

59. *Et etiam si forma corporis ignis invenietur separata et disiuncta* 274b18
adinvicem, ut quidam Antiquorum inquiebant, erit ignis compositus ex omnibus diversis formis infinitis. 274b19

Cum declaravit quod impossibile est invenire corpus simplex infinitum in magnitudine, et quod impossibile est inveniri corpus compositum ex eis unum, et aliquis potest dicere Forte est impossibile invenire unum corpus continuum infinitum discretum autem partium est possibile ut Antiqui dicebant de corporibus indivisibilibus; erunt ergo individua ignis infinita, et unumquodque eorum finitum. Et cum ita sit, dixit respondendo quod contingit impossibile predictum: nam possibile est ponere istas partes congregatas, quapropter contingit magnitudinem unam infinitam esse, et sic continget infinitam gravitatem esse; et iam declaravimus hoc esse impossibile. Et hoc
42ra inten|debat cum dixit *Erit ignis compositus ex omnibus diversis formis infinitis*, idest quod si partes infinite separate abinvicem inveniantur, possibile est continuari adinvicem, et sic proveniet ex eis magnitudo infinita, quod est impossibile; impossibile autem non sequitur ex possibili, illud ergo quod ponebatur possibile est impossibile; et cum contingit quod possibile sit impossibile, continget quod positio propter quam possibile sit impossibile sit possibilis.

60. *Et iam diximus superius quod corpus vadit per omnes mensuras. Si* 274b20
igitur ita sit, quomodo erit possibile quod plura corpora dissimilia sint infinita ita quod unumquodque eorum? 274b21

Intendit per omnes quantitates tres dimensiones, que sunt ad medium et a medio et circa medium, idest quod cum declaratum est quod motus sunt tres, et quod causa illius est quod dimensiones sunt tres, et necesse est quod corpora simplicia sint secundum numerum

---

59. 1 invenietur] -niatur *BJX* 2 inquiebant] -quirebant *B* 3 formis] *om. B* 4 declaravit] -raverimus *B* 7 invenire] -niri *WY* 7-8 partium est] parvum est *JY*: est parvum *Z*: partium forte est *X* 8 possibile] proprie *WXY* 9 erunt] erant *BZ* 12 infinitam[1]] finitam *B* 12-3 et—esse[1]] *om. homoiotel. WXYZ* 13 gravitatem esse] *inv. B* 15 abinvicem] adin- *Y* 18 ergo] autem *WZ* 19 possibile] im- *B* 20 possibilis] im- *JW*$_2$*Z*

60. 3 eorum] ea- *Y* 6-7 sunt tres] *inv. BYZ*

motuum simplicium, et quod dimensiones etiam sint finite, ut declaratum fuit prius: quomodo igitur estimantur corpora dissimilia esse scilicet simplicia diversarum formarum, et infinite? Et quia in demonstratione predicta non complevit declarationem in unoquoque modorum corporum simplicium quod impossibile est scilicet ut sit 42rb infinitum; sed tamen fecit hoc in simplicibus | mobilibus recte, quoniam cum perscrutatus fuit de corpore infinito et divisit ipsum in simplex et compositum, non incepit prius complere divisionem simplicium modorum et destruere unumquodque eorum esse infinitum, deinde post hoc destruere corpus compositum esse infinitum ut oportet in recto ordine, sed dimisit loqui de modis corporum simplicium; et intendebat destruere hoc in corpore composito, quoniam in eo est destructio in hoc corporis simplicis.

Et hoc est cum dixit quod compositum infinitum aut est infinitum quia unumquodque simplicium ipsius est infinitum magnitudine, aut quia sunt finita magnitudine et infinita specie. Videt igitur quod bonum est et rectum facere hoc, sed quia corpus celeste est corporum simplicium et nullum est compositum ex eo, non intrat istam destructionem, scilicet inveniri corpus infinitum corporum simplicium que sunt elementa corporum compositorum; quapropter revertitur hic ad complendum modos corporum simplicium et ad declarandum ex predictis quod impossibile est quod aliquod corpus sit infinitum in magnitudine, sive sit elementum compositi sive non.

42va **61.** *Et etiam si forma corporis sit consimilis, impossibile est etiam ut sit* 274b22 *infinita, primo quod si alius motus a motibus predictis non sit, necessario hoc corpus habebit aliquem motum illorum motuum, et si ita sit, necessario est gravitas infinita et levitas infinita; et iam diximus hoc esse* 274b25 *impossibile, et etiam non assimilatur corpori eunti circulariter in infinitum; impossibile est enim corpus infinitum ire circulariter. Qui ergo*

9 dissimilia] dum similia *B* 10 infinite] -ta *XY* 11 demonstratione] divisione *BXY* 13 tamen] tantum *XY* | recte] -to *Z* 20 destructio] distinctio *Y*: demonstratio *BX* 22 infinitum] + in *JZ* 24 corpus] *om. XY* 25 est compositum] *inv. BX*: compositum *WY* 27 compositorum] *om. WYZ* 29 corpus] illorum *BJX*

61. 2 quod] quia W_2Z: quoniam *J* 3 illorum] horum *BJX* 4 est] *om. B* | hoc esse] *inv. J*: quod omne *B*: + est *J* 6 impossibile] poss- *WYZ* *dicit iterationem circularem esse perpetuum, similis est dicenti celum esse infinitum; et iam declaravimus superius hoc esse impossibile, et etiam quod corpus infinitum impossibile est moveri omnino.* *274b30*

Iam diximus causam propter quam iteravit sermonem de destructione corporis simplicis infiniti in magnitudine; et dixit *Et si forma corporis* etc., idest cum forma corporis fuerit consimilis idest simplex non composita, impossibile est ut forma sit infinita; et hoc dixit quia, cum causa finitatis magnitudinis est forma, necesse est, si magnitudo sit infinita, quod forma sit infinita. Deinde dixit *Quoniam si alius motus a motibus* etc., idest et cum declaratum est quod nullus motus est simplex preter tres motus quos numeravimus, et omne corpus simplex habet motum simplicem, necesse est ut hoc corpus positum infinitum habeat aliquem motum horum motuum. Deinde dixit *Et si ita sit, necessario erit gravitas infinita*, idest et si *42vb* necessarium fuerit ut hoc cor|pus infinitum habeat motum simplicem, aut erit circularis aut rectus: si rectus, aut ad medium et tunc erit grave, aut a medio et tunc erit leve; si autem corpus infinitum fuerit grave, necesse est ut sua gravitas sit infinita, secundum quod prediximus, et similiter de levi; et iam declaravimus superius quod **T 48a** impossibile est inveniri gravitatem infinitam aut levitatem infinitam.

Deinde dixit *Et etiam non assimilatur corpori eunti circulariter*, idest et impossibile est etiam ut motus istius corporis simplicis sit circularis, ut sit simile corpori rotundo idest sue speciei, scilicet corpori cuius motus est circa medium; impossibile est enim corpus infinitum moveri motu circulari, idest quia motus circularis est finite revolutionis, motus autem corporis infiniti est infinitus. Deinde dixit *Qui ergo dixerit itionem circularem esse perpetuam* etc., idest quod sermo dicentis motum circularem esse infinitum est similis in potestate sermoni dicentis corpus mobile circulariter ut celum esse infinitum. Deinde

7 iterationem] itio- *X*: motio- *BJ* 9 est] *om. B* 11 infiniti] -ta *XY* 12 etc.] *om. WZ* | etc.—corporis] *om. homoiotel. BXY* | idest[2]] *om. WYZ* 16 motus a motibus] modus a mobilibus *Z* | et] *om. BX* 17 quos] + narravimus vel *W* | numeravimus] -ramus *Z* 19 positum] com- *B* 19-21 horum—motum] *om. homoiotel. X* 19 horum motuum] omnium motuum horum *B* 22 si rectus] *om. B* 25 prediximus] dix- *Z* 28 etiam] *om. B* | sit] *om. Z* | circularis] infinitus W_2Z 33 dixerit] -xit *B*: -cit *J* | itionem] iteratio- *Z*: itatio- *X*: motio- *J*: motione *B* 34-5 motum—dicentis] *om. homoiotel. B*

dixit *Et iam declaravimus superius hoc esse impossibile* etc., idest quod corpus rotundum impossibile est esse infinitum, cum declaravimus quod corpus infinitum impossibile est moveri neque motu recto neque circulari, aut intendit; et iam declaravimus etiam universalibus hoc, scilicet quod corpus infinitum impossibile est moveri omnino aliquo modo motuum.

43ra 62. *Et etiam si possibile est corpus infinitum moveri, aut naturaliter aut* *274b30* *violente: si violente, omnino habet alterum motum naturalem, habebit ergo alterum locum quo eat naturaliter, sicut in quo it violente, quod est impossibile.* *274b32*

Hec est demonstratio alia a predictis, et est propria corporibus motis recte; et quasi laborat in ea ad perficiendum ex divisione motus illud quod non fuit positum in demonstrationibus predictis: demonstrationes enim predicte fundate sunt super hoc quod corpus infinitum mobile in rectum impossibile est moveri motu naturali: si enim moveatur motu naturali, non movebitur nisi per declinationem que est in eo scilicet gravitatem et levitatem. **T 48b**

Declarat igitur illic quod si moveatur per gravitatem aut levitatem, necessarium est inveniri illic gravitatem aut levitatem infinitam. Deinde ostendebat hoc esse impossibile, propter hoc quod declaratum fuit corpus infinitum impossibile esse moveri naturaliter, quasi igitur aliquis posset dicere Forte movetur violente; propter quod dixit respondendo quod si movetur violente, necessario habet motum naturalem: violentia enim non dicitur nisi in respectu nature, et si res non habuisset motum naturalem, non intelligeretur habere motum violentum. Deinde dixit *Habebit igitur alterum locum* etc., idest secundum quod videtur quod locus in quo movetur idem corpus vio-

36 quod] *om. BX* 37 impossibile—infinitum] infinitum esse impossibile *BX*

62. 1 possibile] im- *XY* 2 violente[1]] -lenter *YZ* | violente[2]] -lenter *Y* | alterum] alium *Z* 3 eat] eant *X*: erat *Z* | it] id *B*: ut *Z* 6 motis recte] *inv. BX* 9-10 si—naturali] *om. homoiotel. B* 10 declinationem] inde- *JW* 11 eo] eis *B* 12 Declarat] -ravit *X* | Declarat—levitatem] *om. homoiotel. Y* | igitur] ergo *JZ* 14 hoc[2]] *om. XY* 15 esse] est *JW* | moveri naturaliter] *inv. B* | quasi] quia *Z* 16 Forte movetur] forme moventur *B* | violente] -lenter *BY* 17 movetur] -veretur *B* 20 Habebit] -bebat *Z*: -beo *X*: -bebitur *B* | alterum] + violentum *B* 21 violente] -lentum *WZ*: -lenter *XY*

lente alius est a loco in quo movetur naturaliter. Et non invenitur ita in corporibus simplicibus, sed videmus quod locus in quo movetur terra violente est idem cum loco in quo movetur naturaliter; sed motus naturalis non differt a violento nisi secundum modum tan- *43rb* tum; motus enim in aere si | fuerit ad superius erit violentus, si ad inferius naturalis; sed tamen visum est quod non intendebat, per locum motus, locum in quo movetur, sed locum quo movetur, quasi igitur diceret Erit igitur alter locus quo movetur naturaliter.

63. *Et dicimus quod corpus finitum impossibile est agere in corpus infi-* *274b33* *nitum et recipere et ea recipere actionem abinvicem omnino, et demonstratio huius est sic: sit ergo corpus infinitum A et finitum B, et sit tem-* *275a1* *pus in quo movetur aut movet C; et expellatur corpus A, et recipiat aliquam formam B in tempore C; et sit D minus quam B, et hoc minus moveatur per alteram magnitudinem minorem in tempori equali primo,* *275a5* *et sit H, quod movetur ex D; dicamus quod proportio D ab B est sicut proportio H ad aliud minus A, et est Z.* *275a7* T

43va

Hec est demonstratio universalis omnibus corporibus rectis et rotundis, et simplicibus et compositis. Omne enim corpus aut est agens aut patiens, aut utroque participans; et cum ita sit, potuit declarare corpus infinitum non esse, nam cum posuerit quod omne corpus aut est activum aut passivum, aut activum et passivum simul, et posuit quod corpus infinitum impossibile est agere aut pati, concluditur ei in secunda figura quod nullum corpus est infinitum,

22 a] in *Z* | ita] *om. Z* 25 tantum] *ante* secundum *BX* 27 non] *om. B* : tamen *X* 28 locum[3]] locus in *B*

63. 2 et recipere] *om. BZ* | ea] eam *Z* 3 ergo] igitur *BJ* 4 corpus] *om. WXYZ* | recipiat] -piet *YZ* 6 per] *om. B* | minorem] + se W_2*Z* 7 dicamus] *om. X* : -cimus *W* 8 A] fine *W* : sine *Y* | Z] + infinito quod tamen est finitum *B* 10 et[2]] *om. B* 13 aut[3]—passivum] *om. homoiotel. B* 14 posuit] -suerit *WY* 15 corpus] *om. Z* : -porum *J* sed quia propositio dicens quod omne corpus aut est activum aut passivum est manifesta per se. Corpora enim celestia videntur agere in ea que sunt sub eis, ista autem videntur agere et pati adinvicem. Et altera propositio non est manifesta scilicet dicens quod infinitum impossibile est agere aut pati; hec autem propositio dissolvitur in duas, quarum una est quod infinitum impossibile est pati, altera autem quod infinitum impossibile est agere in rem finitam.

Incepit declarare propositionem dicentem quod corpus infinitum impossibile est pati a finito, et accepit eadem exempla utriusque *43vb* questionis, deinde divisit demonstrationem et dixit | *Sit ergo corpus infinitum A* etc., idest si aliquis finxerit corpus infinitum reperiri patiens aut agens, sit corpus infinitum linea A, et ponamus aliud corpus finitum et sit B, et ponamus tempus in quo egit A in B aut patitur A a B, et sit tempus C. Deinde incepit demonstrationem alterius istorum duorum, et est ut A infinitum patiatur a B finito in tempore C scilicet secundum totum, et dixit *Et expellatur A et recipiat aliquam formam B*, idest impossibile est ut corpus infinitum patiatur a finito, T 49b quod si sit possibile, ponatur quod A infinitum patiatur secundum totum a B finito in tempore C posito. Et cum posuit propositionem quam intendit destruere, coniungit illi propositionem possibilem et dixit *Et sit D minus B*, idest et coniungamus huic propositioni possibile hoc scilicet ut aliud corpus sit finitum et minus corpore finito quod est B, hoc autem corpus sit D, idest quod ista positio est possibilis, et non contingit ex ea per se aliud impossibile.

Deinde dixit *Et moveatur hoc minus per aliquam magnitudinem minorem in tempore equali primo* etc., idest et ponamus quod corpus *44ra* agens quod posuerimus minus corpore agente quod | est B, scilicet D, patiatur ab eo aliud corpus secundum totum in tempore in quo ex B finito patiatur A infinitum, scilicet in tempore C; et sit H hoc cor-

---

17 est manifesta] *inv. BY* 19 altera] alia *Z* | infinitum] *post* est *(20) B* 21 altera] alia *BJ* 22 autem] est *BJX* | in rem] *om. Z* | finitam] *om. Z* : in- *WY* 25 questionis] passio- *B* | Sit] si *WY* 26 finxerit] fix- *WZ* 27 patiens ... agens] *inv. W* | sit] *om. J* | linea] nam *B* : fit *J* | aliud] illud *B* 28 ponamus] sit *B* 29 sit] si *B* 30 istorum duorum] *inv. BX* | patiatur] *om. BWXY sed suppl.* W_2 31 Et^2] *om. BX* 32 a finito] *om. BWXY sed suppl.* W_2 33 sit possibile] *inv. J* : possibile *B* 35 coniungit] -tingit *BY* 36 possibile] + est *B* 37 sit] *om. WZ* 39 aliud] aliquod *BJX* 42 posuerimus] -suimus *BX* : -nimus *J* 43 aliud] aliquod *BX* 44 patiatur] -titur *BY*

pus quod patitur a minori agente scilicet D; et omne hoc est possibile. Et cum posuit hoc possibile, posuit quod sequitur ex eo, et dixit *Quod proportio D ad B est proportio H ad minus fine scilicet Z*, idest dico ergo quod necessarium est quod proportio minoris agentis scilicet D ad maius agens scilicet B sit sicut proportio agentis ex D minoris scilicet H ad corpus alterum agens minus corpore A infinito; et cum posuit quod hec proportio est necessario ita, incepit declarare hoc et dixit.

64. *Et dicamus etiam quod corpora equalia in magnitudine moventur et transmutantur equaliter; et propter hoc corpus minus movetur minori motu in illo tempore, et corpus maius movetur motu maiori in illo tempore; movetur enim secundum suam magnitudinem magis quam motu minoris temporis.* *275a7* *275a10*

44rb11 Idest quia videmus agentia equalia in patientibus equalibus, patiuntur ab eis in omnibus partibus in tempore equali, et videmus maius agens agere in patientia equalia in eodem tempore magis quam T 50a agens minus, et agens minus minus, et similiter agens maius patitur ab agentibus equalibus minus, et minus magis. Et cum contingit ex hoc quod cum ab agentibus diversis patiantur patientia diversa, quod proportio agentium ad agentia sit sicut proportio patientium ad patentia. Quomodo etiam ex hiis fundamentis positis declaratur quod intendit declarare, apparet sic. Cum enim posuerimus quod agentia equalia agant in patientia equalia eadem actione in eodem tempore, apparet ex hoc quod agentia equalia, quorum proportio adinvicem est sicut proportio patientium adinvicem, agunt actiones equales in temporibus equalibus: agentia enim equalia et patientia equalia sunt huiusmodi scilicet quod proportio agentium adinvicem est sicut proportio patientium adinvicem, sed tamen ista sunt

---

46 posuit[1]] -suerit *YZ* | eo] hoc *Z* 47 fine] *om. BJY* : sive *Z* 49 sit] *om. B* : est *X* 52 et dixit] *om. B*

64. 1 etiam] *om. JX* 2 equaliter] + in tempore uno *WXY* | et] quod *WX* 3 illo[1] tempore] *inv. B* | corpus] *om. B* | maiori] -re *B* 4-5 movetur—temporis] *om. homoiotel. BW* 5 temporis] + Manifestum—comparationem *(= textus [65], infra)* J_2*X* *(cf. Praefatio, p. 47*)* 6 Idest] *om. BY* 10 cum] tunc *BJX* 13 positis] + quod *B* 14 posuerimus] -suimus *BJ* 15 agant] agunt *W* 16 equalia] *om. BX* 17 est—adinvicem] *om. homoiotel. X* | patientium] agentium *Y*

inequali proportione, alia autem in aliis proportionibus. Et cum hoc *44va* apparet in uno modo modorum proportionalium, apparebit | in aliis proportionibus; et cum hec propositio est verificata, scilicet dicens quod actiones omnium agentium, quorum proportio adinvicem est sicut proportio patientium adinvicem in patientia respondentia eis, scilicet actio precedentis agentis in precedens patiens, et consequentis agentis in consequens agens sunt actiones proportionales in temporibus equalibus; et cum ita sit, manifestum est quod querenda est eius *44vb* conversa, et est quod agen|tia et patientia in temporibus equalibus sunt proportionalia.

Dicamus ergo quoniam ita est, verbi gratia ponamus duo agentia quorum alterum sit A, alterum autem B, et duo patientia quorum alterum sit C et alterum D, et ponamus quod totum C patiabur ab A et D a B in eodem tempore, et sit tempus G; dico ergo quod proportio agentis ad agens est sicut proportio patientis ad patiens, scilicet quod proportio A ad B est sicut proportio C ad D et econverso.

44va13 Et demonstratio eius est quia si proportio A ad B non fuerit sicut proportio C ad D, sit ergo maior aut minor; si igitur proportio A ad T 50b B sit minor proportione C ad D, secemus ex A aliquid cuius proportio ad B sit sicut proportio C ad D, sit ergo AH; proportio ergo H ad B sicut proportio C ad D, erit passio D a H et D a B in eodem tempore, sit ergo illud tempus G; sed iam posuimus quod in hoc tempore patitur C ab A, secundum totum cuius H est pars; ergo in eodem tempore patitur idem patiens eadem passione a duobus agen-

21 inequali] -lia *B* : inequa *XY* | proportione] *om. B* 23 hec] *om. WX* 25 patientium] partium *B* 26-7 in—agentis] *om. homoiotel. BY* 26 patiens] agens *WX* 27 sunt] sint *W* 28 cum] + inquam *J*W_2 | sit] *om. W* 28-9 querenda—conversa] conversa eius querenda est *Z* 30 proportionalia] -nales *BWXYZ* 32 A] + et *B* 33 sit] est *Z* | ab] sub *YZ* 34 quod] + ista *B* 35 ad agens] *om. B* 36 econverso] + si enim actiones eedem in eodem tempore W_2*Z* 37 est] *om. Z* : + sic *J* : + sicut *B* 39 minor] maior *JZ* 40 sit[1]] sint *B* 41 erit] + ergo *BJ* | D a H] C ad AH *J* : D ad H *Z* : D ab H *X* | a[2]] ad *JY* 43 cuius] a *W* | est] erit *YZ* : erat Z_2 44 a] *om. B*

tibus quorum unum est maius alterum autem minus; quod est impossibile.

Et propter hoc dixit prius Aristoteles quod maius agit actionem maiorem in eodem tempore quam minus, scilicet in idem patiens; et quasi Aristoteles fuit contentus in hac demonstratione positione pro- *44vb13* positionum ex quibus sequitur hoc. | Omne ergo hoc videtur quasi omissum ab eo quia sequitur ex eis que posuit; sed non omisit hoc in sermone, immo omisit permutationem proportionis et impossibile contingens; cum enim permutaverimus proportionem, continget quod proportio agentis ad patiens primum sit sicut proportio agentis secundi ad patiens secundum, scilicet quod proportio B ad AH, quod est pars infiniti, sit sicut proportio D, quod est agens secundum, ad H quod est patiens secundum. Et cum ita sit, continget ut in eodem tempore et ab eadem re et eadem passione patiantur duo corpora, quorum alterum est finitum scilicet AH, et alterum infinitum scilicet totum A; quod est impossibile.

[65.] *Manifestum est igitur quod corpus infinitum impossibile est movere* *275a10* *aliud corporum finitorum omnino, neque in aliquo tempore. Erit enim aliud corpus parvum quod movebit aliud minus se in illo tempore, secundum modum parvitatis sue, et hoc erit finitum, nam nullam autem habet corpus finitum et infinitum proportionem aut vicinitatem neque comparationem.* *275a14*

Et cum apparet ex hiis propositionibus quod corpus infinitum *45ra* non agit in corpus finitum, dixit | *Manifestum est igitur quod corpus*

---

45 unum] unus *B* 49 quasi] quare *X*: quod *Z* | demonstratione] + et *WZ* 50 hoc[1]] *om. WY* 52 sermone] + uno *B* | proportionis] -nibus *B* 54 primum] *om. BXY* 55 secundum] + patiens *B* 56 est[2]] *om. BY*

[65.] 1-6 Manifestum—comparationem] *om.* J_2X 1-2 Manifestum—omnino] iterum nam ut *(obsc.)* ipsum movetur secundum modum magnitudinis sue et maioritatis sue movetur plus quam moveatur corpus minus. Iam ergo ostensum est nunc quod corpus infinitum non est possibile ut moveatur corpus aliquod ex corporibus finitis BW_2G 1 Manifestum] *om. Z* | igitur] + ex hiis *J* 2 aliud] aliquod *Z* | finitorum] in- *YZ* | omnino] omnia *Y* | neque] enim *Y* | enim] *om. B* 3 aliud[1]] alterum *YZ* | movebit] -vet *YZ* | aliud[2]] alterum *YZ*: + corpus *BJ* | se] *om. Z* 4 secundum—nam] *om. Y* | et—erit] ergo erit hoc iterum corpus *BJG* | finitum nam] infinitum *Z* | nullam] non est *BG* | autem] *om. BJG* 5 habet] inter *BG* | proportionem] -tio *B* | aut] neque *BJG* | vicinitatem] propinquitas *BG*: propinquitatem *J* 6 comparationem] -tio *BG infinitum impossibile est movere aliud corporum finitorum omnino, neque in aliquo tempore*, idest et manifestum est ex hiis que diximus de proportione questio secunda quam intendimus, et est quod est impossibile corpus infinitum agere in corpus finitum in aliquo tempore; quoniam quando duo agentia diversa agunt in idem patiens eadem actione, continget ut sit in temporibus diversis et ut sit proportio agentis ad agens sicut proportio temporis ad tempus.

Si ergo posuerimus alterum agentium finitum ad alterum infinitum, continget ut infinitum agat in nullo tempore; quapropter nullam proportionem habebunt in actione cum nullam habeant in quantitate. Et si posuerimus ipsum moveri in tempore, scilicet infinitum movere finitum, possibile est reperiri aliud corpus finitum quod in illo eodem tempore aget actionem infiniti secundum modum pre- T 51a
dictum; et sic igitur inveniemus duo agentia, quorum alterum est finitum et alterum infinitum, que agunt in eadem actione et in eodem tempore. Et hoc intendebat cum dixit *Erit enim alterum corpus parvum quod movet alterum minus in illo tempore, et erit hoc etiam finitum*, idest quoniam quando posuerimus corpus infinitum movere aliquod corpus finitum in aliquo tempore, possibile est invenire alterum corpus finitum, scilicet minus quam infinitum, movere corpus minus quam quod movet infinitum in illo eodem tempore.

Cum ergo duplicaverimus agens scilicet finitum et similiter patiens finitum parvum, possibile est pervenire ad corpus finitum
45rb quod movebitur in illo eo|dem tempore, corpus scilicet motum ab infinito, propter illam proportionem quam diximus de agentibus et patientibus; igitur in eodem tempore agunt eadem actione et in idem corpus duo agentia eiusdem speciei, quorum alterum est finitum et alterum infinitum. Deinde dixit *Nullam habet corpus finitum et infinitum proportionem aut vicinitatem neque comparationem*, idest et hoc impossibile non contingit nisi confessi fuerimus proportionem habendam; cum enim concesserimus corpus infinitum agere in tem-

---

9 corporum finitorum] corpus finitum *B* 10 neque] + enim *Z* | hiis] eis *WXY* 13 quoniam quando] *inv. B* 17 infinitum] finitum *Y* 19 si posuerimus] cum posuimus *J* 20 reperiri] recipere *BX* 22 igitur] *om. B* : ergo *X* 23 et[1]] *om. W* | in[1]] *om. XY* | in[2]] *om. BX* 25 erit hoc] *inv. BZ* : + erit *Y* 30-1 et—parvum] *om. BWXYZ sed suppl.* W_2 32 scilicet] *om. BWXY sed suppl.* W_2 34 agunt] agent *JZ* 35 est finitum] *inv. W* | et] *om. JZ* 36 finitum ... infinitum[2]] *inv. B* 39 concesserimus] -simus *Z* | infinitum] finitum *BWXY sed corr.* W_2

pore, concederemus habere proportionem ad finitum; sed in rei veritate nullam habent proportionem, sicut dixit; quapropter continget necessario ut actio eius sit in illo tempore.

45va **65.** [66.] *Et etiam corpus infinitum non movet corpus finitum in aliquo* 275a14 *tempore. Sit ergo potentia infinita in linea A et corpus finitum in linea BZ, et sit tempus, in quo sit motus in linea C; D igitur movetur in C minus quam motus BZ, ergo erit proportio totius BZ ad Z sicut proportio motus D ad H; et cum ita sit, contingent quod corpus finitum et infi-* 45vb *nitum moveant idem et in eodem tempore, quod | est impossibile. Iam* 275a20 *enim diximus quod corpus maius movet aliquid in minori tempore quam minus, et hoc erit in omni tempore in quo invenitur actio quam* T 51b *agit per corpus finitum corpus infinitum et corpus finitum. Et impossibile est ut moveat aliquid in tempore infinito, quoniam non habet finem; motus autem est actio que non recipit formas.* 275a24

46ra Cum retulit quod ex propositionibus positis potest declarari secunda, scilicet quod impossibile est ut corpus infinitum transmutet corpus finitum, incepit declarare hoc per lineas, dicendo *Sit ergo potentia infinita in linea A*, idest sit corpus infinitum potentia et magnitudine, cum illud quod est infinitum in magnitudine est infinitum in potentia scilicet corpus agens in linea A. Deinde dixit *Et sit corpus finitum linea BZ* etc., idest et sit corpus finitum patiens ab infinito linea cuius una pars est B et reliqua pars est Z, et sit tempus in quo patitur totum BZ finitum ab A infinito linea C. Deinde dixit

40 concederemus] -demus *JXZ* | finitum] in- *XZ* | veritate] -tem *B* 41 habent proportionem] *inv. W* 42 illo] nullo *JX*

65. [66.] 1 non movet] *inv. Z* | corpus[2]] *om. W* 2 et] + sit *JZ* 3 movetur] -vet *Z*: -vebit *J* 4 ergo—proportio[1]] erit ergo proportio *BJ* : ergo proportio erit *Z* | totius] *om. B* 6 moveant] -veatur *XY* 8 minus] + et isti sunt modi litterarum omnium sicut vides hic W_2*ZG* 9 finitum[2]] in- *Z* | Et[2]] *om. YZ* 12 declarari] + potest *Z* 13 transmutet] -mutat *J*: transivit *B* 14 ergo] igitur *BXY* 15 infinita] finita *B* | sit] si *BX* 17 in[2]] *om. BXY* 19 pars[2] est] *om. W*: pars *XZ*

D igitur movetur in C minus motu B et Z, idest et ponamus aliud corpus scilicet D quod agat in corpus BZ finitum in quo egit A infinitum in illo eodem tempore, scilicet C; necessario igitur erit actio D in ipsum minor actione A, cum D sit finitum, A autem infinitum, scilicet quod D agit in alteram partem BZ, non in utramque, verbi gratia aut in parte A aut in parte Z. Et hoc intendebat cum dixit *D igitur movebit in C minus quam motus B et Z*, idest quod D agens finitum agit in tempore C in patiens BZ actione in parte minori omnium partium Z, scilicet quod agit in hoc tempore in aliqua parte BZ et non in toto.

Deinde dixit *Erit igitur proportio totius BZ ad Z sicut proportio D ad H et A*, idest erit proportio maioris patientis, quod est B et Z, aggregata ad minus patiens quod est B, sicut proportio minoris agentis quod est D ad partem agentem corporis infiniti, et sit AH. Et T 52a
intendit quod cum permutaverimus proportionem, erit proportio agentis, quod est D, ad patiens minus quod est B, scilicet quod est
46rb pars BZ, sicut proportio maioris agentis | quod est pars infiniti scilicet AH, ad maius patiens quod est BZ, congregata. Et cum ita sit, continget necessario ut BZ patiatur ex AH finito in tempore C in quo patiebatur ex toto A infinito; patiens igitur totum BZ patitur in tempore C a duobus agentibus, quorum unum est finitum scilicet AH, et alterum infinitum scilicet A cuius AH est pars; quod est impossibile. Et hoc intendebat cum dixit *Et si ita sit, continget quod corpus infinitum et finitum moveant idem in eodem tempore, quod est impossibile*, idest quod moveant idem eodem motu et in eodem tempore. Et proportio in literis que ceciderunt a quo transtulimus istud, est conversa; posuit enim proportionem maioris patientis ad minus sicut proportionem minoris agentis ad maius, et hoc non potest esse, immo oportet ponere proportionem minoris patientis ad maius patiens sicut proportionem minoris agentis ad maius agens: patiens enim maius est ignotum, quod vult invenire per proportionem.

21 movetur] -vebit *JW* | et[1]] *om. XZ* 22 agat] agit *B* | egit] agat *W* 24 minor] -re *BWY sed corr* W_2 : -ri *Z* 27 et] *om. BJ* 28 parte] -tem *BJ* 29 hoc] *om. WYZ* 31 Erit igitur] *inv. B* 32 et[1]] *om. JZ* | et[2]] *om. BJ* 33 minoris] *om. BWXY* 37 scilicet] quod est *Z* 38 patiens] agens *BXY* | congregata] aggre- *X* 38-9 congregata—BZ] *om. homoiotel. B* 39 ex] ab *W* 41-2 scilicet—A] *om. W* 42 scilicet A] *om. BX* : A *Y* 44 infinitum ... finitum] *inv. Z* 45 quod] *om. B* | idem] + in *B* | tempore] + quod est impossibile *B* 46 ceciderunt] + in literis *WXY* 47 conversa] diversa *Z*: + idest *B* | minus] + patiens *B* 48 proportionem] -tio *B* 48-50 et—maius] *om. homoiotel. B* 51 per] *om. BX*: propter B_2

Deinde dixit *Iam enim diximus quod corpus maius movet idem in tempore minori quam minus*, idest hoc fuit impossibile quia iam diximus quod manifestum est per se quod corpus maius facit actionem quam facit corpus minus in tempore minori quam faciat corpus minus, nedum faciant eandem actionem in eodem tempore agens finitum et agens infinitum. Deinde dixit *Et hoc erit in omni tempore in quo erit actio quam agunt per corpus finitum corpus infinitum et corpus finitum*, idest et contingit secundum hoc ut hoc inveniatur in omni tempore in quo agunt per corpus finitum corpus infinitum et finitum, scilicet ut sit tempus equale.

Deinde dixit *Et etiam impossibile est etiam movere aliquam rem in* *46va* *tempore infinito quoniam non* | *habet finem*, idest et si aliquis posuerit quod infinitum movet in tempore infinito, hoc etiam erit impossibile: quod enim movet in tempore infinito, movet ad nullum finem; et quod movet ad nullum finem, otiose movet; hoc enim non T 52b contingit nisi in hiis que fiunt propter motum situalem et in localibus etiam in motis recte: iam enim declaratum est quod primus movens movet tempore infinito. Nam utilitas huius motionis est continuatio et perpetuitas essendi, quapropter necesse est ut sit infinitum; sed tamen declaratum est quod impossibile est ut tale movens sit corpus, quoniam si esset corpus, moveret in nullo tempore, ut declaratum est in ultimo octavi Phisicorum. Intendebat enim quod impossibile est ponere corpus infinitum movere aliquod finitum in tempore infinito, idest agere in idem numero tempore infinito, quoniam si generandum est, non complebitur sua generatio, et si afficiendum est, non complebitur sua qualitas, et si sit movendum, non perveniet ad finem quem intendebat. Et hoc sequitur in corpore celesti sicut in aliis, scilicet quod impossibile est ut corpus celeste moveat aliquid ex hiis que sunt sub eodem tempore infinito, verbi gratia aliquod esse decem predicamentorum. Deinde dixit *Motus autem est actio que non recipit affectiones*.

---

53 quia] quod *B* 54 facit] habet *Z* 57 infinitum] finitum *Z* 58 finitum] + et *BX* 59 hoc^{2}] *om. Z* 62 etiam impossibile] *inv. Z*: impossibile *BY* | etiam^{2}] *om. JZ* | aliquam rem] *inv. YZ* 63 quoniam] quam *Z* | aliquis] quis *JY* 64 tempore] + sed incipit *B* 64-5 infinito—movet^{1}] sed *X*

64 etiam] *om. YZ* | erit] est *Z*: et *Y* 66 ad] quod *B* 67 hiis] eis *B* | fiunt] sunt *BJX* | propter] preter *XY* 68 etiam] et *Z* 69 movet] + in *W* | Nam] sed quia *Z*: si autem *Y* 72 sit corpus] *inv. Z* 76-7 generatio—sua] *om. homoiotel. Y* 76 afficiendum] affirman- *Z* 77 est] *om. XZ* 78 quem] quam *Z* | hoc] *om. B* 79 sicut] + et *XY* | impossibile est] *inv. JZ* 80 aliquid] -quod *JYZ* | eodem] eo *JX*

46vb15 **66.** [67.] *Dicamus igitur quod corpus infinitum impossibile est movere* 275a24
aliud corpus infinitum neque agere in ipsum aliquod. Sit ergo corpus infinitum agens linea A, et patiens corpus infinitum linea B, et sit tempus in quo agit corpus agens C et D, et moveatur ab A corpus finitum et sit H; A igitur agit in totum B et H, sed hoc impossibile est in eodem tempore, agit enim in minori tempore in minus corpus; A igitur movet
B in D tantum. Dicamus igitur quod sicut est proportio D ad C, ita est 275a30
proportio H ad aliquam partem B que est finita; tunc igitur movet A istam literam scilicet quandam partem in B in tempore C et D, et similiter agit A in corpus maius et minus, non quia actio eius est in tempore
maiori aut minori; et hec est figura. 275b2

47ra

Cum declaravit quod corpus finitum impossibile est ut agat in T 53a
infinitum neque econverso, vult declarare quod neque infinitum in infinitum, et posuit corpus agens infinitum A, et patiens infinitum B, et tempus in quo completur actio CD, quod divisum est in duas partes scilicet C et D; et hoc manifestum est ex sermone eius. Et cum hic posuit hoc sicut diximus et sicut est manifestum ex suis verbis, dixit *Et moveatur ex A corpus finitum quod est H*, idest et ponamus corpus finitum, quod est patiens ab agente posito infinito scilicet A, et sit hoc corpus H. Deinde dixit *Erit ergo corpus A agens in totum B* etc., idest necessarium est ergo ab hac positione ut a corpore agente infinito scilicet A patiantur secundum totum duo agentia quorum unum est infinitum scilicet B et secundum finitum scilicet H; et posuit hoc corpus partem finiti, scilicet posuit corpus

---

66. [67.] 1 igitur] ergo *BX* | est] *om. Z* 2 Sit] si *B*W_2*X* 3 linea[1]] nam *B* | patiens corpus] *inv. BJX* 4 ab A] *om. YZ* | finitum] + scilicet pars W_2*Z* 5 agit] agens *JWX* | hoc] *om. YZ* 7 Dicamus—sicut] *om. J* | igitur] ergo *B* 8-9 que—B] *om. homoiotel. B* 8 movet] -vebit *J* 9 in[1]] *om. JX* 10 actio eius] *inv. Y*: actio non *W*: eius actio non *Z* 13 declarare] determina- *YZ* 16 scilicet] in *YZ* | et[1]] *om. BX* 17 hic] *om. W*: hoc *BXY* | hoc] *om. JZ* | sicut[1]] illud *Y*: quod hoc *Z* | est manifestum] *inv. YZ* 18 moveatur] -ventur *Z* 22 patiantur] -tiatur *YZ*

H partem A verbi gratia partem AH; et impossibile est ut hec duo corpora patiantur ab eo in eodem tempore secundum totum. Principium enim est quod ab eodem agente patitur minus patiens in minori tempore, nedum alterum patientium fuerit finitum et alte- 47rb rum infinitum. A igitur movebit H in D tantum, idest se|quitur ergo ut corpus agens infinitum scilicet A moveat corpus H finitum, quod est pars B patientis infiniti, in tempore quod est pars temporis CD in quo movit infinitum totum; ponamus ergo istam partem T 53b D.

Et cum posuit istam partem, dixit *Et proportio D ad C est sicut proportio H ad quandam partem B, et est finita*, idest et sequitur hoc secundum hoc quod proportio temporis quod est pars temporis totius scilicet D, ad totum tempus scilicet CD, sit sicut proportio patientis finiti scilicet H, quod posuimus partem B infiniti ad illam partem B finitam, cuius corpus H est pars, idest quod proportio H ad ipsam est sicut proportio temporis D ad tempus totum DC. Et ista proportio apparet eo quod concesserimus quod ab eodem agente patiuntur agentia equalia in tempore equali et minus in minori. Et cum ita sit, erit proportio agentis ad agens sicut proportio temporis ad tempus, quoniam si ista fuerint proportionabilia equa proportione, continget necessario ut sit eadem in aliis proportionibus; et iam diximus de hac proportione. Deinde dixit *Et cum igitur A movebit hanc literam scilicet quandam partem B in tempore C et D*, idest 47va necessarium est igitur | ut agens infinitum scilicet A moveat partem B infinitam, cuius proportio ad partem minorem B scilicet H in tempore CD in quo A infinitum movet B infinitum. Deinde dixit *Et similiter agit in corpus maius et minus* etc., idest et contingit ex hoc ut ab eodem agente patiantur maius corpus et minus in eodem tempore, non in temporibus diversis; quod est impossibile.

---

25 verbi—AH] *om. Z* | gratia] + per *J* | partem AH] *om. Y* 26 eo] eodem *BX* 29 movebit] -bitur *BX* 30 moveat] -vebat *Z* 32 movit] -vet *JZ*: -vitur *W* 34 Et[1]] + dixit et *B* | D—proportio] *om. homoiotel. W* 35 finita] in- *BWY* 36-7 temporis[2]—scilicet[1]] totius temporis C *XZ* 37 sicut] sit tempus *Z* 38 patientis] agentis *BWXY sed corr.* W_2 | finiti] *om. B* | illam] aliam *BWX* 41 eo] *om. BWXY sed suppl.* W_2 | eodem] eo *BWXY sed corr.* W_2 42 patiuntur agentia] patiantur patientia *J*: patiuntur *Z* 43 ita sit] *inv. W* 44 quoniam— proportionabilia] *om. Y* | quoniam—proportione] *om. Z* | si ista] *om. BWX* | fuerint] fiunt *B* | proportionabilia] -tionalia *BWX* | equa] equalia *BX* | proportione] -ni *Y* 46 diximus] pre- *BX* | Et cum] *om. X*: tunc *JZ* 49 infinitam] finitam *JZ* 52 ab—minus] *om. B* 53 in—diversis] diversis temporibus *Z*

67. [68.] *Manifestum est igitur quod impossibile est ut corpus infinitum* 275b2 *moveat corpus infinitum in tempore finito; et si hoc est impossibile, possibile est igitur ut sit in tempore infinito; et iam diximus quod tempus infinitum non habet finem; corpus autem mobile habet finem.* 275b4

Iam manifestum est quod corpus infinitum impossibile est agere T 54a in corpus infinitum in tempore finito; sed quia aliquis potest dicere quod, si ageret, erit in tempore infinito, dixit *Et si hoc est impossibile, possibile est igitur ut sit in tempore infinito*, idest et si aliquis dixerit quod si hoc est impossibile in tempore finito, forte est possibile in tempore infinito. Et hoc quod dixit habet sufficientiam, quoniam cum posuerit patiens infinitum, continget necessario ut eius passio sit infinita; et si posuerit patiens infinitum et passionem finitam, est quasi positio contradicens sibi. Et dicendum est ab ali- 47vb quo quod ponens hoc petit prin|cipium, estimandum est enim quod nulla differentia est sive ponatur primo finitum quod est contradicens questio, sive ponatur passio eius finita. Et cum ita sit, dixit respondendo *Et iam declaravimus quod tempus infinitum* etc., idest et iam diximus in Phisicis quod tempus est infinitum et quod non habet finem; et cum posuerimus hoc infinitum pati in tempore infinito, tunc eius actio non habebit finem ad quem perveniret; erit ergo aliquod quod semper generabitur et numquam complebitur; et declaratum est huiusmodi contrarium in sexto Phisicorum.

68. [69.] *Et etiam omne corpus sensibile habet potentiam activam et* 275b5 *passivam insimul; et cum ita sit, impossibile est corpus sensibile esse infinitum.* 275b6

Cum declaravit quod impossibile est in corporibus infinitis neque agere neque pati, ponit secundam propositionem ut concludatur quod

---

67. [68.] 1 igitur] *om. Y*: + quia *B* 2 moveat—infinitum] *post* finito *Z* 5-6 impossibile—infinitum] *om. homoiotel. Z* 6 aliquis] *om. YZ* 8 possibile] im- *B* 9 finito] in- *BX* | possibile] im- *B* 10 hoc] + est *Z* 13 positio] ponitur *B* 14 est enim] *inv. BJX* 17 respondendo] dicendo *Z* | declaravimus] dixi- *JX* 18 tempus] + quod *BJ* 19 posuerimus] -suimus *JXZ* 20 infinito] *om. B* | eius actio] *inv. BX* | perveniret] -niet *WY* 22 huiusmodi] *om. J*: eius *B*

68. [69.] 1 etiam omne] *inv. Z* 4 est] + ut *JZ* 5 concludatur] -dat *BX*

nullum corpus est infinitum; et hec quidem proposito est quod omne corpus naturale aut est activum aut passivum aut ambo. Hoc autem videtur sensui, corpus enim celeste videtur agere in ea que sunt apud nos, alia autem que sunt sub eo videntur agere et pati insimul. Et componitur sic in secunda figura: omne corpus aut est activum aut T 54b passivum, et nullum corpus infinitum est activum aut passivum, ergo nullum corpus est infinitum. Et intendit per sensibile sensibile tactu; et tacuit corpus celeste quoniam omnes Antiqui non credebant ipsum esse non patiens, immo omnes dixerunt ipsum esse aut ignem aut compositum in quo dominatur ignis, et maxime de stellis.

48ra11 **69.** [70.] *Et etiam omne corpus quod est in loco est sensibile; et cum ita* *275b6* *sit, nullum corpus infinitum est extra celum, quoniam illic neque est locus illic neque corpus finitum. Manifestum est igitur quod nullum corpus extra celum est omnino. Si ergo aliquis dixerit quod possibile est* *275b10* *quod extra celum sit aliud corpus spirituale, necesse est ei dicere quod illud corpus est in loco: extra enim et intra notant locum; quapropter erit sensibile etiam; et nullum corpus sensibile est nisi in loco.* *275b11*

Cum posuit quod omne corpus aut est activum aut passivum, et hec propositio est accepta a sensu, potest aliquis dicere quod corpus infinitum possibile est esse extra mundum et quod sit neque agens neque patiens; et propter hoc videtur mihi quod inducit hunc sermonem post illam demonstrationem; et fundavit ipsum super duas propositiones, quarum altera est quod omne corpus quod est in loco est sensibile, secunda est quod corpus quod est extra mundum est in loco; ergo corpus quod est extra mundum est sensibile.

Deinde ponit quod omne corpus sensibile est finitum, quia est *48rb* activum aut passi|vum, ergo non est extra mundum corpus infinitum; et secundum hoc explanavit hunc locum Themistius. Ipse

---

8 sensui] -su *WXY* 13 Antiqui] + qui *BX* 14 non—esse] *om. homoiotel. B*

69. [70.] 2 celum] + omnino *Z* 2-4 quoniam—est[1]] *om. Y* 2 neque] nullus *J* : non *X* : si non *Z* 2-3 est locus] *inv. J* : locus *Z* 3 illic] *om. B* | corpus[1]] + aliud *Z* | finitum] + omnino *Z* 4 est[1]] *ante* extra *X* | ergo] *om. J* : enim *B* | quod] *om. BJ* | possibile] *om. B* : im- *Z* | est[2]] *ante* possibile *X* : *om. B* 5 quod[1]] ut *WXY* 7 erit] est *Z* 10 possibile] im- *B* | extra mundum] *om. B* 14 corpus] -pora *BX* 16 ponit] -suit *JZ* | est[1]] + non *JZ* | finitum] in- *JY* | est[2]] + aut *BX* autem Aristoteles videtur secundum hanc conclusionem non propalasse nisi duas propositiones, quarum altera est quod omne corpus quod est in loco est sensibile; et est maior propositio in sillogismo quem dixit Themistius, sed non manifestavit secundam dicentem quod omne corpus quod est extra mundum est in loco, neque propositionem dicentem quod omne corpus sensibile est finitum; sed videmus ipsum, postquam dixit propositionem dicentem quod omne T 55a corpus quod est in loco est sensibile, quod dixit *Et cum ita sit, nullum igitur corpus infinitum est extra mundum*, et est conclusio.

Deinde induxit aliam propositionem quam posuit causam istius propositionis, et dixit *Quoniam nullus locus est illic*, et ista propositio non habet cognitionem in aliquo cum prima, que dixit quod omne corpus quod est in loco est sensibile, nisi intelligatur conversa eius scilicet quod omne corpus sensibile est in loco; quoniam quando adiungebatur huic quod est extra mundum non est locus, concludetur quod extra mundum non est corpus sensibile; et cum huic coniungeretur, quod si extra mundum est sensibile, et omne corpus est in loco, concludetur quod si extra mundum est corpus, est in loco. Deinde destruet quod illuc non est aliquis locus, et concludet quod nullum corpus est illic; et hoc apparet esse quod intendebat.

Et signum quod intendebat conversionem istius propositionis dicentis quod omne corpus quod est in loco est sensibile, est quia post hoc dixit *Si aliquis igitur dixerit Forte possibile est quod extra celum sit aliud corpus spirituale*, idest si igitur concessum non fuerit
48va hoc nobis scilicet | quod si extra mundum non sit corpus sensibile, quod omne sensibile est in loco, et dixit aliquis Forte illic est corpus

---

19 autem] *om. B* 22 quem] que *X*: quam *Z* | non] *om. WYZ* 23 omne corpus] illud *WX*: nullum *B* 23-6 extra—est[1]] *om. homoiotel. Y* | neque—sensibile] *om. homoiotel. Z* 23 est[2]] *post* loco *JZ* 24-6 finitum—est[1]] *om. homoiotel. W* 26 sensibile] equale *Y* | quod[2]] *om. XY* 27 conclusio] econverso *Z* 28 induxit] -ducit *BX* | aliam] -quam *B* 29 est] *dittogr. W* 30 cognitionem] contradictio- *B*: coniunctio- *X*: communitatem *J* | que] qua *YZ*: quam *W* 31 intelligatur] + sermo *BX* | conversa eius] *inv. BXY* 32 sensibile] + corpus *B* | quando] *om. B* 33 adiungebatur] -getur *BX* | est[1]] *om. BJ* | concludetur] -ditur *J* 33-4 concludetur—corpus] *om. B* 34 extra—est] non est extra mundum *J* | corpus] *om. WY* | et] + si *Z* | huic] hac *Z* 35 coniungeretur] -getur *BJX* 37 quod] quo *B* | est aliquis] *inv. B*: est *X* 39 Et] + dat *JZ* | Et—intendebat] *om. homoiotel. B* 40 dicentis] dicens *Z* | est[3]] *om. JWYZ* 41 igitur] *om. J*: enim *Z* | dixerit] + igitur *W* | Forte] *om. J* : *post* est *Z* | quod] quo *B* 42 aliud corpus] *inv. JZ* 43 si] *om. JZ* | corpus] + est *BX* 44 quod] quia *J*W_2*Z* | aliquis Forte] *inv. Z*

spirituale non sensibile propter quod non contingit ut sit in loco, respondebimus ei quod necesse est ei dicere quod illud corpus est in loco, quoniam extra et intra significant locum, ut declaratum est in Phisicis; et cum est in loco, est sensibile; et ideo posuit primo hanc propositionem. Deinde dixit *Et non est corpus sensibile nisi in loco*, et ipse post declarabit proportionem dicentem quod extra mundum non est locus.

70. [71.] *Et etiam volumus perscrutari ad verificandum hoc quod diximus, et dicemus in hoc sermonem rationabilem quod corpus infinitum impossibile est moveri circulariter quoniam non habet omnino medium; et etiam quia corpus rotundum movetur circa medium. Et etiam impossibile est hoc corpus infinitum moveri motu equali, quoniam si moveretur motu equali, indigeret alio loco infinito et sine fine, et indigeret loco* *48vb* *ad suum mo|tum accidentalem secundum substantiam loci sui motus naturalis.* *275b12* *275b15* *275b18*

Hec est alia demonstratio a predictis, et est logica scilicet quia est ex veris propositionibus sed non propriis, immo generalibus. Et com- **T 55b** pendium sermonis sui est quod si hoc est corpus infinitum, aut est circulare aut rectum, et qualecumque sit, impossibile est moveri. Si enim fuerit rotundum infinitum, non habebit medium circa quod movetur, aut quia omne habens medium habet extrema, aut quia circa medium revolvitur corpus rotundum; et quod revolvitur circa finitum est finitum; et hec est causa magis propria quam prima. Et si fuerit rectum et infinitum, necesse est ut locus in quo movetur sit maior eo; erit ergo infinitum maius infinito. Et hoc intendebat cum dixit *Indiget loco infinito et sine fine*, et etiam non tantum indiget uno loco sed etiam duobus infinitis, scilicet loco in quo movetur naturaliter et loco in quo movetur violente, et utroque infinito. Et hoc

---

46 necesse] -arium *Z* | ei[2]] *om. JZ* 47 quoniam] + per *Z* | significant] notant *JZ* 48 ideo] inde *YZ* | primo] post *Z*

70. [71.] 3 medium] modum *Y* 5 moveri] *om. Y* : *ante* hoc *BJX* | motu equali] *inv. Z* | moveretur] -vetur *WYZ* 6 equali] + et *W* | indigeret[1]] -get *Z* : -gent *Y* | alio] *om. J* : alio modo *Z* : aliquo *B* : autem *Y* | indigeret[2]] -gent *YZ* 9 scilicet] *om. Y* 11 sermonis sui] *inv. W* | quod] quia *JZ* | hoc] hic *B* 14 aut[1]] autem *B* | habens] *om. BX* 16 est causa] *inv. B* : causa *X* | quam prima] *om. XZ* 18 ergo] igitur *YZ* 19 loco] + in *B* 21 et[1]] + in *Z* | et[2]] + in *B* intendebat cum dixit *Secundum substantiam loci sui motus naturalis*, idest eiusdem modi infinitatis; erunt ergo duo loca infinita, quod est impossibile. Et non intendebat quod locus motus corporis naturalis est alius a loco sui motus accidentalis, quoniam locus in quo lapis movetur naturaliter idest aqua et aer, est idem cum eo in quo movetur violente, immo intendit quod locus ad quem lapis movetur naturaliter est alius a loco in quo movetur violente: hoc est manifestum de motibus corporum simplicium, sicut est manifestum quod locus in quo corpus quod recte movetur quiescit naturaliter, est alius a loco in quo movetur; et iam declaravimus hoc ante.

49ra16 71. [72.] *Et etiam si motus corporis infiniti fuerit equalis, erit aut naturalis aut accidentalis, et erit potentia movens ipsum istis duobus motibus* *275b18*
infinita. Potentia enim corporis infiniti est infinita, et corpus potentie *275b20*
infinite est infinitum, et omnino movens infinitum. Et iam diximus in libro de Motibus quod impossibile est quod in aliquo corpore infinito sit potentia finita; si ergo corpus infinitum potest moveri naturaliter et acci- **T 56a**
dentaliter, erunt res infinite secundum duas potentias. *275b24*

Compendium huius sermonis est quod si corpus rectarum dimensionum est infinitum, habebit duos motus, naturalem et accidentalem, et continget ut due potentie moventes secundum istos duos motus sint infinite; movens enim corpus infinitum oportet esse potentiam infinitam. Declaratum est enim quod potentie corporum
49rb infi|nitorum sunt infinite, et etiam econverso quod potentie corporales infinite sunt corporum infinitorum; et cum ita sit, continget quod corpora infinita sint bina, scilicet corpus habens potentiam

---

22 loci sui] *inv.* *Z* 23 infinitatis] -finitis *XYZ* 25-6 lapis movetur] *inv.* *YZ* 26 idest] vel *JZ* | et] vel *JZ* 27 violente] naturaliter et loco in quo movebitur violente et utroque infinito *Y* 28 in] *om.* *BX* | est^{2}] + enim *W* 30 a loco] locus *Z* 31 declaravimus] determina- *BXZ*

71. [72.] 1 equalis] naturalis *B* 2 erit potentia] potentia est *Z* 3 Potentia—corpus] *om. homoiarch.* *W* | Potentia enim] *inv.* *J*: potentia *YZ* | enim corporis] *inv.* *Z* | infinita] *om.* *J*: + potentia *Z* 4 et^{1}] erit *B* | infinitum^{2}] finitum *WY* 5 sit] est *Z* 8 quod] *om.* *Z* 9 motus] + scilicet *BJ* 11 motus] modos *YZ* | oportet] + ei *Z* 13 etiam econverso] *inv.* *B*: econverso *Y* 14 continget] -git *YZ* 15 sint] sunt *W*: fuerint *Y* | bina] bona *Y*: duo *J* | scilicet corpus] *inv.* *WXY* | corpus habens] *inv.* *B*

infinitam moventem naturaliter, et corpus habens potentiam infinitam moventem accidentaliter; et hoc intendebat cum dixit *Erunt res infinite secundum duas potentias*, scilicet secundum duas naturas, naturam que movetur naturaliter et naturam moventem aliam extrinsecus violente; necesse est enim quod movens violente sit movens aliud a movente naturaliter, et quod deferens alteram potentiarum sit aliud a deferente aliam. Et manifestum est quod hec demonstratio est alia a demonstrationibus predictis, quamvis utatur in ea una propositionum declaratum in Phisicis scilicet quod corpus infinitum impossibile est habere potentiam finitam. Et intendit per librum de Motu librum de Phisicis; et iam declaravimus hoc in hoc tractatu scilicet quod impossibile est reperire potentiam finitam in corpore infinito.

72. [73.] *Et etiam quod movet corpus infinitum aut est animatum aut* *275b25*
non. Et si est animatum, quomodo potest animal esse infinitum? Et si T 56b
movens est aliud non animatum infinitum, erunt duo infinita, movens scilicet et motum, sed differunt in forma propter potentiam. *275b29*

49va Idest quod corpus quod movet corpus infinitum aut est animatum aut non; et quia necessario est ut sit infinitum, cum eius potentia sit infinita, oportet, si fuerit animatum, ut sit animal infinitum, quod est impossibile. Et si fuerit inanimatum, erunt duo corpora utrumque infinitum, movens et motum, et cum hec differunt secundum naturam; et hoc intendebat cum dixit *Sed differunt in forma*; et innuebat per hoc aliud impossibile, et est quod si natura alicuius forme fuerit innata esse in corpore infinito, necesse est ergo ut natura

16 corpus] *om. B* | infinitam[2]] *om. Z* 19 que] *om. Z* | movetur] -vet *BJ* : -ventem *Z* | naturam moventem] *inv. Z* 20 est enim] *inv. BJ* 21 aliud] alio *W* | movente] violente *B* | potentiarum] -tiam *B* 22 aliam] alteram *WX* : altera *Y* 23 una] via *B* 25 finitam] in- *BW* 26 hoc[1]] *om. WY*

72. [73.] 2 Et[1]] *om. XYZ* | est] *om. WXYZ* | animal] animatum *B* | animal esse] *inv. X* : animal *BJ* 3 animatum] *om. XY* | infinitum] *om. BJ* | duo infinita] *inv. Z* 4 sed] non *B* | differunt] deferens *Y* : diversantur *B* | propter] per *XY* | potentiam] virtutem *B* 5 Idest] *om. XY* 6 et] est *B* | necessario] -arium *Z* : -cesse *J* 8 inanimatum] anima- *BX* 9-10 secundum—differunt] *om. homoiotel. B* 12 forme fuerit] *inv. B* | innata] infinita *Z* | esse] erit *XYZ* | ergo] *om. BWXY*

istius forme sit eadem; forme enim diversarum naturarum innate sunt habere materias diverse quantitatis, cum diversitas formarum sit causa diversitatis quantitatum et magnitudinum recipientium formas. Et est manifestum quod si fuerit mobile per se si igitur fuerit rectum, continget ut sit aut animatum aut inanimatum, et si erit animatum, erit animal infinitum, et si fuerit inanimatum et fuerit rectum, contingent ei dispositiones predicte; et ideo tacuit complere istas divisiones.

49vb5 73. [74.] *Et etiam si corpus totius non fuerit continuum, et fuerit discre-* *275b29*
tum separatarum partium, ut Democritus et Leucippus dicebant, necessa- *275b30*
rio erunt motus omnium partium totius consimiles. Separatio enim et
discretio earum fuit per figuras earum, natura autem est eadem, sicut *276a1*
aurum separatum in partibus diversarum figurarum et omnes partes sunt
auree. Si ergo natura omnium partium est eadem, erunt motus etiam
eorum omnes idem necessario, sicut diximus superius, quod quo movetur
gleba et tota terra et quo movetur scintilla et totus ignis. Et in illo loco
declaratum est etiam quod si hec corpora omnia fuerint discreta gravia, *276a5*
nihil erit in eis leve, et si omnia fuerint levia, nihil erit in eis grave. *276a6*

Cum declaravit quod corpus infinitum impossibile est quoniam si T 57a
fuerit, aut erit simplex aut compositum, et declaravit quod si compositum fuerit, necessario erit in eo simplex infinitum in magnitudine, aut unum aut plura, si fuerit compositum ex simplicibus finitis in specie; si autem finita fuerint in magnitudine, erunt necessario infinita in specie, et si non, nullus modus esset propter quem compositum esset infinitum. Et quia aliquis posset dicere quod possibile
50ra est compositum esse infinitum ex corporibus finitis in ma|gnitudine
et finitis in specie, ita quod sint infinita in individuo aut partes indi-

13 istius] illius *Z* | forme[2]] forma *Z* 15 quantitatum] -tatis *W* | et] *om. BX* 17 continget] -geret *JW* 17-8 aut[1]—erit] *om. BXY* | aut[2]—inanimatum] *om. W sed suppl.* W_2 18 animal] aliud *Y* | infinitum—fuerit[1]] *om. X* | inanimatum et] *om. BY*: animatum et si *X* | fuerit[2]] *om. BY* 19 contingent] -get *BX*: -git *Y*

73. [74.] 1 fuerit[2]] fuit *Y* 3 consimiles] -lis *XYZ* 4 earum[1]] *om. Z* | earum[2]] eo- *BZ* 6 motus etiam] *inv. W* 8 quo] + modo *B* | illo loco] *inv. Z* 11 declaravit] declaratum est *W* | est] + esse *JZ* 12 simplex] simul *Y* 15 finita fuerint] *inv. BX* 15-6 necessario infinita] *inv. W* 16 infinita] finita *Z* | esset] *om. WYZ* 19 aut] et *BX*: ut *JZ*

visibiles quas facit Democritus: iste enim ponit eas eiusdem nature et infinitas secundum individuum, immo etiam ponit eas discretas non contingentes se. Et ideo incepit contradicere huic sermoni, ut compleatur sermo de destructione infiniti, et dixit *Et etiam si corpus totius* etc., et est si aliquis dixerit quod corpus totius est infinitum, non quia est continuum sed quia partes eius sunt infinite secundum individuum separate in loco, ut ponit Democritus et Leucippus, necessario erunt motus istarum omnium partium consimiles idest aut ad superius aut ad inferius. Et cum hoc non sequitur nisi sermonem eorum, notificavit hoc dicendo *Discretio enim et separatio earum est per figuras, natura autem earum est eadem*, idest et contingit ut motus earum sit idem in specie; quoniam facientes hec corpora ponunt ea eiusdem nature in specie, et non differunt individua eorum nisi secundum figuram, ut frustra auri diversarum figurarum; omnia enim sunt eiusdem nature scilicet auree.

Deinde dixit *Si ergo natura omnium partium sit eadem* etc., idest ut declaravit in primo istius tractatus, quia ab eadem natura non provenit nisi unus motus, ideoque quo movetur gleba et tota terra si esset imaginata extra suum locum. Et cum declaravit quod necessarium est T 57b
quod motus istorum corporum sit idem in specie, dixit *Manifestum est igitur* etc., idest manifestum est quod si motus istorum coporum fuerint idem, necessario erunt alterius duarum naturarum scilicet aut omnia erunt gravia aut levia: si gravia, impossibile est corpus leve esse, si levia impossibile est corpus grave esse. Et hoc quod dixit non
50rb sequitur positionem quod ista corpora | sint infinita, sicut sequitur positionem quod sint elementa totius, propter quod estimatur quod non proprie sequitur. Et hoc sequitur ex positione eorum quod sunt infinita, quia cum fuerint infinita, nihil reperitur extra ea; et propter hoc dixit Themistius in hoc loco quod non sequitur ex hoc sermone

22 ideo] + nunc *JZ* | ut] *om. W* 24 etc.—totius] *om. homoiotel. B* 27 consimiles] -lis *X*: -libus *BZ* 28 nisi] ut *YZ*: + secundum *X* 29 hoc] *om. WYZ* | enim] autem *B* | earum] eo- *XY* | est] *om. WY* 30 earum] eo- *BXY* 36 istius] huius *BJX* | quia] quod *WXY* 39 sit] sint *BX* 41 fuerint] -rit *X*: finitum *Y* | alterius] aliud *B* 42 gravia[1] ... levia] *inv. W* | est] erit *WXY* | corpus leve] *inv. J* 42-3 leve—corpus] *om. homoiotel. B* 43 non] *om. BWXY* 44 sint] sunt *WY* 45 positionem] privatio- *Z* | sint] sunt *YZ* 46 quod] que *BX* | sunt] *om. W* 47 quia] *om. Y*: et *Z* | cum—infinita] *om. WYZ* 48 hoc sermone] *inv. Y*

conclusio quam intendebat concludere, sed sequitur quod natura partium indivisibilium non est una.

74. [75.] *Et etiam si corpus infinitum aut est grave aut leve, est ergo aut* 276a6 *in orizonte totius aut in eius medio; quod est impossibile, quoniam est corpus infinitum. Et omne corpus carens medio caret orizonte omnino, quidam ergo eius erit superius et quidam inferius. Et cum ita sit, nullus locus erit motus corporum, et cum non habent locum non habent motum* 276a10 *omnino; corpora enim moventur aut naturaliter aut accidentaliter necessario, et sunt in suis locis terminata et in aliis locis etiam.* 276a12

50va Manifestatio istius sermonis est quod vult declarare quod si corpus infinitum fuerit tangibile, aut est grave aut leve; si grave, erit in medio, et illius quod est in medio est extra aliud corpus, et extra quod est aliud corpus terminatum est, et omne terminatum est finitum; si autem fuerit leve, erit in orizonte totius, et quod est in orizonte totius est terminatum ab orizonte continente ipsum, etiam et quod est huiusmodi est finitum; et cum ita sit, nullum corpus infinitum est tangibile, quasi ergo diceret omne corpus grave aut leve est contentum ab extrinseco: corpus autem infinitum non est conten- T 58a tum ab extrinseco.

Deinde dixit *Et omne corpus carens medio caret orizonte omnino*, et manifestatio huius est quia est alia declaratio scilicet quia non sequitur a positione corporis tangibilis esse infinitum ut sit infinitum propter locum, immo et per positionem eius, quod est corpus esse infinitum, negatur esse locus omnino: locus enim aut est medium aut orizon. Et cum posuerimus corpus esse infinitum secundum tres dimensiones, necessario est infinitum secundum omnes dimensiones,

74. [75.] 1 si] *om. WXY* | est[1]] *om. B* | grave ... leve] *inv. W* | est ergo] *om. WYZ*: est igitur *B* | aut[3]] + est *Y* 3 omnino] ideo *B*: ideo neque *J* 4 quidam[1]] quoddam *YZ* | ergo] *om. J* | eius erit] *inv. WY* | quidam[2]] quoddam *Z* 7 locis etiam] *om. Y* 8 quod[1]] quia *BX* 9 est] erit *B* 10 illius] illud *BJX* | aliud] *om. X*: aliquod *B* | extra[2]] *om. B* 11 aliud corpus] *inv. Z* | est[2]] *om. WY* 12 erit] *om. BZ*: et *WXY* | orizonte[1]] + est *Z* 12-3 et—totius] *om. homoiotel. XY* 13 etiam] *om. JYZ* 15 diceret] –cit *J*: -cetur *B* | omne] + ergo *Z* 19 est[2]] *om. BX*: *post* declaratio *J* | scilicet] *om. B* 20 corporis] *om. X* | tangibilis] -bile *X*: sensibilis *Y* 22 esse locus] *inv. Z* | aut est] *inv. BJ*: aut *X* 24 necessario—dimensiones] *om. homoiotel. X* 24-5 necessario—dimensiones] *om. homoiotel. BWZ*

et cum fuerit infinitum secundum tres dimensiones, non habebit medium, et cum non habuerit medium non habebit orizonta, et cum non habuerit medium neque orizonta, non erit locus, et cum non fuerit locus non erit motus neque naturalis neque accidentalis: cum ergo posuerimus corpus infinitum, non erit motus omnino.

50vb7 75. [76.] *Et dicamus quod si aliquod istorum corporum separatorum* *276a12*
fuerit fixum, aut vadens ad aliquem locum accidentaliter, hoc accidens erit naturale alii corpori. Et verificatio illius est quoniam iste sermo
convertitur: dicamus igitur quod omnia corpora non sunt tantum levia *276a15*
aut gravia, sed quedam gravia et quedam levia. Manifestum est igitur ex propositionibus et sillogismis quos diximus, quod corpus totius habet
finem et quod impossibile est esse infinitum. *276a17*

Videtur quod sermo iste est ad affirmandum quasdam propositiones quas dixit ante. Accepit enim ante quod omnis motus aut est a medio aut ad medium: et aliquis potest concedere hoc in motibus naturalibus et negare in accidentalibus, et scilicet concedere quod superius scilicet et inferius sunt loca motus naturalis non accidentalis, quoniam negando superius et inferius negantur motus naturales et accidentales. Propter quod dixit *Dicamus quod si aliquod corpus* etc., idest et quod verificat quod loca motuum accidentalium sunt eadem cum locis motuum naturalium est quia, cum posuerimus ali-
quod corpus quiescens in loco accidentaliter aut etiam motum acci- **T 58b**
dentaliter, necessario ille motus aut quies accidentalis sunt alii cor-
51ra pori in illo loco na|turales loca ergo motuum accidentalium sunt loca
motuum naturalium, et econverso scilicet quod si posuerimus corpus aliquod quiescens aut motum in aliquo loco naturaliter, necesse est ut quies aut motus in illo loco sint alii corpori innaturales.

25-6 habebit—non[1]] *om. homoiotel. Z* 26-7 non[2]—medium] *om. homoiotel. Z* 26 orizonta] + ergo non habebit orizontem *J* 27 et cum] *inv. B* 28 naturalis ... accidentalis] *inv. WY*

75. [76.] 1 quod] *om. YZ* 2 fixum] finitum *B* 3 alii] illi *B* : alteri *J* 4 igitur] ergo *Z* | tantum levia] *inv. BJ* 5 aut] et *BJX* | ex] *om. BWX* 6 quod] et *W* 11 scilicet] + quod *W* | scilicet concedere] *inv. J* 12 scilicet et] *inv. B* : et *X* | sunt] sint *JY* 13-4 naturales ... accidentales] –ralis ... –talis *JY* 16 est] et *BX* 19 naturales] -rale *B* : -ralis *J* | ergo] enim *B* 19-20 loca motuum] *om. B* 20-1 corpus aliquod] *inv. BX* 22 sint] sunt *B*

Et cum verificatum fuerit nobis ista conversio, verificabitur quod loca motuum naturalium eadem sunt cum locis motuum innaturalium; cum enim verificatum fuerit nobis quod ubicumque fuerit inventus motus accidentalis invenitur naturalis, et econverso verificabitur quod loca motuum naturalium et accidentalium sunt eadem. Ideoque, videtur mihi, dixit *Et quod verificatio illius est quod iste sermo convertitur, dicamus igitur quod omnia corpora* etc., idest quod hoc est contrarium ei quod sequitur a positione eius, quod est corpus esse infinitum, et est ut omnia corpora sint eiusdem speciei; aut enim sunt diverse speciei aut non sunt loca diversa: quasi ergo voluit per hoc dicere quod ponens hoc non potest ponere corpus simplex esse infinitum, negat enim multitudinem locorum et multitudinem corporum. Et cum declaravit hanc questionem proprie et generaliter, dixit suam primam intentionem propter quam declaravit hoc, et est quod corpus istius mundi congregatum ex sensibilibus corporibus, impossibile est ut sit infinitum in magnitudine, et dixit *Manifestum est igitur ex propositionibus et sillogismis factis ex illis propositionibus quod mundus est finitus in magnitudine et quod impossibile est esse infinitum in magnitudine.*

51rb24 **76.** [77.] *Modo igitur volumus perscrutari utrum celi sunt plures, ut pre-* *276a18*
misimus. Omnia igitur corpora aut moventur aut stant; si moventur, aut **T59a**
naturaliter aut violente, si stant similiter aut naturaliter aut violente. Et
etiam dicamus ad locum in quo corpus stat non violente, movetur non
violente, et in loco in quo movetur non violente stat non violente; et *276a25*
etiam si iste motus fuerit violentus, contrarium eius erit naturale. *276a26*

Cum declaravit mundum esse infinitum in magnitudine, incepit perscrutari utrum sit unus aut plures, et hoc intendebat per celos.

---

23 verificabitur] -catur *YZ* 25 enim] *om. YZ* | verificatum fuerit] *inv. Z* 26 invenitur] -nietur *WX* 28 Et] *om. Z* | quod[1]] *om. J* 31 omnia] tota *W*: tunc *Y* 31-2 aut—speciei] *om. homoiotel. BWXY* 32 aut] + ut *BZ* | sunt[2]] sint *BXY* | loca diversa] *inv. Z* | quasi] quod *B* 34 locorum—multitudinem] *om. homoiotel. X* 36 suam] + positionem *Z* | primam] + et *Z* | propter quam] propter quod *Y*: quapropter *BX* 37 congregatum] + esse *WXY* 39 illis] istis *W* 40 et quod] *om. Z*

76. [77.] 1 igitur] ergo *YZ* | sunt] sint *XY* | premisimus] promis- *JXY* 2 igitur] enim *BJ* 4 ad locum] quod locus *WYZ* | violente] *om. YZ* 6 iste] *om. Z* 7 esse] *om. B*

51va Ce|lum enim dicitur grece tribus modis, dicitur enim de toto mundo, et dicitur de parte eius que circulariter movetur, et etiam dicitur de orbe stellarum fixarum. Et manifestum est quod recto ordine fuerit hec perscrutatio post alteram, quoniam non desiderat anima scire hoc nisi post scientiam ipsius esse finitum in magnitudine, quamvis sit questio nobilior prima et difficilior.

Et ipse proponit propositiones propter hoc, quarum una est quod corpora naturalia aut sunt mota aut quiescentia, et hec propositio manifesta est per se; secunda est quod illud quod movetur ex eis aut movetur violente aut naturaliter, et hec etiam manifesta est per se, similiter de quiescentibus scilicet quod aut naturaliter quiescunt aut violente, et hoc etiam est manifestum per se. Deinde addidit aliam propositionem, et dixit *Et etiam ad locum in quo corpus stat* etc., idest et manifestum est etiam quod ad locum in quo corpus quiescit naturaliter non violente movetur illud corpus, quando extra ipsum T59b
fuerit naturaliter non violente; et est etiam manifesta per se quoniam cum status terre naturaliter est in medio, necesse est ut motus eius sit naturaliter ad medium; et similiter de igne. Et conversio istius propositionibus manifesta est per se etiam, et est quod omne quod movetur ad aliquem locum naturaliter quiescit in eo naturaliter. Deinde induxit etiam aliam propositionem et dixit *Et etiam si iste motus fuerit violentus, contrarium eius erit naturale*, idest et etiam manifestum est per se quod cum aliquis motus fuerit violentus alicuius corporum simplicium, erit motus contrarius illi motui naturalis illi corpori, verbi gratia quia motus ad superius est innaturalis terre, erit motus ad inferius ei naturalis.

---

9 enim[1]] igitur *Z* 12 fuerit] fuit *JZ* 13 ipsius] + que est ipsum *JW*$_2$*Z* 18 movetur violente] *inv. BX*: violente *WY* | violente ... naturaliter] *inv. Z* | etiam] *om. XZ* 19 scilicet quod] *inv. B* | quod aut] *inv. Z* 20 hoc] *om. JZ* | est manifestum] manifestum *W*: manifesta est *JZ*: est manifesta *B* 21 corpus] *om. BX* 21-2 stat—corpus] *om. homoiotel. WYZ* 23 violente movetur] *inv. W* 24 fuerit] *ante* extra *(23) B* : *om. X* 25-6 sit naturaliter] *inv. YZ* 26 conversio] econverso *B* 29 induxit] -ducit *Y*: adduxit *J*: dixit *B* | etiam[1]] *om. JYZ* | aliam] alteram *JZ* | etiam[2]] + et *Z* 30 naturale] -ralis *BWY* 32 erit] est *Z* 33 quia] quod *WZ* | motus] movetur *YZ* | est] + non *JWZ* | innaturalis] naturalis *BJ*

51vb11 77. [78.] *Et etiam si terra que est extra istum mundum movetur ad* *276a27* *medium istius mundi violente, movebitur a medio istius mundi ad illud medium naturaliter; et si motus eius a medio istius mundi ad medium illius mundi fuerit violentus, erit motus eius a medio illius mundi ad medium istius naturalis.* *276a30*

Cum posuit quod omne corpus naturale est aut motum aut quiescens, et quod omne motum aut naturaliter movetur aut violente, et cum posuit quod omne quod violente movetur movetur naturaliter secundum contrarium istius motus, vult declarare impossibile consequens a positione duorum mundorum aut plurium, et est quod si posuerimus duos mundos esse, et posuerimus quod terra secundi mundi moveri possit ad medium istius mundi sicut movetur in illo mundo ad suum medium, necesse est ut motus eius a medio illius ad medium istius sit aut naturalis aut violentus, cum posuerimus iam quod nullus motus fugit istos duos motus. Si ergo motus eius a **T 60a** medio istius mundi ad medium istius fuerit violentus, quia igitur impossibile est moveri a medio illius ad medium istius antequam *52ra* moveatur ad circumferentiam mundi in | quo est, et notum est quod motus eius ad circumferentiam est violentus, necesse est ut motus eius in isto mundo ad medium sit violentus, idem est enim cum motu a medio illius mundi ad circumferentiam eius; et iam diximus quod ille motus est violentus, necessario ergo erit motus eius ad medium violentus, et cum ita sit, continget ut aliqua terra movebitur ad medium violente; quod est impossibile.

Et si concesserimus quod motus eius ad istud medium est violentus, erit motus eius ab isto medio ad illud medium naturalis, motus enim a medio contrarius est illi qui est ad medium; et iam diximus

---

77. [78.] 1 istum] illum *Z* | movetur] + a medio illius mundi *Z* 1-3 ad—medium[1]] *om. Y* 2 istius[1]] huius *Z* | violente] + ipsa *Z* 3 medium[1]] *om. Z* 6 motum] -tus *B* 7 motum] -tus *B* 8 cum] *om. B* | omne] + corpus *JZ* 9 declarare] + quod *X* | impossibile] + est *X* 10 a positione] positionem *J* 12 moveri possit] *inv. BX* | possit] posset *JZ* 13 illius] eius *B* 15 motus eius] *inv. Z* 16 istius[1]] illius *JX* | istius[2]] illius *Z* | igitur] *om. Z* 17 est] *om. W* | illius] istius *W* 19-20 necesse—violentus] *om. homoiotel. BYZ* 20 isto] hoc *J* | sit] est *WX* | est enim] *inv. YZ* 21 motu—medio] a medio motu eius *B* | illius] istius *BX* | mundi] + motus *JZ* 22 ille] iste *BJX* 22-3 necessario—violentus] *om. homoiotel. YZ* 23 movebitur] -veatur *Z* 25 istud] illud *WZ*: aliud *J* 26-7 motus enim] *inv. Y*

quod si aliquis motus fuerit violentus alicui corpori, necesse est ut contrarium illius motus sit naturale illi corpori; continget ergo quod aliqua terra moveatur a medio naturaliter. Et similiter si posuerimus motum eius a medio istius mundi ad medium illius violente, erit motus eius a medio illius mundi ad medium istius naturalis: hoc ergo impossibile contingit si posuerimus mundos esse plures. Et quia hoc non contingit nisi post verificationem quod natura elementorum existentium in alio mundo est eadem nature istorum elementorum secundum speciem; et quod loca mundorum etiam sunt eadem secundum speciem, incepit verificare hoc.

52rb14 **78.** [79.] *Et etiam si celi et mundi fuerint plures, sunt ex partibus consi-* *276a30*
milibus in natura necessario. Et potentia cuiuslibet corporis eorum assimi-
latur adinvicem scilicet ignis, terre, et sunt inter ea; potentia enim illius *276b1*
ignis assimilatur potentie istius. Si ergo partes mundorum plurium non
fuerint huiusmodi, erunt partes corporum illorum mundorum similes par-
tibus istius mundi totius, nomine tantum non forma. Si igitur ita sit, erit
totum ergo simile scilicet mundus iste similis illis nomine non in forma.
Et si corpora mundorum sunt similia adinvicem, sicut diximus superius **T 60b**
quod ille ignis assimilatur isti et similiter de aliis corporibus, manifestum
est quod quedam eorum moventur ad medium et quedam a medio. Et *276b5*
52va *iam declaravimus hoc cum posuimus motus corporum et | determinavi-*
mus ipsos, et diximus necessario quod motus eorum sunt finiti. *276b9*

Cum posuit quod si posuerit duos mundos, continget ut corpora naturalia moveantur naturaliter duobus motibus contrariis insimul, quoniam terra inquantum movetur ab illo mundo ad medium istius mundi, continget ut moveatur a medio et ad idem medium a medio

---

29 corpori] + et *BWX* 31 violente] -lentum *B* : -lentus *JX* 32 illius] istius *WYZ* | mundi] *om. BX* | mundi—istius] *om. W* | istius] illius *YZ* 36 et] sed *W*: hoc est *Z* 36-7 et—speciem] *om. homoiotel. X*

78. [79.] 3 ignis] + et *Z* | et] + eorum que *J* : + earum que *Z* 4 istius] + et potentia terre illius assimilatur potentie istius *BEJ* 5 similes] + erunt *Z* 6 non] *om. B* : + in *WXY* | ita sit] *inv. B* 7 totum ergo] *inv. JY* | iste] *om. XY*: illi *J*: istius *Z* 7-8 iste—mundorum] *om. B* 7 illis] *om. J* : illi *X* : illius *Y* 8 sunt similia] *inv. W* | sicut] *om. B* 9 ignis] + non *Y* 12 motus] movetur *B* : + omnium *JX* 13 continget] -git *Z* 15 quoniam] + si *J* | inquantum] + terra *J* 16 continget] -git *WY* | moveatur] -veantur *Z* 16-7 a^2—medium] *om. homoiotel. Z*

quidem illius mundi quia movetur ab illo mundo ad medium istius, ad medium autem quod est in illo mundo quia movetur a medio quod est hic ad illud medium. Et quia hoc non sequitur nisi cum posuerimus quod natura elementorum mundi positi eadem est cum natura elementorum istius mundi, incepit demonstrare hoc.

Et intendebat per celos corpora rotunda et per mundos corpora residua; et cum dixit *Sunt ex partibus consimilibus in natura*, intendit ex partibus convenientibus in specie; et hec propositio non fuit manifesta per se, sed acquiritur demonstratione: cum enim fuerit demonstratum quod impossibile est invenire corpus naturale simplex nisi moveatur ad medium vel a medio vel circa medium, manifestum est quod medium non est nisi per corpus spericum, et corpora sperica non sunt nisi per corpora existentia in medio; et cum ita sit, manifestum est quod si ponatur alius mundus esse ut sit ex corporibus simplicibus, necessario est ex motibus a medio et ad medium et circa medium. Et quia impossibile est estimare quod sit alius mundus non conveniens isti nisi in nomine tantum, sicut est inopinabile estimare quod fit alius homo non conveniens isti homini nisi in nomine
52vb tantum, quoniam qui tale ponit non po|nit aliquid terminatum T 61a
neque diffinitum, quapropter sequitur ut si aliquis posuerit alium mundum, ut ponat ipsum univoce cum hoc, et quia hoc est manifestius et probabilius, accepit hoc Aristoteles in hoc loco, quoniam illud quod diximus collocatum est in potentia in predictis, et ipse innuet ipsum post. Et dixit *Si igitur corpora mundorum plurium* etc., idest si individua plurium mundorum non fuerint sicut individuum istius mundi, erunt partes individuorum illorum mundorum convenientes cum individuo istius mundi in nomine tantum. Deinde dixit *Si igitur ita sit, erit etiam totum* etc., idest et cum partes mundorum que sunt extra istum mundum non conveniunt cum partibus istius

---

17 quia] quod *W* 18 quod] qui *BX* 22-3 per[2]—et] *om. homoiotel. X* 22 corpora[2]] + rotunda *Y* 24 non] *om. W* | fuit] est *WXY* 26 impossibile est] *inv. Z* | simplex] similiter *YZ* 27 vel[2]] aut *WXY* | medium[2]] + et *BX* 28 est[1]] + per se *B* 29 existentia] entia *BWZ* | medio] -dia *BX* 31 ex] a *Z* | motibus] mobili- *BY* 32 circa] contra *Z* | impossibile] inopina- *JYZ* | estimare quod] ut *B* 32-4 sit—quod] *om. homoiotel. WXYZ* 33 tantum] *ante* in *J* 33-4 sicut—in] *om. Z* 34 fit] sit *XY*: si *B* | homo] mundus *X* | nomine] hom- *WZ* 35 qui] si *W* | tale] -liter *B* 37 hoc[2]—manifestius] manifestius est hoc *Z* 38 hoc Aristoteles] *inv. Z* 40 ipsum] *om. W*: ipsam *YZ* 41 sicut] secundum *BXYZ* 44 sit] est *WY* 45 extra] + ipsum *B*

mundi nisi in nomine tantum, continget quod individua istorum mundorum non conveniant cum individuo istius mundi nisi in nomine tantum, in formis autem differunt, quod est inopinabile et impossibile. Et cum posuit quod si alius mundus sit, necesse est ut sit univocus cum isto, dixit *Et si mundi sunt similes adinvicem* etc., idest et cum posuerimus quod partes mundorum necesse est ut sint eiusdem speciei, manifestum est quod partium mundi positi extra mundum istum quedam moventur a medio quedam autem ad medium, erit ergo illic corpus celeste et elementa quatuor.

Deinde dixit *Et iam declaravimus hoc cum posuimus* etc., idest et hoc apparet ex eis que declaravimus prius de numero partium istius mundi et numero motuum cum declaravimus quod isti motus recti sunt finiti et ad loca terminata et ex locis terminatis. Et cum ita sit, et est necessarium si alius mundus sit ut sit conveniens isti mundo, *53ra* necesse est ut partes eius et partes istius sint eedem specie, | cum T 61b manifestum est quod esse istius mundi ex istis partibus est necessarium, et impossibile est aliter. Et debes scire quod qui fuerit assuetus verbis loquentium difficile, aut omnino non poterit concedere istam propositionem et sibi similes, isti enim opinantur quod omnia sint possibilia, et nihil est necessarium, et primus Philosophus debet verba eorum destruere sicut omnium Sophisticorum, quorum sapientia est negare prima principia et destruere sapientiam.

79. [80.] *Si igitur unum elementorum dicatur habere aliquem motum,* *276b9* *dicamus quod si motus corporum omnium mundorum fuerint idem, erunt elementa omnium mundorum eadem necessario. Et si ita sit, erit terra illius mundi mobilis ad medium istius mundi, et ignis illius mundi*

---

46 continget] -git W_2Z 46-8 continget—tantum] *om. homoiotel. XY* 46 istorum] illo- *J* 47 conveniant] -nient *B* : communicat *J* 48-9 inopinabile ... impossibile] *inv. BXZ* 50 sunt similes] similes sint *B* 52-3 mundum istum] ipsum istum mundum *B* 53 autem] + moventur *Z* 54 ergo] igitur *BX* 55 declaravimus] + prius de numero *B* 56 que] + iam *Z* 56-7 prius—declaravimus] *om. homoiotel. WYZ* 58 et[2]] *om. B* 60 eedem] ea- *B* : + in *JZ* | cum] et *JYZ* : et cum *X* 62 est] + esse *YZ* | debes] -bemus *BJ* : -bet *Z* 64 sint] sunt *BJX* 66 verba—destruere] destruere verba *J* | sicut] sunt *Y* : *om. J* | Sophisticorum] -carum *JXY*

79. [80.] 1 unum] *om. B* | motum] -tuum *B* : + motuum *X* 3 elementa omnium] omnia elementa *Z* 4 illius[1, 2]] istius *B* | mobilis—mundi[2]] *om. homoiotel. BY* | illius[2] mundi] *inv. W* : mobilis *Z* *ad superius istius mundi mobilis; et hoc est impossibile quoniam si motus* 53rb *ignis et terre, qui sunt in illo mundo, sint sicut narravi|mus, tunc movebitur terra ad superius necessario, quia movebitur desiderans suum mun-* 276b15 *dum, et ignis movebitur ad medium, et similiter erit motus istius terre ab isto medio ad illud medium motu naturali; media enim et extrema mundorum similia sunt adinvicem.* 276b18

Cum declaravit quod necesse est ut partes mundi positi sint similes partibus istius mundi, revertitur et declarat hoc esse necessarium ex predictis; et est modus declarationis quem invenimus superius. Et dixit *Si igitur unum elementorum* etc., idest et si necessarium est quod quodcumque elementum sit positum in illo mundo, debet habere unum motuum istius mundi scilicet a medio aut ad medium aut circa medium, *Dicamus quod si motus corporum omnium mundorum* etc.; et hoc quod dixit manifestum est quoniam si motus corporum plurium mundorum fuerint idem in specie scilicet quod in quolibet mundo sint tres motus, a medio aut ad medium aut circa medium, necessarium est quod partes omnium mundorum sint T 62a eedem in specie.

Deinde revertebatur dicere impossibile quod sequitur ex hoc, et iam utebatur ipso prius antequam posuisset propositiones ex quibus sequitur, et dixit *Et si ita sit, erit terra illius mundi* etc., et hoc quod dixit necessarium est, scilicet quod si natura terre illius mundi posite eadem fuerit cum natura istius terre, necessarium est ut moveatur ad medium quo movetur ista terra; et similiter de igne scilicet quod si fuerit cum isto idem in specie, necessarium est ut moveatur ad orizonta istius mundi. Et non potest aliquis dicere quod natura ignis qui 53va est in medio illius mundi sit mo|veri ad superius illius mundi, et qui est in isto mundo moveri ad orizonta istius mundi; et similiter de terra que est illic et terra que est hic. Continget enim ex hoc quod natura ignis in illis mundis esset natura terre istius mundi, quoniam

5 istius] huius *Z* | mobilis] *om.* *Z* | est] *om.* *WYZ* 6 sint] sit *WXZ* 7 movebitur] –vetur *J* 8 terre] + et *B* 11 sint] sunt *W* 12 istius] *om.* *Y*: huius *JW sed corr.* W_2 13 quem] quam *W* | invenimus superius] *inv.* *JYZ* 14 unum] + quodque *JZ* 17 corporum] *om.* *XY* 19 plurium] omnium *Z* 20 sint] sunt *JWY* 24 ex] *om.* *Y*: + se *WX* 27 moveatur] -vetur *B* 29 orizonta] –tem *J* 31 illius[2]] istius *WZ* 32 istius] illius *YZ* | et] *om.* *B* 33 illic] hic *Z* | hic] illic *Z*

quod movetur ad orizonta illius mundi, et quod movetur ad medium istius mundi movetur ad idem, et similiter quod movetur ad medium illius mundi movetur ad medium istius mundi; contingit ergo quod motus istorum elementorum sit accidentalis, aut quod eadem natura moveatur motibus contrariis.

Et cum declaravit quod necessarium est quod elementum grave illius mundi moveatur ad medium istius mundi, et similiter quod ignis illius mundi moveatur ad orizonta istius mundi, dixit *Et hoc est impossibile* etc., idest et si necessarium est quod terra illius mundi moveatur ad medium istius mundi, quia eadem est natura illius terre, continget ut terra moveatur in illo mundo ad superius necessario; non enim potest venire ab illo mundo ad istum antequam moveatur in illo mundo ad suum orizonta. Deinde dixit *Quia movetur desiderans ad suum mundum*, idest cum fuerit nature istius terre, et terra que est hic movetur desiderio ad medium istius mundi, continget similiter ut illa terra moveatur ad medium istius mundi. Et dixit *Ad* **T 62b** *suum mundum*, et potest aliquis dicere quod suus mundus non est nisi mundus in quo est, quia cum posuerimus ipsum de natura istius terre, erit suus mundus iste mundus necessario, scilicet mundus terre que est hic est iste mundus necessario.

Et ex hox apparet quod impossibile est ut elementa habeant nisi unum mundum, quoniam si plures mundi essent, non esset superius *53vb* naturale | neque inferius, sed in sola comparatione; et si ita esset, non diversarentur nature elementorum propter diversitatem motus, immo esset de motibus naturalibus sicut de motibus animalium, scilicet quia animal movetur ad loca opposita. Deinde dixit *Et similiter erit motus istius terre* etc., idest et similiter continget quod terra istius mundi moveatur a suo medio ad medium illius mundi, mediorum enim mundi nature necessario debent esse eedem in movendo terram

---

36 mundi] *om.* *YZ* | movetur[1]] -ventur *Y*: -veretur *B* 36-7 et—mundi[2]] *om.* *W* 37 mundi[1]] + necessario *J* 41-2 medium—ad] *om. homoiotel.* *B* 42 illius] istius *YZ* 44 eadem] *om.* *BWXY sed suppl.* W_2 | est] *om.* *Z* | natura] + huius et *J*W_2*Z* | illius] istius *BX* 45 continget] -git *JW* 47 Quia] quod *W* | movetur] -vebitur *Z* 48 nature] -ra *BX* 49-50 continget—mundi] *om. homoiotel.* *B* 51 mundum] medium *Z* | et] quia *J*W_2 53 scilicet] cum *WX*: est *Y* 53-4 scilicet—necessario] *om. homoiotel.* *B* 54 hic] + et *JZ* | necessario] *om.* *Z* 55 ut] quod *Z* 57 esset] + tunc *J* 58 diversarentur] -retur *JZ* | nature] -ra *BJZ* 59 animalium] *post* naturalibus *Z*: illis *Y* 60 quia] quod *W* 61 et] *om.* *JZ* | istius[2]] illius *YZ*

ad ea cum fuerit extra ea; cum igitur ista terra fuerit in medio istius mundi, erit extra medium illius mundi, et cum fuerit extra medium, quodcumque illud medium fuerit, continget ut moveatur ad ipsum, et similiter continget quod terra que est in medio illius mundi, cum sit extra illud medium, ut ad ipsum moveatur.

80. [81.] *Si igitur mundi non fuerint huiusmodi, non erunt nature cor-* 276b18 *porum simplicium, que sunt in celis et in pluribus mundis, eadem. Si igitur fuerint huiusmodi, erit natura corporis eadem, et concedunt dicentes mundos esse plures hoc necessario, si posuerint medium et orizonta esse unum, scilicet quod medium omnium mundorum et orizontes* 276b20 *omnium mundorum sint eadem; et cum ita sit, impossibile est quod mundi sint plures uno.* 276b21

54ra Vult in isto capitulo coniungere propositiones demonstrationis cum conclusione ut sint presentes coram considerante, ut sint apud intellectum sicut sensum, et incepit dicere *Si igitur mundi non fuerint huiusmodi* etc., idest et manifestum est, si omnes mundi positi non T 63a fuerint secundum modum quem narravimus quod in omnibus illis sint mobilia a medio et ad medium et circa medium, continget quod natura partium mundorum non sit eadem. Deinde dixit *Si igitur fuerint huiusmodi* etc., idest si necessarium fuerit in eis ut sint mobilia istis motibus, erit necessarium quod natura corporum ex quibus constituuntur plures mundi sit eadem natura. Deinde dixit *Et concedunt dicentes plures mundos esse hoc necessario*, idest et continget dicentes plures mundos esse concedere hoc necessario scilicet quod corpora ex quibus partes mundi componuntur sint eiusdem nature.

---

64 fuerit[2]] *om. Y*: *ante* ista *J* 68 illud] istud *BJ* | ad ipsum] *post* moveatur *BX*: *om. Y*

80. [81.] 1 Si] isti *Z* | fuerint] -erunt *Z* 2 in[2]] *om. XY* 3 corporis] -porum *WXY* 4 et] in *B* 5-6 et—mundorum] *om. homoiotel. B* 5 orizontes] -tis *Y*: orizon *J* 8 isto] hoc *W* 9 ut[1] sint] cum fuerint *Z* | considerante] + scilicet *BJ*W_2 12 quod] quoniam *W* 13 sint] sunt *JW* | et[1]] *om. B* 15 etc.] *om. BX* 17 constituuntur] *ante* sit *JZ* | mundi] + partes *Z* | sit] *om. YZ* | Deinde dixit] *om. Z* 18-9 hoc—esse] *om. homoiotel. BY* | idest—necessario] *om. homoiotel. X* 19 hoc] hic *YZ* 20 partes] plures *WX* | sint] sunt *JY* | sint eiusdem] *inv. Z*

Deinde dixit *Si posuerint medium et orizonta esse unum*, idest et continget eis, cum posuerint corpora simplicia ex quibus mundi constituuntur, quod orizon sit unus et medium unum, scilicet secundum quod mihi videtur; quia quo totum movetur et quiescit, illic moventur partes et quiescunt; aut intendit quod non contingit eis cum non posuerint unum orizonta et unum medium, et posuerint naturam elementorum in pluribus mundis esse unam, ut elementa moveantur contra suam naturam. Et dixit *Et concedunt dicentes* etc., idest et concedunt mundum esse unum, et ideo contingit concessioni eius quod est mundum esse eiusdem nature, scilicet aliam concessionem scilicet quod orizon sit unus et medium unum; propter quod dixit *Si posuerint medium et orizonta esse unum*, idest si concesserint
54rb quod sequitur a po|sitione partium mundorum esse eiusdem nature, ut orizon et medium sit unum; et dixit hoc quod qui concesserit has duas propositiones, necessario concedet mundum esse unum. Et propter hoc dixit post *Et cum ita sit, impossibile est ut mundi sint plures*, idest et cum medium sit unum et orizon sit unus, impossibile est ut mundi sint plures. Hic autem induxit tres propositiones quibus nullus potest contradicere: quarum prima est quod partes mundorum necessario sunt eiusdem nature, et secunda est quod orizon et medium partium que sunt eiusdem nature est unum numero, et tertia est quod cum medium fuerit unum numero et similiter orizon, mundus erit unus numero; et omnes iste demonstrate sunt hic.

81. [82.] *Si ergo aliquis dixerit quod natura corporum simplicium, que* 276b22 T
sunt in pluribus mundis, est diversa secundum eorum remotionem a locis eorum secundum magis et minus, dicemus quod hic sermo est extra normam. Corpus enim cum removeatur a suo loco parum vel multum, non
diversatur in sua forma omnino, sed solummodo in quantitate remotio- 276b25

21 dixit] + et *Z* | et^{2}] *om. Z* 23 constituuntur] + concedere *J*W_2*Z* | scilicet] idest *BXY* 24 mihi videtur] *inv. Z* | quia] quod *WX* 25 moventur] -vent *Z* 26 posuerint^{2}] -suerunt *WY* 27 ut] *om. X*: et *Y*: + etiam *Z* 28-9 dicentes—concedunt] *om. homoiotel. W* 30 eiusdem] eius *BXY* 31 quod^{1}] quia *BX* 32 dixit] -ximus *BZ* | posuerint] -suerimus *Z* 34 orizon ... medium] *inv. B* | sit] sint *JZ* 37 medium] + cum *Z* | sit^{2}] *om. BX* 38 sint] *om. Z* 41-3 et—numero] *om. homoiotel. X* 42 cum] tantum *Z* 43 mundus] + igitur *Z*

81. [82.] 1 Si] et *praem. BX* | ergo] *om. BX*: igitur *Y* 2 secundum eorum] ergo secundum *Z* 3 hic] iste *Z* 4 removeatur] moveatur *W*

54va *nis, in forma autem non, sive appropinquet | suo loco sive removeatur; sed si appropinquaverit aut removeatur, omnino movebuntur, et motus eorum sunt manifesti, et hoc non indiget disputatione.* 276b27

Cum posuit quod causa diversitatis motuum corporum mobilium (scilicet motus eorum a locis oppositis) est diversitas naturarum, et posuit quod si inveniretur in eis ex locis oppositis plus quam unum, ut contingeret ut eadem natura moveretur duobus motibus oppositis, quapropter contingeret quod natura simplex non diversa esset diversa, non diversa propter positionem, diversa propter hoc quod movetur ad contraria scilicet a medio et ad medium, secundum quod dictum fuit quod continget si media fuerint plura uno.

Et cum hoc fecit, induxit questionem contradicentem huic propositioni dicenti quod ab hiis naturis simplicibus non provenit nisi unus motus tantum, et dixit *Si ergo aliquis dixerit* etc., idest si aliquis dixerit quod non contingit quod terra illius mundi moveatur ad medium istius neque econverso, quamvis habeant eandem naturam in duobus mundis, similiter de aliis simplicibus: corpus enim idem ex eis non equaliter est propinquum in duobus locis consimilibus que sunt in duobus mundis; et quamvis igitur sit unum tamen semper movetur ad propinquiorem duorum locorum consimilium, verbi gratia terra istius mundi propinquior est suo medio quam medio alterius, quapropter movetur ad istud medium non ad illud; et si esset in illo mundo, moveretur ad suum medium; moveretur igitur duobus motibus contrariis, quamvis sit eiusdem nature propter propinquitatem et remotionem ab altero duorum locorum consimilium in T 64a
54vb specie, oppo|sitorum secundum ubi, verbi gratia quod motus a medio istius ad medium illius, quamvis sit oppositus motui a medio illius ad medium istius. Et corpus simplex, secundum quod est simplex, impossibile est moveri duobus motibus oppositis, tamen hoc possibile est secundum appropinquationem et remotionem; appro-

6 appropinquet] propinquet *B* 7 removeatur] -veantur *XY* | movebuntur] re- *B* 8 indiget] + in *BY* 9 diversitatis] –tatum *W* 12 ut[1]] *om. JZ*: per *BWX* | contingeret] -veniret *Z* | ut[2]] quod *BJ*: ex *Y* | moveretur] -veatur *Z* 16 fuit] fuerit *Z* 21 habeant] -bent *X*: -beat *JZ* 22 similiter] + et *B* | aliis] + corporibus *Z* 24 tamen] tantum *B* 25 locorum] motuum *Z* 27 esset] non *W* 28 illo] alio *BX* | moveretur[2]] -vetur *BX* 33 illius] istius *Y* 35 appropinquationem] propinquitatio- *Z*

pinquatio enim et remotio aliquid addunt supra simplicia. Continget ergo simplici ut sit compositum ex natura sua, et ex eo quod ei contingit ex propinquatione et remotione, ut compositum habeat duos motus oppositos in duabus horis.

Et Aristoteles dixit respondendo *Quod iste sermo est extra normam*, idest quia diversitas motuum corporum non est nisi propter diversitatem formarum substantialium, non propter diversitatem in relatione aut quantitate neque in alio predicamento; diversitas autem secundum appropinquationem et remotionem non est in substantia. Et debes scire quod appropinquatio et remotio non operantur nisi in motibus rerum quarum movens est extrinsecum secundum appropinquationem et remotionem a movente; quapropter necesse est verificare in hoc loco quod motus elementorum non est ab extrinseco: quamvis manifestum sit per se, tamen ipse induxit sermones contradicentes sermonibus Antiquorum de motu et quiete elementorum, et maxime de terra. Antiqui enim ponunt causas existentes in quiete terre et in motu eius attractionem que est apud universam terram et locum universe terre et manifestum est quod non movetur terra ad universam terram ubicumque fuerit, quoniam si pars eius moveretur ad universam terram, contingeret ut esset sicut motus ferri ad magnetem, et si ita esset, possibile esset ut terra moveretur natura ad superius.
55ra | Et universaliter si motus terre non fuerit secundum attractionem aut propter naturam loci, aut propter naturam corporis quod est in illo loco, neque secundum expulsionem propter motum celi, manifestum est quod hoc quod dixit Aristoteles in hac demonstratione est verum. Et scire hoc, scilicet quod terra et elementa non moventur neque attractione neque expulsione, est propinquum principiis naturalibus; et hoc erit quando sciverimus causam attractionis **T 64b**

36 enim] igitur *Z*: illis *B* | addunt] -dit *B*: -det *X* 37 natura sua] *inv. BJX* 38 propinquatione] ap- *J* 40 iste sermo] *inv. YZ* | normam] moram *B* 46 appropinquationem] propinqua- *XYZ* 47 necesse] *om. J*: -ario *B* 49 sit] *post* se *B* | induxit] -ducit *B*: -ducet *X*: duxit *W* 50 Antiquorum] aliquo- *YZ* 51 causas] -sam *JWZ* | existentes] *om. JY*: extrahens *X* 52 in] *om. Z* 52-3 in—et] *om. homoiotel. BWXY sed suppl.* W_2 53 quod] + si *Z* 54-5 ubicumque—terram] *om. homoiotel. WYZ* 55 terram] *om. X* 56 possibile esset] *inv. JZ*: possibile *B* 57 attractionem] accidentem *B* 58 loci—naturam[2]] *om. homoiotel. W* 59 illo loco] *inv. Z*: loco *J* 60 quod[2]] + hoc *B* 61 verum] manifestum *Z* 62 attractione] abstrac- *WY* 63 quando] *om. X*: quoniam *W*: quoniam cum *Z* | attractionis] abstrac- *W*

in aliis corporibus attractivis, et viderimus ipsum fugere res in quibus estimatur quod attrahant elementa; et similiter manifestum est etiam quod neque secundum impulsionem; impossibile est enim invenire corpus naturale cuius motus non sit naturalis; et universaliter manifestum est quod motus eorum est per se. Et ideo dixit *Sed si appropinquaverint locis eorum aut removeantur, omnino movebuntur; et motus eorum sunt manifesti et non indiget disputatione*, idest quod non indigemus disputatione ad hoc quod motus eorum sit per se et naturalis.

82. [83.] *Volumus modo perscrutari utrum motus corporum non sit vio-* 276b27
55rb *lentus, aut motus eorum naturales sint violenti etiam.* | *Dicamus igitur*
quod corpus quod impossibile est moveri omnino, impossibile est moveri
violente. Si igitur motus corporum mundorum sit naturalis, erit motus 276b30
partium consimilium formarum necessario ad eundem locum, scilicet aut
ad aliquod medium aut ad aliquem finem. 276b32

Quia fundamentum huius demonstrationis est super hoc quod elementa habent motus naturales, incepit perscrutari utrum motus omnium corporum sit violentus, aut omne corpus habeat motum naturalem. Deinde dixit *Dicamus ergo quod corpus* etc., idest quod corpus quod est impossibile moveri omnino naturaliter impossibile est moveri violente. Et dixit hoc quia quod violente movetur, movetur contra suam naturam; quod ergo caret motu naturali caret violento. Deinde dixit *Si ergo motus corporum mundorum* etc., idest et T 65a
cum posuerimus quod motus corporum simplicem sit naturalis eis, continget quod corpora simplicia convenientia in specie, que sunt in pluribus mundis, moveantur ad eadem loca, scilicet quod gravia

64 attractivis] abstractis *Z* | viderimus] -demus *BJ* | ipsum] *om. B* | fugere] fingere *Z* 65 attrahant] -hunt *W*: trahat *J* 66 est enim] *inv. BYZ* 68 appropinquaverint] -verit *BJX* 70 disputatione] *om. W* 70-1 idest—disputatione] *om. homoiotel. B* 71-2 sit—naturalis] *om. Y*

82. [83.] 1 non] *om. Z* 2 naturales] *om. JX* 3 omnino—moveri] *om. homoiotel. BW sed suppl.* W_2 | omnino] + quod W_2Z 4 erit] erunt *JWY* 5 formarum] forma *B* | scilicet] *om. Z* | aut] sive *Z* 6 medium aut] *inv. Z* 9 sit] est *J* | habeat] –bet *J* 11 est impossibile] *inv. JW* 12 violente movetur] *inv. Z*: violente *Y* 13 naturam] + scilicet contra motum qui est secundum naturam W_2Z | violento] -lente *WZ* 14 ergo] igitur *B* eorum moveantur ad media, levia autem ad fines; et intendit Et cum ita sit, continget impossibile predictum scilicet quod idem corpus movebitur duobus motibus contrariis, a medio scilicet et ad medium.

55va3 **83.** [84.] *Si aliquis ergo dixerit quod corpora que sunt in mundis, scili-* 276b32
cet terra et alia elementa, sint eadem in forma, in numero autem plura, quoniam non assimilantur adinvicem neque universum universo, sed assimilantur adinvicem, scilicet quod partes istius mundi sunt similes partibus illius, et similiter partes ceterorum mundorum, quapropter non diversantur in forma sed in numero tantum. Ignis enim istius mundi est 277a5
alius ab igne illius mundi numero; et cum corpora mundorum differunt in numero, differunt etiam in locis, quoniam partes corporis istius sunt similes partibus corporis illius mundi, et eedem in forma, verbi gratia quod partes istius terre assimilantur terre alterius mundi. 277a8

Hec est questio secunda, quod mundi sunt plures; potest enim aliquis dicere quod sicut videmus in omni quod est unum in specie plura individua, ita est de mundo et suis partibus; sed cum posuerimus sicut dicitis quod partes mundorum sunt similes adinvicem, necesse est quod partes mundi et partes eorum sint idem in specie, et similiter de locis; et cum ita sit, necesse est ut mundi sint plures individuo; et cum posuit hanc questionem, incepit dissolvere eam.

55vb5 **84.** [85.] *Si igitur diversitas corporum in numero esset causa diversitatis* 277a9
locorum, oporteret ut partes istius terre diverse in numero et gravitate, quod quodlibet corpus corporum eius haberet medium quo vadit; quod est impossibile. 277a10

---

18 et[1]] + hoc *J* 19 sit] est *BX* | corpus] *om. Z* 20 movebitur] -vetur *Z* | duobus] + modis *B*

83. [84.] 1 ergo] *om. WY*: igitur *Z*: *ante* aliquis *J* 2 sint] sunt *BJ* 3 neque] + in *B* | universum] -sa *XY* 5 illius] istius *BXY* 7 mundi] + in *J* | numero] movetur *Y* 8 istius] + mundi *WX* 10 assimilantur] -latur *B* 12 quod[1]] quoniam *B*: quia *WY* | omni] *om. Y*: homine *JZ* 13 sed] et *WY*: quia *BX* 14 sicut] illud W_2YZ | dicitis] -centi W_2Z 15 partes[1]] *om. JXYZ* | sint] sunt *WXY*

84. [85.] 2 gravitate] -tem *Z* 3 quodlibet corpus] *inv. Z* | corporum] *om. YZ* | eius] *om. Z*

Dicit et responsio huius questionis est si causa multiplicationis **T 65b** locorum, scilicet esse plura media in numero et plures orizontes essent plura elementa in numero verbi gratia essent plures ignes in numero, scilicet quia ignis istius mundi sit alius ab igne ceterorum mundorum in numero, et similiter in aliis elementis, contingeret ut medium in numero diversaretur in eodem mundo propter diversitatem partium elementorum in numero, verbi gratia si causa, in essendo duo media in numero in duobus mundis, esset quia terra unius mundi est alia a terra alterius mundi in numero, oporteret ut quelibet pars terre in eodem mundo haberet proprium medium: essent ergo media secundi numerum individuorum partium terre, quod est impossibile; videtur enim quod locus cuiuslibet partium cuiuslibet horum elementorum sit idem numero, immo idem est numero locus totius et partis; impossibile est ergo quod esse plura elementa numero sit causa in essendo plura loca numero, scilicet prima loca que sunt medium et orizon.

56ra4 **85.** [86.] *Si ergo contradicens non poterit contradicere istis sermonibus et* *277a10* *rationibus, necessario erit unum medium et unus orizon. Si igitur medium fuerit unum et orizon unus, et celum etiam erit unum per istos sermones et necessitates.* *277a12*

Vult affirmare in hoc sermone, et quiescere in hoc quoniam dignum est, quod cum nullus poterit contradicere sermonibus affirmantibus medium esse unum numero et similiter orizonta, concedere celum esse unum. Et iste locus est probabilis, scilicet verificare rem cum non inveniatur ei contradicens secundum suam consuetudinem,

5 Dicit] *om. J* | responsio] -spondet *JYZ* | huius questionis] huic questioni *JZ* 7 essent[1]] erunt *WY*: omne *B* | in[1]] esse *WY* 8-9 scilicet—numero] *om. homoiotel. Y* 8 quia] quod *BJX* 9 elementis contingeret] *inv. Z* 10 propter] per *X* 10-1 propter—numero] *om. B* 11 elementorum] + in eo *W* | in[1]] + eodem *X* | numero] + in eo *Y* 12 essendo] + in eo *Z* 13 unius] istius *JZ* | est] et *W* | alterius] istius *Z* 14 proprium medium] *inv. BX* 15 secundi] -dum *BX* | individuorum] -duarum *B* 17 cuiuslibet] *om. JZ*: eius *B* 18 impossibile—ergo] *om. WYZ* | quod] + est *Z* 19 sit] non est *Z* 20 prima] puta *J*

85. [86.] 2 necessario] -arium *B* | erit] est *JZ* 3 et[1]—erit] *om. X* | etiam] *om. B* | unum[2]] + etiam *BJX* 4 necessitates] -tem *B* 6 dignum est] *inv. Z* 7 orizonta concedere] *inv. B* 8 rem] iterum *Z* 9 inveniatur] -nietur *B*: -niretur *X* | contradicens] -trarietas *W* | suam] + Ari. *BWXY* qui semper inducit sermones probabiles post demonstrationes. Aristoteles enim utitur eis duobus modis, aliquando preinducens eos ante demonstrationem et ad exercitandum et dubitandum, et aliquando post ut faciat convenire.

56rb11 **86.** [87.] *Et etiam terra vadit ad aliquem locum et ignis vadit ad ali-* 277a12 *quem locum, et hoc manifestum est. Omne enim mobile alteratur ab ali-* **T 66a** *quo in aliquid, alterationes autem diversantur secundum formam; forma* 277a15 *enim ex qua movetur corpus ad aliam diversatur; et omnis alteratio habet finem, verbi gratia sanabile enim alteratur ex infirmitate ad sanitatem, et similiter augmentabile alteratur ex parvitate ad magnitudinem, et similiter mobile in loco alteratur ex quibusdam locis diversis in forma ad quendam, et vadit ad locum qui est finis, non ad alterum, verbi gratia quia corpus sanabile movetur neque secundum transmutationem neque ubi-* 277a20 *cumque voluerit movens, sed ubi est finis motus scilicet sanitas. Et cum ita sit, contingit quod motus terre et ignis non sint in infinitum sed ad finem, scilicet ad loca diversa illis locis ex quibus moventur: locus enim superior est diversus ab inferiori; ista autem duo loca sunt orizontes motuum.* 277a23

56va Cum accepit, in demonstratione quod mundus est unus, quod motus simplices habent loca propria, hoc autem est fundatum super hoc quod motus simplices sunt finiti, incepit affirmare hoc et dixit *Et etiam terra movetur* etc., idest et etiam una propositionum receptarum in hac demonstratione est quod terra et ignis moventur ad loca terminata, et similiter aqua et aer, et nullum horum movetur in infinitum neque ad quemcumque locum.

Deinde incepit declarare hoc et dixit *Omne enim mobile alteratur ab aliquo in aliquid*, et intendit per alteretur transmutari ab aliquo terminato in aliquod terminatum idest a contrario in contrarium; et

10 inducit] -duxit *Z* 11 preinducens] -indicans *YZ* 12 demonstrationem] -nes *B* | et[1]] *om. BX*

86. [87.] 1 vadit[1]] *om. BWXY*: movetur *J* | ignis] + etiam *JY* 2 enim] igitur *Z* | alteratur] *om. Y* 4 qua] quo *WYZ* 5 enim] igitur *Z* | et] *om. Z* 6 ad] in *WY* 7 locis diversis] *inv. Z* 7-8 in[2]—alterum] *om. X* 8 non] *om. Z* | quia] quod *Z* 9 movetur] *post* transmutationem *Z* 10 sed] scilicet *B* 12 enim] *om. BWXZ sed suppl.* W_2 13 autem] *om. J* 14 unus] + et *Z* 15 est fundatum] *inv. Z* 16 finiti] + et *BX* 18 moventur] -vetur *B* 19 et[1]] *om. BZ* | et[2]] *om. Z* 22 in] ad *Y* | ab[2]] in *WYZ* hoc declaratum est in Phisicis in motibus rectis. Deinde dixit *Alterationes autem diversantur secundum formas*, idest et hoc est causa diversitatis modorum transmutationum scilicet quod quedam eorum sunt in ubi, quedam in quantum, quedam in quale; forma enim ad quam pervenit transmutatio diversatur in unaquaque transmutatione; et hoc voluit notificare quia non quodcumque sit, transmutatur in quodcumque. Deinde dixit *Forma enim ex qua moventur corpora* ad aliam formam diversantur, et hoc est causa alterationum que sunt **T 66b** finite. Et cum declaravit ratione quod transmutationes sunt finite, secundum quod sunt ex contrariis et in contraria, incepit declarare hoc per inductionem, scilicet declarare esse contrarias aut esse finitas, et dixit *Omnis alteratio habet finem*, et intendit per alterationem transmutationem tribus motibus, quoniam unusquisque istorum motuum habet finem terminatum, et est ex contrario in contrarium, verbi gratia quod transmutatio in qualitate est ex sanitate in infirmitatem, et in quantum ex parvo in magnum aut econtrario.

56vb Et quia invenire contrarietatem motu locali est magis occultum quam in alteratione, dedit exemplum in motu alterationis ut videat considerans quod ita debet esse in motu locali, et dixit *Et similiter mobile localiter*, idest et similiter est de mobili naturaliter in loco ex corporibus simplicibus, quoniam transmutantur ex loco contrario ad locum contrarium idest ex superiori ad inferius aut econverso. Deinde dixit *Et non vadit nisi ad locum qui est finis*, idest et manifestum est quod mobile in loco non movetur nisi ad locum diversum a loco ex quo movetur, sicut est de alteratione; et universaliter necesse est in omni transmutatione, ut a quo est sit diversum ad quod est, nisi transmutatio sit otiosa. Deinde dixit *Movetur neque per transmutationem* etc., idest secundum quod mihi videtur, nisi in verbis ceciderit error, quod manifestum est ex sanatione, et sibi similiter quod

---

25 autem] enim *Z* | formas] -mam *BJ* 26 transmutationum] transumptio- *Y*: transumptionem *Z* 27 quedam in quantum] *om. W* 28 unaquaque transmutatione] unamquam- -nem *W* 29 hoc] hic *XY* | sit] *om. JZ*: fit *B* 31 que] quod *WXY* 33 ex] in *WYZ* 35 et dixit] *om. Y* 36-7 istorum motuum] *inv. W* 39 parvo] magno *Y* | in^2] *om. WY*: ad *J* | magnum aut] parvum et *Y* | econtrario] -verso *BJY* 40 invenire] -niri *WXY* | contrarietatem] + in *XY* | est] *post* occultum *Z* 43 idest] *om. W* 44 transmutantur] -tatur *BX* 45 contrarium] -tinuum *Z* 46 nisi] natura *YZ* 48 loco] quo *B* | sicut] illud *Y* | est] *om. XY* 49 transmutatione] transumptio- *Y* | diversum] + ab eo *J*W_2: + a loco *Z* 50 sit] esset *WXY* 52 similiter] -libus *XY*

non transmutantur ut transmutentur, neque transmutantur ad
quamcumque transmutationem transmutans voluerit, ut in habenti-
bus animam sed ad finem naturalem. Et iam declaravit quod motus
locales recti sunt finiti in pluribus locis; et cum verificavit proposi-
tionem dicentem quod omnis transmutatio est finita, dedit conclu- **T 67a**
sionem et dixit *Et si ita sit, erit motus ignis et terre non in infinitum*,
et ita componitur sillogismus: ignis et terra et alia simplicia sunt
transmutabilia; et omne transmutabile transmutatur ex contrario in
contrarium; ergo ignis et terra transmutantur ex contrario in contra-
rium; sed omnia transmutabilia ex contrario in contrarium sunt
finite transmutationis, ergo ignis et terra et alia sunt finite transmu-
tationis, ergo habent loca naturalia ad que transmutantur secundum
57ra naturam et | ex quibus transmutantur, et oportet ut sint contraria.

Et cum ita sit et apparet quod omnes partes uniuscuiusque corpo-
ris eorum moventur ad eandem partem, manifestum est quod locus
omnium individuorum terre est unus numero, et similiter de aliis
elementis. Et hoc intendebat cum dixit *Et ista duo loca sunt orizontes
motuum*, idest quod superius et inferius sunt ista duo loca que appa-
rent esse finita; duo enim elementa levia moventur ad superius, alte-
rum quidem simpliciter, alterum autem in respectu, et similiter gra-
via ad inferius, aut simpliciter aut in respectu.

87. [88.] *Et etiam secundum aliquem modum est in motu circulari* *277a23*
*contrarietas per diversitatem, scilicet motus existentis in diametro circuli;
motus autem totius circuli non habet contrarium omnino. Manifestum* *277a25*
*est igitur quod motus partium revolutionis sunt ad loca diversa finita.
Verificatum est modo sermonibus necessariis quod motus omnium corpo-
rum habent finem, et quod non vadunt ad infinitum.* *277a27*

---

53 transmutantur[1, 2]] -tatur *X* | ut] vel *B* | ut—transmutantur] *om. homoiotel. JZ* | neque] in *W* 54 transmutans] *om. Y* 55 motus] + naturales *Z* 58 si] cum *W* 59 alia] + sunt *B* 61-3 transmutantur—terra] *om. homoiotel. WYZ* 62 sed] et *X* 63 finite[1]] in- *B* 63-4 ignis—ergo] *om. homoiotel. B* 65 ut] quod *Z* 66 et[2]] *om. B* 67 moventur] -vetur *XYZ* 70 ista] illa *JZ*

87. [88.] 1 etiam] quod *YZ* : que *W* | aliquem modum] *inv. WYZ* 2 existentis] -tentes *WXZ* : -tens *B* 5 Verificatum] -catio *Z* 6 ad] in *WYZ*

57rb Cum posuit quod omnis motus necessario, secundum quod est motus, est ex contrario in contrarium, et hoc universaliter in omni motu (et iam declaratum fuit in motu circu|lari quod non habet contrarietatem neque est ex aliquo ad aliquod, et hic facit questionem), incepit dissolvere ipsam demonstrando quod in motu circulari est contrarietas secundum aliquem modum, secundum autem alterum non, scilicet quod in partibus eius invenitur contrarietas secun- T 67b dum aliquem modum et secundum alium non; cum autem movebitur eadem pars eius ex aliquo extremo diametrorum circuli ad extremum circuli, erit contrarietas; cum autem movebitur eadem pars ab aliquo puncto quousque revertitur ad idem punctum, non erit contrarietas. Et hoc intendebat cum dixit *Et etiam secundum aliquem modum est in motu circulari contrarietas per diversitatem*, idest quod in motibus partium revolutionis erit contrarietas, idest ex quo et ad quod, inquantum pars incepit ex extremo diametri, et pervenit ad extremum secundum, quapropter stella in uno extremo dicitur ascendens et in altero occidens, et ascendens dicitur contrarium occidenti. Deinde dixit *Motus autem totius circuli non habet contrarium*, idest motus autem partium quarum motus est revolutio completa non habet contarietatem omnino, quoniam ex puncto ex quo incepit, ad punctum ex quo incepit, pervenit.

Deinde dixit *Manifestum est igitur quod motus partium revolutionis etiam sunt ad loca diversa*, idest manifestum est modo quod motus partium circuli sunt ad loca diversa ab eis ex quibus sunt, cum non compleverint revolutionem sed fuerint ab extremo diametri usque ad extremum oppositum. Et hec contrarietas non invenitur in circulo nisi propter diametrum: diametrum enim est linea recta, et contrarietas non invenitur nisi in duobus extremis longitudinis recte, ut in tertio Phisicorum declaratum est, et in principio huius libri.

---

8 est] esse *BXY* 10 aliquod] aliquid *WXY* 11 ipsam] ipsum *X*: + in *J* 14 aliquem] alium *Z* | autem] enim *J* 15 eadem] *om. BX* 15-7 eius—pars] *om. homoiotel. Y* 15 eius ex] ab *Z* | extremo] + diametri *BX* 16 erit] est *Z* | contrarietas] + secundum *Z* 17 pars] *om. W*: + cadit *Z* | revertitur] -tatur *BX* 18 etiam] *om. W* 19 per] propter *B*: secundum *J* 22 dicitur] *om. Z* 24 Motus] totus *B* 26 ex[2]] in *Z* 29-30 idest—diversa] *om. homoiotel. BY* 29 modo] + secundum *Z* 30 partium] *om. Z* | circuli] -lares *Z* | loca diversa] *inv. X* 31 compleverint] -pleverit *BXZ*: -pleavit *Y* | sed] si *Z* | fuerint] -erit *BZ*: fiunt *W* 32 invenitur] + nisi Y 32-34 in—nisi] *om. W* 32 circulo] -cularibus *J* 33 diametrum[2]] diameter *JZ* 34 invenitur] -nietur *Z* | nisi] *om. Y* 35 tertio] quinto *J* | et] *om. XY* | et—libri] *om. B*

57va14 **88.** [89.] *Et signum quod movens non movetur in infinitum est terra: terra enim quantocumque appropinquaverit medio, citius movetur, similiter ignis quanto magis ad superius, tanto citius movetur. Si ergo motus esset in infinitum, velocitas esset in infinitum; si ergo velocitas esset secundum hoc, gravitas et levitas essent secundum hoc, scilicet in infinitum. Et sicut eorum que ad inferius moventur quedam moventur citius secundum diversitatem in gravitate, ita esset si additio gravitatis esset in infinitum.* 277a27 277a30 277a33

Hoc est signum aliud a predictis, quod motus elementorum sunt finiti, quamvis utatur in hoc quibusdam propositionibus predictis: ponit igitur primo quod manifestum est per se quod motus terre et ignis et aliorum elementorum videntur, cum appropinquaverint locis propriis, citius moventur; et cum ita sit, necesse est ut medium sit finitum similiter superius. Quomodo autem sequitur | 57vb hoc ex predictis patet et notum est per se, quoniam si motus eorum esset in infinitum, contingeret ut velocitas eorum cresceret in infinitum, quoniam superius et inferius ponuntur infinita; et si velocitas esset infinita, gravitas et levitas essent infinita, quod est impossibile secundum quod declaravimus. Cum ergo dixit *Si igitur motus eorum esset in infinitum, velocitas esset infinita*, intendit Et cum posuerimus quod quanto magis intraverit illam partem ad quam movetur, tanto movetur citius, et posuerimus quod illa pars in quam movetur est infinita, continget quod velocitas transeat in infinitum, et si inveniatur velocitas infinita, invenietur gravitas aut levitas infinita. Et hoc intendebat cum dixit *Si igitur velocitas est secundum hoc, gravitas et levitas erunt secundum hoc*, scilicet in infinitum.

Sed potest aliquis querere quod hoc sequitur a positione quod grave et leve moventur in infinitum, scilicet gravitatem esse infini-

---

88. [89.] 1 infinitum] + et *B* 2 citius] cuius *XY* 3 ad] advicinatur *W_2ZG* | tanto—movetur] *om. BWXY sed suppl. W_2* 4 velocitas[1]] -citer *B*: -cius *W* | in[2]—esset] *om. homoiotel. Y* 5 essent] *om. Y*: *post* hoc[2] *BJ* 7 gravitatis] -tas *B* | infinitum] + esset additio velocitatis in (ad *G*) infinitum *JW_2ZG* 8 predictis] preteritis *B* 10 igitur] ergo *J* 12 moventur] -veri *BJ* 13 superius] *om. B* | hoc] *om. YZ* 14 patet] *om. XY* | notum] motum *Z* | eorum] *om. WYZ* 15 contingeret] -get *Y* | contingeret—infinitum] *om. homoiotel. B* 18 ergo] igitur *JZ* | igitur] ergo *WY* 19 Et] *om. Y*: quod *Z* 20 magis] + motus *B* | intraverit] + ad *BX*: + in *Y* 21 quam] qua *BZ* 22 contingett] -git *Z* 24 levitas] velocitas *BX* 25 in] *om. WY* 26 quod[1]] quare *Z* | quod[2]] *om. B* tam, quamvis non ponatur quod quanto magis appropinquaverit motum, tanto citius movebitur, quid ergo prodest propositionis istius positio? Dicamus igitur quod ipse voluit notificare quod, cum declaratum est quod si motus equalis ponatur infinitus, continget ut virtus movens sit infinita, necesse est et dignius, cum ponatur motus transiens in velocitate in infinitum, ut sit a potentia infinita. Et ideo dixit *Si igitur velocitas secundum hoc, gravitas et levitas essent secundum* T68b *hoc*, scilicet in infinitum.

Deinde dixit *Et sicut eorum que moventur ad inferius* etc., idest et monstrat per hunc sermonem causam propter quam, si velocitas fuerit infinita, quod gravitas et levitas sint infinite, et est quia si cause diversitatis rerum in velocitate est diversitas eorum in declinatione *58ra* | idest in gravitate et levitate, sequitur quod quanto magis fuerit grave aut leve, tanto magis erit velox, et manifestum est quod hoc convertitur scilicet quod quanto magis fuerit velox, tanto magis erit grave et leve; et cum ita sit et fuerit velocitas infinita, necessario erit gravitas et levitas infinita.

89. [90.] *Et etiam corpus vadens ad superius et corpus vadens ad inferius non vadunt ex aliquo movente violente, sicut inquiunt quidam homines quod elementa moventur per constrictionem. Et si ita esset, motus ignis magni esset tardus ad superius et motus terre magne ad inferius; sed nos videmus contrarium, ignis enim magnus citius movetur ad suum locum. Et etiam si motus elementorum esset constrictione, non esset motus terre et ignis, cum appropinquaverint suis locis, velocior: omne enim quod a suo expellente removetur, tarde movetur, et dixi|mus quod ad locum ex quo elementa moventur violente, moventur naturaliter.* *277b1* *277b5* *58rb* *277b8*

Quia una propositionum ex quibus componitur demonstratio est quod corpora simplicia moventur per se naturaliter, scilicet a princi-

---

28 appropinquaverit] -querit *W*: + ad *X* 29 tanto citius] *inv. B* 31 continget] -git *Z* 33 a] alia *Z* 34 igitur] ergo *Z* | velocitas] + esset *JZ* 35 in] *om. B* 37 monstrat] -stravit *JY* 38 cause] omne *BX* 40 sequitur] scilicet *WXY*: secundum *B* 41-2 velox—erit] *om. homoiotel. B* 41 et] *om. WY* 42 erit] *om. WXY* | et] vel *BX*: aut *Y*

89. [90.] 1 superius—ad] *om. homoiotel. Y* 2 non] *om. BWXY sed suppl.* W_2 | aliquo] alio *BJX* 3 constrictionem] + quia *Z* 5 nos] *om. BJ* 6 esset[1]] + cum *BJ* 7 omne] esse *XY* 8 diximus] -cemus *J* 11 scilicet] *om. WZ*

piis existentibus in eis, non ab extrinsecis, potest aliquis dicere quod non moventur ad superius et ad inferius nisi ex causis extrinsecis violentis, ut plures Antiquorum credebant de terra quod violente stat in medio et violente movetur ad medium propter motum circularem,
incepit contradicere huic ostendens quod non dignum est ut sit que- **T69a**
stio de motu per se, cum manifestum est quod ista moventur per se scilicet corpora simplicia.

Et posuit primam rationem super hoc, scilicet quod si terra movetur ad inferius violente et a movente extrinseco, oporteret quod terra maior moveretur tardius; et nos videmus quod quanto magis terra magnificatur, tanto citius movetur ad inferius, et hoc sequitur quando movens violente ponitur idem, scilicet quia idem movens, quando moverit duo corpora quorum alterum est maius alterum autem minus, movebitur motu tardiori. Et dicentes has scientia concedunt quod constringens elementa est idem, quoniam si ponatur quod sit motus totius, erit idem, et si ponatur quod quedam constringunt quedam, tamen causa earum erit motus totius.

Deinde dedit secundam rationem et dixit *Et etiam si motus elementorum* etc., iste sermo componitur in prima figura et secunda, in prima quidem sic: motum violente quanto magis appropinquaverit suo fini, tanto minus est velox, et quanto est minus velox, tanto
58va magis tardatur | in suo motu, ergo tanto tardius movetur quanto propinquior suo fini. Deinde accipitur hec conclusio et componitur in secunda figura sic: omnia elementa quanto magis appropinquaverint suo fini, tanto magis velociter movebuntur, et mobilia violente non velocificantur cum appropinquaverint suo fini, ergo elementa non moventur violente. Ipse autem inducit ipsum in modum ipothetici,

14 credebant] puta- *BX* 16 quod] *om. Y*: quia *BX* | non] *om. X*: numquam *JY* | est] et *B* 17 de motu] *om. WXY sed suppl.* W_2 | motu] *om. B* 19 si terra] *inv. Z* 23 quando] quoniam *Y* | scilicet] et *Z* | quia idem] *om. Z*: cum idem *J* 25 minus] + quod maius est *JZ* | scientia] sententiam *JZ* 28 quedam] + non quedam *Z* | earum] *om. YZ*: eo- *J* | totius] + est eadem W_2*Z* 30 secunda] prima *W* 31 motum] -tus *B* | appropinquaverit] tardabitur in *B* 32 fini] motu *B* 32-3 tanto[1]—movetur] ergo motus violentus tanto est minus velox *Y* 32 minus[1]] magis *X* 32-6 minus[1]—tanto] *om. homoiotel. W* 32 est[1]] erit *B* 32-3 est[1]—movetur] tardabitur in motu et quanto magis tardabitur in suo motu, tanto erit minus velox cum appropinquaverit suo fini *X* | et—movetur] *om. B* 32 est minus] *inv. Z* 33 propinquior] + est *Z* 34 accipitur] -cepit *X* 36 magis] minus W_2 *(cf. supra, app. crit. ad 32-6)* | violente] -lenter *J* 37 velocificantur] -citantur *Z* 38 ipsum in] *inv. Z* | modum] *om. B*: modo *Y* | ipothetici] -tico *B*

sed continuatio declaratur ex hoc quod diximus; et ipse dedit causam propter quam motus violentus quanto magis appropinquaverit suo fini tanto tardius movetur, que est remotio a movente extrinseco. Deinde dixit *Et dicebamus quod ad locum ex quo elementa moventur violente, moventur naturaliter*, idest quod talibus rebus non indigemus quod talia indigeantur declarari ratione, quoniam manifestum est per se quod ad omnem locum ad quem elementum movetur violente movetur eius contrarium naturaliter, verbi gratia quia terra movetur ad superius violente, ignis movetur ad superius naturaliter.

58vb **90.** [91.] *Oportet ergo aspicientem in istis rebus et querentem eas conce-* *277b8*
dere hoc quod diximus ex istis opinionibus et propositionibus concessione
non dubia, et quod sit sufficiens et contentus sermonibus Phisicis Primis *277b10*
quos scripsimus, quoniam iam scripsimus hec et diximus quod motus circularis unus est sine principio sive sit in hoc mundo sive in aliis mundis;
deinde invenimus ex isto motu quod celum est unum et mundus unus, et **T69b**
non plures ut alii inquiunt. *277b13*

Cum declaravit quod mundus est unus ex propositionibus naturalibus, et hoc etiam apparet ex propositionibus divinis, incepit declarare quomodo apparet hoc ex duabus scientiis, et quod oportet propter hoc credere hoc quesitum, et dixit *Oportet ergo aspicientem* etc., idest et dixit hoc quia plures homines non possunt credere principia prima propter crementum eorum in opinionibus falsis, ex quibus infiguntur in eis propositiones probabiles, contrarie primis propositionibus manifestis per se, sicut opinantur per se in lege nostra quod Deus posset creare mundos infinitos nisi esset diminutus. Quasi ergo dixit Oportet ergo aspicientem in istis propositionibus credere eas et

41 movente] violente *Z* 42 dicebamus] -cemus *JX* | locum] + extra *Z* 43 indigemus] + scilicet *Z* 44 quod] quia *BJ* | indigeantur] -geant *Z* : -gent *J* 45 est] + hoc *Z* | se] + et *Z* | ad[1]] *om. B* | quem] + aliquod *WX* | movetur] + non *Z* | violente] + est *Z* 46 movetur] motus *YZ* | contrarium] -rius *Z*

90. [91.] 1 et] *om. BWXY* 2 opinionibus et] *om. Z* | propositionibus] + et probationibus *Z* 3 Phisicis] philosophis *BX* 4 scripsimus[1]] + hoc *B* | quoniam—scripsimus] *om. homoiotel. XY* | iam scripsimus] *inv. J* 5 sive[2]] + sit *Z* 6 ex] in *YZ* | et[1]] quod *Z* | mundus] + est *BJX* 10 hoc] hee *Y* | et] *om. YZ* | quod] *om. W* 11 credere hoc] *om. WYZ* 12 idest et] *om. X* : et *W* 16 posset] -sit *W* 17 propositionibus] + positis *J*W_2*XZ* scire que sunt necesse in hac scientia, et non dubitet in eis propter propositiones probabiles contrarias eis, aut propter estimationem quod principia istius artis debent esse in fide, sicut principia geometrie: ista enim duo faciunt dubitationem in propositionibus naturalibus.

59ra Et cum dixit quod hoc apparet ex rebus dictis in Scientia Divina, incepit dicere hoc et dixit *Quoniam scripsimus iam illic et diximus quoniam motus circularis* etc., idest iam declaravimus, in sermonibus positis quasi fundamenta in Scientia Divina, quod motus circularis debet esse unus sine principio et sine fine, et quod alii motus reducuntur ad istum motum scilicet ad motum circularem, sive ponamus istos motus accidentales in isto mundo sive in alio mundo, scilicet quod necessarium est ut reducantur in unoquoque mundorum ad talem motum. Et hoc quod dixit declaratum est in octavo Phisicorum, quoniam hoc non declaratur in Scientia Divina, sed ponitur. Et secundum hoc debemus intelligere hoc quod dixit scilicet *Iam scripsimus*, idest iam scripsimus in Scientia Divinali, et declaravimus in **T 70a** Scientia Naturali scilicet quod motus naturalis debet esse in hoc mundo sine principio et sine fine, et similiter in quolibet mundo.

Et cum dixit hoc, dixit *Deinde invenimus ex isto motu quod celum est unum et mundus unus necessario*, idest et cum verificatum est nobis hoc fundamentum positum in Scientia Divinali, declaravit quod motus localis circularis debet esse unus eternus, et quod movens et faciens ipsum non debet esse in materia. Et declaratum est illic quod omne quod non est in materia impossibile est ut inveniatur ex eo plus quam unum ens numero, scilicet quod impossibile est ut inveniantur ex eo duo, et declaratum est illic quod ab uno movente non invenitur nisi unum motum. Et cum ita sit, necesse est ut celum sit unum numero, ergo necesse est ut mundus sit unus numero; et hoc intendebat cum dixit *Deinde invenimus ex hoc motu quod celum est*

18 sunt] est *YZ* | necesse] -arie *BJX* | dubitet] -tare *Z* 19 propositiones] -portiones *B* | quod] quia *Z* 20 istius] huius *Z* 23 scripsimus iam] *inv. YZ* | illic] illis *B* 25 fundamenta] -tum *YZ* 26 alii motus] *inv. Z* 28 motus] modos *Y* | sive—mundo] *om. homoiotel. Y* 29 ut] + quod *B* 31 Et] *om. B* 33 idest—scripsimus] *om. homoiotel. BJ* | scripsimus] de- *X* | Divinali] divina *BJZ* 34 motus naturalis] *inv. B* 36 isto] primo *Z* 37 mundus] + est *BJX* 38 Divinali] + et *BJ* 40 Et] ut *Z*: + cum *J* 41 inveniatur] -niantur *W* 42 inveniantur] -niatur *XYZ* 44 invenitur] -niatur *JY*: -niuntur *W* 45 necesse est] *om. W*: et W_2 | sit] *om. WZ*

59rb *unum*, idest quod invenimus ex uno motu quod movens est | unum et quod unum movens non facit nisi unum motum.

Et potest aliquis dicere, cum declaratum est quod movens non movet celum nisi desiderio existente in celo propter intellectum eius, quid prohibebit ergo quod non sint plures celi? Et unumquodque eorum desiderat hoc primum movens eiusdem modi intellectus, sicut in nobis unus intellectus movet plura imaginata, verbi gratia cibum esse utilem movet omne famelicum ad cibum, et forte sic est de celis. Dicamus igitur quod iam declaratum est quod imaginatum illic est idem imaginatum ex omnibus modis; et cum ita sit, imaginatio est eadem sicut imaginatum est idem, et cum imaginatio sit eadem, non invenietur nisi in eodem imaginato. In hominibus autem imaginatio non est imaginatio eadem ex omnibus modis, quapropter possibile est imaginari idem imaginatum pluribus imaginationibus et a pluribus imaginantibus; secundum hoc igitur debemus intelligere imaginationes corporum celestium, et iam declaravimus in eo quod scripsimus in ea scientia.

59va20 **91.** [92.] *Modo autem dicamus quod elementa particularia sunt tria,* 277b13 **T 70b**
ergo et loca eorum sunt tria. Quorum unus est locus corporis descenden- 277b15
tis ad inferius, secundus est corporis rotundi et est ultimum, tertius est
medius inter illa duo loca, et est corpus leve existens in isto loco necessa-
rio; quoniam si non fuerit in hoc loco, erit in alio loco extra loca pre-
dicta, quod est impossibile. Si ergo locus medius est accidentalis corpori 277b20
levi, erit naturalis corpori; sed nullum corpus erit preter ista corpora; et
cum ita sit, erit locus istius corporis medius locorum necessario; et nos
59vb *dicemus post diversitatem istius corporis. Iam ergo de|claravimus quali-*
tatem elementorum particularium et etiam quantitatem locorum et

---

47 movens est] *inv. BJ* : movens *X* 49 quod] + hoc est *B* : + hoc *X* 51 quid] quod *Z* | prohibebit] -hibet *Z* | sint] sunt *Z* 52-3 sicut—intellectus] *om. homoiotel. B* 53 in] *om. Z* 54 esse] *om. Z* | ad cibum] *om. Z* | sic] sicut *B* | est] erit *JZ* : esse *B* 55 illic est] *inv. BJ* 56 cum—sit] est ita sic *Y* | imaginatio] -tione *B* | est] *om. B* 57 est] *om. YZ* 59 eadem] *om. WXY* 60 et] *om. Z* | a] *om. B* 61 imaginationes] -tionis *B* 63 ea] hac *Z*

91 [92.] 2 ergo—tria] *om. homoiotel. Y* 4 medius] motus *Z* 5 si] + ita *W* | hoc] isto *BJ* | erit] est *Z* | loca] *om. BJ* 7 naturalis] + alii *J* | corpori] + alteri *Z* | erit^{2}] est *BJX* | ista] + tria *BJ* 8 corporis medius] *inv. B* | necessario] *om. B* 10 particularium] + et quantitatem eorum et locum cuiuscuiusque ipsorum *ZG* *numerum, et cum perfecte et sufficienter. Volumus ergo dicere modo* 277b25 *etiam quod mundus est unus, et non est possibile quod sint multi mundi; et ostendam iterum quod sempiternus est, et non est factus ex aliqua re, et non cadit sub corruptione.* 277b29

Intendit per elementa particularia corpora simplicia que non sunt ex corporibus prioribus eis, et sunt tria, grave, leve et neutrum; et cum manifestum est quod corpora simplicia sunt tria, necesse est ut loca sint tria. Et hoc non contingit nisi ei qui concesserit quod omne corpus est in loco essentialiter, et Alexander noluit hoc; et etiam Aristoteles dixit quod celum non est in loco essentialiter, quomodo ergo dixit quod loca sunt tria? Dicamus ergo quod cum dixit in Phisicis quod celum non est in loco, fundavit quod nihil est extra celum, et hoc fuit acceptum illic pro constanti, et in hoc loco declaratum est. Qui ergo posuerit quod mundus est plus quam unus, necesse est ei dicere quod celum est in loco; et cum declaratum fuerit quod mundus sit unus, declarabitur quod extra ipsum nullum corpus est, et cum huic coniungatur quod declarabitur in Phisicis, idest quod cor- T 71a pus quod non est in loco est extra quod non est aliquid, declarabitur quod celum non est in loco. Videtur ergo quod dixerit hoc quod sequitur dicentes mundos esse plures, scilicet quod celum est in loco *60ra* et quod ex istis tribus locis non inveniantur | plura numero; et si ex eis inveniantur plura numero, continget quod pars et totum moventur ad eundem locum numero elementorum motorum, recte igitur hec tria loca sunt eadem numero, quoniam duorum utrumque fuerit unum numero scilicet locus gravis et levis, continget quod locus tertius sit idem numero, et est locus corporis neutrius; et cum mundus

---

11 cum] *om. J*: est *Y*: etiam *X* | perfecte] -tione W_2Z | ergo] *om. BJ* | dicere modo] *inv. WXY* 12-4 quod—corruptione] *om. BWXY (sed cf. infra, c. 91 [92], v. 77-84)* 14 corruptione] + sed (sed autem *Y*: sed tamen *X*) volumus, antequam declaremus hoc, querere et perscrutari de hoc *BJWXY (cf. infra, t. 92 [93] et app. crit. ad loc.)* 16 tria] scilicet *YZ* | grave] + et *BX* 18 nisi ei] *inv. J*: ei *Z* | concesserit] -cesserat *Z* 21 dixit—loca] *om. B* 22 fundavit] + illud super hoc JW_2Z 23 acceptum illic] *inv. YZ* | et] *om. B* | est] *om. WYZ* 26 sit] est *BX* | declarabitur] + ei *Z* 27 declarabitur] -batur *BJ* 28 non[1, 2] *om. BWXY sed suppl.* W_2 29 ergo] igitur *BJX* 31 quod] quia W_2Z | non] *om. WXY sed suppl.* W_2 | inveniantur plura] *inv. YZ* | et[2]] sed *W* 31-2 et[2]—numero] *om. homoiotel. YZ* 32 eis] hiis *J* | continget] + enim *Z* | totum] + non *J* | moventur] -veantur *WXY* 34 tria loca] *inv. W*: *post* sunt *Z* 35 et] aut *JZ*

non inveniatur nisi in hiis tribus locis, et sunt eadem numero, ergo mundus existens in eis est idem numero.

Deinde dixit *Quorum unus est locus corporis descendentis ad inferius quod est in medio*, idest locus terre et aque; et quoniam hec duo elementa numerantur in gravi, et iste locus in rei veritate est ultimus aeris. Deinde dixit *Et locus secundus* etc., idest longitudo in qua imaginatur corpus esse celeste, et potest intelligere per locum tertium hic illud quod est corpus ultimum corporum celestium, quoniam omnia que sunt extra illud sunt essentialiter in loco. Deinde dixit *Et est ultimum locorum*, potest intelligi locorum imaginatorum vel locorum partium celi preter ultimam partem, verbi gratia preter speram stellarum fixarum. Deinde dixit *Et locus tertius* etc., idest locus duorum corporum levium scilicet ignis et aeris, que sunt media inter gravia et neutra. Deinde dixit *Quoniam si non fuerit in hoc loco* etc., idest necessarium est ut corpus leve sit in medio inter duo extrema scilicet inter corpus grave et corpus neutrum: non enim est locus quartus; et cum dixit quod hoc est impossibile, et potest aliquis dicere Forte iste locus est ei accidentalis, dixit *Si igitur locus medius fuerit accidentalis corpori levi, erit naturalis alii corpori; sed nullum corpus est preter ista*,
60rb idest impossibile est ut iste locus medius sit accidentalis huic | cor- T 71b
pori, quoniam si ei fuerit accidentalis, continget ut alii sit naturalis; sed non est corpus quartum cui sit ille locus naturalis, locus enim naturalis non habet nisi unum corpus naturale et econverso, et ideo si esset ei accidentalis, contingeret ut haberet alium locum naturalem, et tunc contingeret quod loca essent quatuor, quod est impossibile. Deinde dixit *Et nos dicemus post diversitatem istius corporis*, scilicet corporis levis, idest quia dividitur in corpus calidum siccum et hoc est ignis, et in corpus calidum et humidum quod dicitur aer; et dixit hoc de levi tantum quoniam divisio gravis in duo manifesta est per

---

38 est] erit *J* 40 et[2]] *om. YZ* | hec duo] *inv. Z* 41 rei] *om. YZ* 42 idest] + et *XZ* | longitudo] locus *Z* | qua] quo *Z* 44 est] *om. Z* : + preter *BX* | quoniam] quia *JZ* 45 que] *om. XY* 47 preter] + ultimam *Z* 48 fixarum] pre- *Y* 49 gravia] + et levia *Z* 52 locus quartus] *inv. JZ* 53 hoc] *om. Z* | iste] ille *WXY* 54 est] *om. BX* | accidentalis] + vel acciderit *B* | dixit] dicere *X* : dicit *Z* 55 alii corpori] *inv. W* 56 idest] + et *BXY* 58 sed—naturalis] *om. homoiotel. BWYZ sed suppl.* B_2W_2 | quartum] quantum B_2X | cui] enim B_2 | sit] *om. BX* | ille] iste B_2 | locus[2]] + naturalis *Y* 59 ideo] non *W* 61 contingeret] *post* quatuor *B* 62-3 scilicet corporis] *om. Y* 63 idest] + et *B* | calidum] + et est quod diximus *XY* | siccum] *om. BWXYσ* 63-4 et hoc est] *om. W* : et sic est *J* : et est hoc *Z* : et hoc et *BX* 64 et[1]] *om. X* 65 hoc] quod *B* | manifesta] -tum *WX*

se, divisio autem levis in duo est comprehensum per rationem. Quapropter Alexander dixit quod corpus quod est illic non dicitur ignis cum hoc nisi equivoce; et iam perscrutati sumus de hoc in determinationibus Meteororum et in mineris et in determinationibus Generationis et Corruptionis.

Deinde dixit *Iam ergo declaravimus* etc., idest et hoc quod hic accepimus pro fundamento in declaratione istius questionis de cognitione qualitatum elementorum et numeri eorum et locorum iam declaratum est a nobis prius. Et voluit rememorare per hec omnia que declaravit de partibus mundi usque ad hunc locum; et ideo debemus intelligere ex sermone eius et quantitate eorum idest esse finita in magnitudine; et propter hoc dixit *Volumus etiam modo rememorare et dicere quod mundus sit unus*; et cum dixerimus hec omnia, volumus complere sermonem perpetuum aut aliquid eternum, quod mundus sit unus per declarationem inducentem maiorem dubitationem, deinde solvemus eam, deinde declarabimus mundum esse unum et eternum qui semper fuit et semper erit; et hoc intendit cum *60va* dixit *Et declarabimus ipsum esse etiam eternum perpe|tuum et non generatum neque cadentem sub corruptione*; et ipse perscrutabatur de hoc quoniam quidam Antiquorum opinabantur quod aliquid potest generari, deinde post remanere perpetuum, aut aliquid esse eternum quod semper fuit et potest corrumpi tamen post.

59vb10 **92.** [93.] *Sed volumus, antequam declaremus hoc, querere et perscrutari de* *277b29*
60vb16 *hoc. Potest enim aliquis querere et dicere quod impossibile est mundum esse unum, quoniam in quodlibet quod fit ex natura aut arte, dispositio forme solius est alia a sua dispositione cum materia verbi gratia orbis: forma enim*

---

66 levis] *om. BWXY sed suppl.* W_2 | in] *om. BW* 68 determinationibus] -tione libri *W*: + in libro *Z* 69-71 Meteororum—declaravimus] *om. X* 69 Meteororum] metaphysicorum *B*: methodorum *WY* | et^1—determinationibus] *om. Z* 72 declaratione] + est *Z* | questionis] + et *B* | cognitione] mixtione *W* 73 numeri] -ris *J* | numeri eorum] numerorum *B* | locorum] + et *YZ* 76 et] in *JZ* 76-7 esse finita] *inv. W* 77 etiam] *post* rememorare *BX* 78 et^2] *om. Z* 79 volumus] + modo *XY* | sermonem] illud movere *Z* | quod] et *Z* 82 et^1] *om. BXZ sed suppl.* Z_2: quasi *Y* 83 eternum] + et *BX* 85 aliquid] -quis *BXY* 86 esse] *om. BXY*

92. [93.] 1-2 Sed—hoc *VZ*] *post* corruptione *(t. 91 [92], v. 14) transp. BJWXY* 1 Sed] + autem *Y*: + tamen *X* 3 quoniam—ex] omnia enim quae sunt *J (cf. infra, c. 92 [93], v. 38)* | quodlibet] quo- *XY* | fit] sit *BX* | dispositio] + in eis *J* *orbis secundum suam dispositionem est alia cum fuit cum cupro et auro, et* 278a1 *forma circulationis est alia a cuprea aut aurea. Cum enim diximus rotunditatem, non collocatur in ea cuprum aut aurum quoniam non est substantie auri neque cupri; si ergo dixerimus rotunditatem auream aut* 278a5 *cupream, tunc inducemus formam cum materia insimul; si autem dixerimus tantum rotunditatem, non manifestabitur materia omnino, orbeitatem enim possibile est esse in diversis materiis. Cum igitur dixerimus orbei-* 61ra *tatem aut rotunditatem, non intelligemus nisi formam | tantum, et cum dixerimus rotundum intelligemus formam cum materia; et sic debet esse in quolibet. Et cum ita sit, et celum sit sensibile, ergo est pars partium mundi;* 278a10 *et omne sensibile est sensibile visu; Si ergo celum est unum particularium, cum ergo dixerimus celum simpliciter est aliud quam cum dixerimus hoc celum. Celum ergo simpliciter est forma, celum enim simpliciter est sola forma, hoc autem celum est forma cum materia. Et omne habens formam* 278a15 *et figuram aut est plura aut possibile esse plura. Et iste sermo sequitur sermonem de exemplari mundi, sicut quidam dicunt quod mundus habet exemplar eternum apud creatorem. Et iste sermo intelligitur cum mundus non habuerit exemplar; videmus enim quod omne quod est forma in materia habet plures partes consimiles infinitas; manifestum est igitur ex* 278a20 *hoc sermone quod celi aut sunt plures aut possibile est esse plures. Ex hiis igitur opinionibus et propositionibus potest aliquis estimare.* 278a21

Hec est fortissima questionum opinantis mundos esse plures, et propter difficultatem huius questionis dubitavit Galienus. Et fundamentum huius sermonis est super propositiones quarum una est 61rb quod omne corpus naturale componitur ex materia et forma: | celum est naturale, ergo est ex materia et forma; et cum huic coniungatur quod de eis que sunt ex materia et forma, aut plura uno individuo T 72b

5 fuit] fuerit *BJ* | cum²] + in *Z* | et¹] aut *Z*   6 aut] vel *JZ* | diximus] -xerimus *BXY*   7 est] + enim *Z*   8 ergo] enim *Z*: igitur *WY*   10 materia omnino] *inv. BJ*   12 intelligemus] -ligimus *Z*   13 intelligemus] -ligimus *Z* | esse] *post* quolibet *(14) Z*   14 ergo est] *inv. Z*   15 visu] per materiam *JG* | ergo] igitur *BX*   17-8 celum³—forma¹] sola *BJX*   18 celum] *om. Z* | Et omne] Omne enim *J*   19 possibile] + est *BY*   21 Et] + etiam testificatur *J (cf. infra, c. 92 [93], v. 100, et app. crit. ad loc.)* | intelligitur] *om. J*: -ligi *B*: videtur *(obsc.) X*   22 habuerit] -buit *JZ* | videmus enim] verbi gratia quia nos videmus *J*: videmus igitur *Z*   25 estimare] + et dicere quod celi *(dei Z)* sunt multi aut possibile est ut sint multi *JZG*   26 fortissima] forma *B* | questionum] -tio propositionis *W* | esse] *om. XY*: *post* plures *JZ*   27 huius] istius *W*: *post* questionis *BJ*   30 naturale] + celum *Z* | cum] *om. Z*   31 que] *om. WXY*

inveniuntur aut hoc est possibile. Forma enim innata est predicari de pluribus et inveniri in pluribus materiis, aut numero aut specie; et cum ita sit, contingit, inquantum mundus componitur ex materia et forma, ut inveniantur plura uno individuo, aut hoc sit possibile. Is enim est intellectus cuiuslibet forme; et ideo dicitur in diffinitione universalis quod sit natum dici de pluribus: forma enim est subiectum universalitatis. Et dixit *Omnia enim sunt ex natura aut arte*, idest omnia que sunt aut ex natura aut ex arte habent formam et materiam. Deinde dixit *Et cum dispositione forme* etc., idest et diximus quod omne naturale aut artificiale componitur ex materia et forma, et cum videmus quod in hoc subiecto est aliquid cuius dispositio abstracta apud intellectum est alia a sua dispositione extra intellectum, scilicet in materia verbi gratia rotundum; cum enim nos acceperimus ipsum intellectum abstractum a materia, dicitur rotunditas, cum materia autem verbi gratia in auro aut cupro dicitur rotundum. Et hoc intendebat cum dixit *Verbi gratia orbis* etc., idest forma enim scilicet rotunditas intellecta per se idest secundum quod est forma alterius dispositionis quam cum sit in cupro aut in auro.

Deinde dixit *Et forma rotunditatis est alia a rotundo et cupro*, idest cum declaratum est quod forma rotunditatis est alia a rotundo et a cupro, manifestum est quod hoc individuum componitur ex duobus, materia scilicet et forma. Deinde dixit *Cum enim dixerimus rotunditatem* etc., idest et signum eius est quia cum nos diffinierimus rotunditatem, non recipiemus aliquam materiam omnino: natura enim
61va rotunditatis | non est substantie auree neque alicuius materie, sed manifestum est quod substantia forme est alia ab essentia materie. T 73a
Deinde dixit *Si igitur dixerimus rotunditatem auream* etc., idest ex

32 possibile] im- *B* 33 aut[2]] + in *B* 35 inveniantur] -niatur *YZ* 37 universalis] *om. X*: nisi *B* | natum] in- *W* | enim] autem *BX* 38 enim] + que *JZ* | aut] + ex *B* 39 materiam] naturam *B* 40 Et[1]] *om. WY* | cum] *om. JZ* | dispositione] –tio *JY* 42 et] *om. BX* | cum] + ita sit *J*W_2 42-3 abstracta apud] *inv. Z*: abstracta *WY*: abstracta in *BX* 43 intellectum[1]] -lectu *W* 44 materia] essentia W_2*YZ*: esse *J* | enim] *om. W* | acceperimus] -peramus *B* 45 ipsum] + in *B* 46 autem] *om. Z*: aut *Y* | gratia] + non videtur debere dampnari *B* 47 enim] *om. J*: tantum *W* 48 rotunditas] + dicitur *Z* | forma] + est *XY* 49 sit] sint *Z*: fuit *Y*: fuerit *BX* 50 dixit] *om. BJ* | idest] + et *B* 50-2 idest—cupro] *om. homoiotel. XY* 51 a[2]] *om. B* 53 enim] igitur *B* 54 eius] huius *X*: hoc *B* | quia] *om. Z*: quod *W* | diffinierimus] -niamus *Y*: -nimus *Z* 55 recipiemus] -perimus *Z* 56 sed] scilicet *XY* 57 forme est] *inv. WY* | est alia] *inv. J*

quo declaratum est quod cum nos dixerimus rotunditatem auream, notabimus per hunc sermonem formam et materiam, et si rotunditatem non demonstramus nisi formam: formam enim orbeitatis possibile est inveniri in diversis materiis; et sic est de omnibus scilicet que componuntur ex materia et forma.

Deinde dixit *Et si ita sit, et celum est sensibile, ergo est pars partium mundi; et omne sensibile non est sensibile nisi per materiam*, idest et cum manifestum est quod omne sensibile componitur ex materia et forma, impossibile est enim sensibile esse per formam, quia est intelligibile secundum formam et sensibile secundum materiam, et similiter est de omnibus partibus mundi; et quia celum est sensibile et una partium mundi, necessario componitur ex materia et forma. Deinde incepit dicere quod modus est ex quo apparet in partibus mundi et celo, ne aliquis dicat quod in hac inductione non fuerint inducta omnia particularia, et dixit *Si igitur celum sit unum particularium* etc., idest si igitur est unum individuorum sensibilium, necessarium est ut componatur ex materia et forma: celitas enim est aliud quam dicere hoc celum, celitas enim dicitur de pluribus individuis; et sicut hoc invenitur in pluribus partibus huius celi, ita non est longe quin inveniantur plura uno individuorum totius celi. Et cum dixit *Celum enim absolute est* sola forma, intendit quod illud quod intelligitur de celo inquantum celum est sola forma: per sensum autem comprehenditur forma cum materia.

Et cum affirmavit quod celum componitur ex materia et forma, *61vb* | dixit *Omne enim habens formam et figuram* etc., idest et omnium habentium formam, aut in actu forma dicitur de pluribus numero, aut hoc erit possibile, quamvis illa pluralitas non fuerit in actu; et T73b contradictio est quia forte hoc possibile est secundum formam, impossibile secundum materiam, cum possibile sit quod tota materia sit inclusa in hac forma secundum Aristotelem, ut dicet post in dis-

59 quo] hoc *JX* | est] + aliud *B* | quod] + quia *B* 60 notabimus] voca- *Z* 61-3 enim—componuntur] *om. Y* 61 orbeitatis] orbita- *J* : orbieta- *W* 62 materiis] + numeris *B* | que] quia *BXZ* 64 si] cum *BJ* | et^{2}] *om. Z* | est^{2}] *om. WYZ* 66 quod] + et *B* 67 est enim] *inv. JXYZ* 67-8 quia—formam] *om. homoiotel. Y* 69 et^{1}] *om. W* 71 quod modus] modum *J* | est] *om. BJZ* | apparet] + hoc *Z* 72 dicat] -ceret *Z* 75 aliud] apud *Y* 78 inveniantur] -niatur *B* 79-80 intendit—forma] *om. homoiotel. WYZ* 81 autem] aut *Z* 85 fuerit] + maior *Y* 87 cum] est *YZ* | possibile] im- *B* : + enim *Z* | sit] est *Z* 88 ut] *om. BWXY sed suppl.* W_2 : et *Z* | dicet] -xit *Y* : -cit *J* : docet *W*

solutione huius questionis. Si ergo demonstratum est quod impossibile est mundos esse plures, necessarium est opinari quod materia eius conclusa est in forma. Deinde dixit *Et iste sermo sequitur sermonem* etc., idest et testatur huic sermoni, scilicet mundos esse plures, qui ponit in creatore mundi exemplar secundum quod fecit mundum: dicens enim exemplar esse, contingit ei necessario dicere ut mundus habeat plura individua. Et quia quod contingit ab aliquo est sequens ipsum, dixit *Et iste sermo sequitur sermonem* etc., et innuit Platonem: cum enim dixit examplar eternum esse secundum quod creator fecit mundum et quod mundus est novus, continget ei necessario possibile esse creare plures mundos.

Deinde dixit *Et etiam testificatur iste sermo cum mundus non erit exemplar*, idest et iste sermo sequitur hoc quod diximus de compositione mundi ex materia et forma, scilicet ut aut sint plures, aut hoc sit saltem possibile quamvis non dicamus exemplaria Platonis; et dixit hoc ne aliquis dicat ei Quare ducis rationem ex eo quod non opinaris cum multotiens destruis exemplar esse? Deinde dixit *Verbi gratia quod nos videmus quod omne quod est forma in materia* etc., idest bene potest quilibet concedere inductione ex omni habenti for-
62ra mam in materia, in essentia in|veniuntur plura individua aut infinita.

Deinde dixit *Manifestum est igitur per hunc sermonem quod celi sunt plures principaliter*, idest quod iste sermo non prestat omnino esse plures, sed tantum dat alterum duorum aut plures esse actu aut potentia; et sufficit in hoc sermone, quia est inductio in qua non inducuntur omnes species habiture istud iudicium *Ex istis ergo opinionibus et propositionibus potest aliquis estimare*, idest intendit, secundum quod mihi videtur, per opiniones opinionem Platonis de exem- T 74a

---

91 conclusa] in- *JZ* 93 ponit] -nitur *Z* 94 esse] *om.* *YZ* | dicere] *om.* *Z* 95 quia] *om.* *YZ* | aliquo] alio *Z* 97 eternum esse] est eternum *Z* 98 creator—mundum] creati sunt mundi *Z* | continget] -git *BX* 100 testificatur] rectifi- *BY*: reedifi- *Z* | mundus] -do *WXY* | non] + esse aliud *B* 103 sit] *om.* *Z* | saltem] -tim *B* 104 hoc] *om.* *Z* | Quare] quamvis *WY* : quoniam *Z* | ducis] in- *J*: tu dicis W_2Z | rationem] *post* eo *B* 105 destruis] -struit *Z* | exemplar] -ria *JY* 106 quod[1]] quia *BJ* | quod omne] *inv.* *X*: omne *B* 107 potest] + vel quis *B* | quilibet] quis *X* | omni habenti] *inv.* *Y* | habenti] -te *JW* 108 in materia] *om.* *Y*: quia *Z* | in essentia] *om.* *BWX* | aut] *om.* *BXY* 109 celi] + aut *JZ* 111 tantum] tamen *BJX*: terra *Z* | esse[2]] *om.* *Z* 112 qua non] quantum *Z* 113 istud] illud *Z* | iudicium] indi- *B* | ergo] *om.* *Z*

plaribus, et per propositiones hanc inductionem quam affirmavit et rationem per quam induxit causam esse predicati in subiecto. Cum enim declaratum fuerit predicatum esse in subiecto inductione, et sustentatum fuerit hoc per causam esse predicati in subiecto, erit magis sufficiens: hec autem est causa que dicitur, quia causa essendi plura individua ex hiis compositis est quia res que componitur ex materia et forma non est innata habere hanc materiam solam, ideoque iste sermo est valde sufficiens. Et virtus imaginativa valde iuvat in hoc: possumus enim imaginari extra mundum vacuitatem continentem ipsum, et quod in illa vacuitate est possibile alium mundum esse sicut dispositio de omnibus individuis que sunt hic, quoniam quodlibet eorum existit in vacuo et extra ipsum inveniatur aliud individuum: aer enim non comprehenditur visu, et hoc induxit plures dicere esse vacuum. Et cum imaginatio assimilatur intellectui, estimabant loquentes nostre legis vacuum esse intellectum primum; et est necessarium dicenti quod mundus fuit postquam non fuit scilicet vacuum dicere esse.

62rb12 **93.** [94.] *Necesse est ergo nobis cum fuerit sollicitudo nostra circa hanc* *278a22*
intentionem querere de hiis opinionibus donec sciamus que est earum rectum et que non. Dicamus ergo quod bene dixit dicens quod aliquod secundum suam formam tantum est aliud quam secundum suam formam et materiam, et iste est sermo rectus sine contradictione. Sed tamen *278a25*
celi non sunt plures propter hoc; neque possibile est ut sint multi propter hunc sermonem; celum enim factum est ex tota materia. *278a27*

116 per] *om. Z* 116-7 et[2]—quam] *om. BX*: per *Y* 117 induxit] indicium *BXY* | predicati] *om. BXY* | subiecto] causato *B*: creato *Y*: tanto *X* 119 hoc] *om. WYZ* 120 est causa] *inv. JXYZ* 121 res que] *om. XY* 123 ideoque] ideo *J*: ideo quod *WY* | valde] *om. Z* 124 hoc] hac *YZ* | possumus enim] *inv. Y*: quia possumus *J* | vacuitatem] + et *B* 125 et] ut *Y* 127 inveniatur] -nitur *BJX* 128 enim non] *inv. Z* 129 cum] *om. Z* | intellectui] *om. WY* 130 estimabant] + autem *Z* 131 et] *om. B* | necessarium] + hoc *JZ* 131-2 scilicet vacuum] idest vacuum scilicet *JZ*

93. [94.] 1 ergo] *om. YZ* | fuerit] fuit *Z*: est *J* | sollicitudo] locutio *Z* 1-2 hanc intentionem] *inv. Z* 2 est earum] eorum est *BJX* 3 rectum] -ta W_2ZG | quod[1]] quia *BJX* 4 suam[1]] *om. YZ* | tantum—formam] *om. homoiotel. Y* 5 est sermo] *inv. W* 6 possibile] principaliter *BY* | est] *om. XY*: *ante* possibile *Z*: et *B* 6-7 ut—sermonem] *om. BWXY*

Vult demonstrare quid est in isto sermone predicto de veritate et quid de falsitate, et incepit dicere quod ille cuius sollicitudo est perscrutari de hoc debet perscrutari de hac questione dubitante, et dixit *Oportet igitur nos cum sollicitudo nostra est* etc., intendit per intentionem hoc quod declaravit quod mundus est unus, et intendit per opi- T 74b niones propositiones positas in hoc sermone. Et quia iste sermo continet in se universaliter duas propositiones, quarum altera est *62va* quod omne sensibile | componitur ex materia et forma cum forma tantum aliud est ab habente formam scilicet quod coniungitur ex materia et forma, altera autem est quod in eo quod componitur ex materia et forma inveniuntur plura uno individuo aut hoc est possibile, incepit dicere quod prima propositio sit vera et dixit *Dicamus modo* etc., idest dicamus modo quod faciens questionem bene dixit in hoc quod dixit quod forma cuiuslibet est aliud a composito ex forma et materia, unde declaratum fuit ipsum esse compositum ex materia et forma.

Et quamvis ista propositio est vera, tamen conclusio quam concludit faciens questionem non est vera, quia altera propositio est falsa, et dixit *Sed tamen celi non sunt plures* etc., idest quamvis hec propositio sit vera, tamen conclusio non sequitur; secunda enim est falsa, et est quod ex omni facto ex materia et forma inveniuntur plura individua aut est possibile. Hoc enim non est verum nisi in eo quod factum est ex quadam parte materie ei convenientis, quod autem fit ex tota materia impossibile est reperiri duo individua in actu; et hoc quod dixit intellectum est per se.

8 quid est] *post* veritate *W* 9 est] *om. BX* 10 hoc] + et *Z* | debet] dixit *WYZ* | dubitante] -tare *Z*: -tatione *J*: -ficante *BX* 11 Oportet igitur] si igitur oportet *Z* | igitur nos] *inv. B* | etc.] + et *BX* 12 et] *om. WYZ* 15-7 cum—forma] *om. homoiotel. WX sed suppl.* W_2 16 formam] -mas *YZ* | coniungitur] -gatur *B* 17 materia] forma *B* | forma] + sed *Z* | autem] *om. Z* | in eo] *om. Y*: ex omni *BX* | quod[2]] *om. WY* 18 plura] *om. W* | possibile] im- *Z* 19 sit] est *WXY* | Dicamus] + ergo *XY* 20 bene] unde *Z* 21 quod[2]] et *Z* 21-2 ex—materia] *om. B*: ex materia et forma *J* 22-3 unde—forma] *om. YZ* 24 est] *om. B*: sit *J* 25 quia] quoniam *B* | propositio] -tionis *Z* 27 sequitur] + vera *Z* | enim] + sequitur *Y* 28 materia ... forma] *inv. BX* 29 aut] + hoc *BX* | est possibile] *inv. B* | verum] *om. B* 30 materie ei] matricis *Z* | convenientis] querentis *YZ* | fit] *om. B*: sit *Z* 31 reperiri] -perere *BJ* 32 est] *om. XYZ*

62vb4 **94.** [95.] *Et nos inducemus exemplum ut sermo noster sit magis decli-* 278a28 *nans et propinquior manifestationi. Dicamus ergo quod si simitas est profunditas in naso, ergo simitas est materia in carne; si igitur ex tota* 278a30 *carne formetur una caro in qua est simitas, nullum aliud membrum erit membrorum corporis simum nisi illa caro tantum, et impossibile est esse in alio loco omnino; et similiter si materia hominis fuerit caro et os, tunc homo debet esse totaliter ex carne et ossibus; si igitur fuerit impossibile ut caro et ossa non dissolvantur neque diminuantur, impossibile est etiam* 278a35 *alium hominem esse omnino, et similiter de omnibus aliis. Et universa-* T75a *liter omnis substantia composita ex materia impossibile est inveniri ex ea sibi simile nisi fuerit alia materia similis materie ex qua composita fuit.* 278b3

Cum narravit quod impossibile est inveniri ex eo quod factum est ex tota sua materia aliud individuum existens cum eo in actu, vult declarare hoc per exemplum ut concessio sit manifestior, et dixit *Et nos inducemus exemplum* etc., idest ut sermo noster sit magis declinans ad declarationem. Deinde dixit *Si simitas est profunditas in naso aut in carne, ergo simitas est materia in carne*, idest verbi gratia quia diffinitio 63ra simitatis est profunditas in | naso: nasus autem est carnis, ergo materia simitatis est caro. Deinde dixit *Si igitur formetur ex tota carne* etc., idest si igitur posuerimus quod iam formetur ex tota carne nisi simi unus nasus simus, impossibile erit alium hominem esse cuius nasus sit simus: simitas enim non existit nisi in carne propria, sed totam illam carnem posuimus in illum nasum simum. Et cum dixit *Nullum aliud membrorum corporis erit simum*, intendit ut mihi videtur quod nullus nasus nasorum corporum hominum erit simus nisi ille nasus in quo conclusa est illa caro recipiens simitatem. Et cum dicit *Impossibile est esse in alio loco omnino*, potest intelligi quod simitatem est impossibile esse in aliqua materia nisi in propria; et potest intelligi quod

---

94. [95.] 2 et] *om. X* | quod] *om. YZ* 3 naso] + aut in carne *EJZG* | ergo] + est *Z* 4-5 erit—simum] corporis simum erit modorum simi corpori *Z* 5 est] + etiam *BX* 6 materia] natura *B* 7 ex carne] *om. Y* | fuerit] sint *WZ* | impossibile] + est *YZ* 8 non] neque *WXY* 9 omnibus aliis] *inv. Z*: omnibus *W* | universaliter] + est *J* 10-1 ea sibi] casibus *B* 12 quod[2]] quo *B* 14 hoc] *om. J* | Et[2]] *om. WYZ* 16-8 in[2]—autem] *om. XYZ* 17 materia] forma *J* 19 igitur] ergo *Z* | formetur] -matur *B* 19-20 etc.—carne] *om. homoiotel. Y* 20 igitur] *om. BX*: iam *Z* | iam] *om. Z* 21 erit] est *JZ* 23 aliud] + membrum *WY* 24 mihi videtur] *om. Z* 25 nisi] aliusque *JZ* 26 cum] *om. WYZ* | dicit] -xit *BXY* 27-8 simitatem—quod] *om. homoiotel. Z* 27 est impossibile] *inv. JY* simitas non erit in alio membro nisi in naso; cum enim simitas non potest esse nisi in naso, et posuerimus quod tota caro nasi simi est conclusa in uno naso, non invenietur nasus simus nisi unus tantum.

Et cum dedit exemplum de naso simo, dedit etiam de homine, et demonstravit etiam conversionem huius sermonis et est quod, in illo ex quo non invenitur nisi unum individuum tantum, materia est conclusa necessario, sicut est necessarium ex illo in quo materia est conclusa non invenitur nisi unum individuum. Et dixit *Et similiter si* T 75b
materia hominis fuerit caro et os etc., idest et similiter si materia hominis fuerit caro et os, et posuerimus quod nullum ens est nisi unus homo tantum, necesse est ut ille homo sit ex tota carne et ossibus, sicut est necesse si aliquis homo fuerit creatus ex omnibus ossibus et carne quod sit unicus. Deinde dixit *Si igitur impossibile fuerit ut caro et os dissolvantur* etc., idest et si posuerimus quod iste homo, quem
63rb posuimus esse creatum ex tota mate|ria hominis, impossibile est corrumpi et dissolvi in ea ex quibus componebatur, impossibile est fieri alium hominem omnino; et totum hoc manifestum est.

Deinde dixit *Et similiter contingit in generatione aliarum rerum*, idest quia cum posuerimus aliquam materiam conclusam in aliquo ente, impossibile erit ex eo esse plus quam unum individuum, et cum fuerit non conclusa, in eo possibile est invenire plus quam unum secundum quod possibilis est in illa materia partitio. Et ideo cum fuerit necessarium individua esse infinita generabilia et corruptibilia, materie autem impossibile est ut sint infinite in actu, ut sit hoc secundum successionem idest quod in una materia succedant plura individua, immo infinita. Et ideo, quia materie divisibiles sunt in actu in partes finitas et est possibile ut succedant in qualibet materia individua infinita, necessarium est ut sit dispositio individuorum

29 in^2] + uno *Z* 30 et] + cum *BJ* | posuerimus] -suimus *JY* | nasi] *om. JZ* 32 exemplum—dedit] *om. homoiotel. Z* 34 tantum] *ante* unum *W*: modo *X* 36 non] nisi *B* | invenitur] -niatur *B*: -nietur *XYZ* 38 posuerimus] -suimus *JY* 39 necesse est] *inv. JZ* | homo^2] *om. Z* 40 necesse] + ut *B* 41 ut] quod *Z* 45 totum] + ergo *BX* 46 dixit] + et similiter generatio in omnibus aliis idest *BX* 47 posuerimus] + quod *B* 48 erit] + inveniri *BX* 48-9 individuum—unum] *om. homoiotel. YZ* 49 non conclusa] inconclusa *W* | invenire] -niri *WX* 50 in illa] nulla *B* | partitio] -ticipatio *W* 52 materie] modo *Z* 53 quod] quia *Z* 54 immo] numero *W* 55 finitas] in- W_2Z | est possibile] *inv. Z*: possibile *Y* 56 individua infinita] *inv. B*

generatorum et corruptorum in qualibet scilicet specie, scilicet inve-
niri ex quolibet generato plura individua finita in actu, infinita
secundum successionem eorum. Et quia istud non complebitur istis
individuis nisi inveniantur esse partem alterius individui quod est
totus mundus, contingit in hoc individuo ut sit unum numero, quo-
niam si fuerit plus uno, erit pars alterius mundi; et ille alter mundus
si esset plus uno esset pars alterius, et sic procederetur in infinitum
aut inveniretur unicus mundus numero ingenerabilis et incorruptibi-
lis; quapropter contingit ut mundus sit ingenerabilis et incorruptibi-
lis et quod eius materia sit conclusa in eo. Et propter hoc etiam dixit
post *Et universaliter ex omni substantia composita ex materia impossi-*
63va *bile est ut sit alia sibi similis nisi fuerit alia materia similis il|li ex qua*
componebatur, idest et universaliter diximus quod omne individuum
compositum ex materia et forma cum non fuerit extra ipsum mate-
ria, sicut materia ex qua componebatur illud individuum, impossibile
est ut in eo inveniatur aliud individuum sibi simile; et si ipse fuerit
ingenerabilis et incorruptibilis, impossibile et inveniri ex eo aliud
individuum simplex scilicet neque in alia hora.

95. [96.] *Revertamur igitur et dicamus quod celum est unum composi-* 278b3
tum ex sua materia sicut alia corpora composita. Si igitur non fuerit
compositum ex medietate sue materie, sed ex tota, tunc erit unum non 278b5
plures; et cum dicitur celum absolute, est aliud quam cum dicitur hoc
celum; impossibile est igitur secundum hunc sermonem ut sint plures celi
neque sint plures omnino. Hoc enim celum continet totam materiam ex
qua fit; remanet ergo nobis declarare de celo cum sua essentia sit compo-
sita ex toto corpore naturali sensibili. 278b9

---

57 scilicet[1]] *om. BJX* 59 istud] illud *JZ* | complebitur] + in *JZ* 60 inveniantur] –niatur *Y* 61 hoc] toto *B* | ut] *om. B* 62 uno] + individuo *B* 63 procederetur] -ceditur *Z* 64 numero] *om. W* | incorruptibilis] corrup- *XY* 66 etiam] *om. BJX* 66-7 dixit post] *inv. Z* 68 sit alia] *inv. Z* | sibi similis] *inv. B* | alia[2]] illa *Y* | materia] *om. Z* 69 componebatur] -nitur *Z* 72 in] cum *J*W_2*XZ* : cum in *Y* | inveniatur] -nietur *B* 73 incorruptibilis] corr- *Z* | inveniri] *post* eo *W* 74 simplex] -pliciter *JX* | alia] aliqua *XY*

95. [96.] 1 igitur] *om. Z* : ergo *X* : modo *WY* 2 corpora composita] *inv. B* | Si igitur] Et si *J* 3 compositum] -situs *B* | sue materie] *inv. BX* | unum] + et *Z* 4 celum] *om. Z* 6 neque sint] *dittogr. Y* : neque sunt *X* 7 fit] + et *Z* | ergo] *om. XZ* | de celo] et *Z* | cum] quod *JZ*

63vb Cum declaravit quod impossibile est inveniri duo individua insimul ex omni composita ex tota sua materia propria, vult declarare quod mundus est compositus ex tota sua materia. Et ipse vult anteponere quod mundus est compositus sicut alia corpora composita, et dixit *Revertamur ergo et dicamus quod celum est compositum ex sua materia sicut et alia corpora composita*, et intellige per celum totum mundum, idest manifestum est quod totus mundus est compositus ex forma et materia sicut alia corpora composita; et videtur (si intellexit per celum in hoc sermone totum mundum) quod sit illud quod est in eo quasi forma corpus neutrum, et quod est in eo quasi materia corpora gravia et levia: corpus enim neutrum non habet materiam ex qua componitur, quoniam si haberet, haberet contrarium et esset generabile et corruptibile; et in principio istius tractatus probavit quod corpus neutrum caret subiecto et contrarietate.

Sed habet questionem utrum res intellecta ex hoc corpore est hoc sensatum aut aliud: si erit hoc sensatum, tunc erit res intellecta eadem cum sensata, si aliud, continget ut sit composita ex materia et forma; ergo habet potentiam ut corrumpatur; dicamus ergo quod ista natura neutra non est existens per se in actu sed est materia forme corporis celestis que est in actu. Si ergo ista materia non fuerit composita, non habebit formam ita ut sit intellecta, sed est una numero et individuata per se, et est intellecta in respectu ad aliud, sicut est de illa materia scilicet quod non operatur per corpus celeste nisi inquantum est subiectum forme operationis intellectus, sicut intelligitur *64ra* materia inquantum est sub|iecta forme tamen prima materia est in potentia et ista in actu; et ideo in hac nulla potentia est que denudari possit a sua forma, et non habet nisi potentiam ad ubi. Debemus ergo intelligere de corpore celesti ipsum esse neutrum ad hanc naturam que est secundum intellectus, non quod est aliquid habens formam et materiam, quoniam si ita esset, grave esset aut leve.

---

10 composita] -sito *JW* | sua materia] *inv. Z* 13-4 dixit—et[2]] *om. homoiotel. YZ* 13 ex] in *BWX* 14 et[1]] *om. BX* 16 forma ... materia] *inv. W* 20 haberet] *haplogr. Y* 21 istius] huius *Z* | probavit] -palavit *BX* 23 Sed] et *W* 25 sensata] -sato *YZ* 25-6 materia ... forma] *inv. BX* 27 non est] est et non *J*: non *BWY* | est[2]] ex *BX* | materia] + et *B* | forme] *om. WY*: forma *B* 28 in] *om. Z* 29-30 sed—intellecta] *om. homoiotel. X* 29 sed] + ut *YZ* | est una] *inv. B* 30 et[1]] *om. J*: est *Z* 31 illa] naturale autentico illa *B*: communi *J*: contrarie *W*: quiete *YZ* 33 tamen] cum *Z* 34 denudari] individuari *YZ* 36 ad] *om. WYZ* 36-8 naturam—et] *om. X* 37 secundum] -dus *Z*: subiectum *J* | est[2]] habet *Z* | aliquid] *om. Y*: -quam *Z* | habens] *om. Z* 38 esset[2]] *post* leve *JZ*

Deinde dixit *Si igitur non fuerit composita ex medietate sue materie* etc., idest si igitur mundus non fuerit compositus ex quadam parte sue materie, impossibile est inveniri alium mundum; et dixit medium proprie quia si ponatur ipsum generatum ex plus quam medio, impossibile est inveniri alium mundum, quoniam materie plurimorum individuorum terminate sunt quantitatis, scilicet quod necesse est ut sit quantitas materie Sortis equalis naturaliter quantitati materie Platonis; et si fuerit excessus inter eas, erit terminatus, quoniam sicut non ex qualibet qualitate materie fit quodlibet ens, sic est de quantitate. Deinde dixit *Et secundum hoc cum celum dicitur absolute est aliud quam cum dicitur hoc celum*: modo revertitur ad hoc ex quo declaravit quod celum componitur ex aliquo quasi materia et ex aliquo quasi forma.

Et cum dixit Aristoteles quod si mundus fuerit compositus ex tota materia necesse est ut sit unicus, voluit declarare hoc etiam in mundo ut declaretur ex hoc ipsum esse unum, et dixit *Remanet nobis ultimo declarare de celo* etc., idest et cum declaratum est quod celum idest mundus compositus est ex materia et forma, et declaratum est etiam quod illud quod compositum est ex tota sua materia est unicum, remanet nobis declarare quod mundus compositus est ex tota materia. Et differt | *64rb* hec demonstratio a demonstratione predicta in hoc scilicet quod ista demonstratio propter quid, prima autem quia, aut demonstratio ducens ad inconveniens: demonstrationes autem ducentes ad impossibile sunt generis signorum; sed declaratio eius quod est causam esse acceptam pro medio termino in hac demonstratione, accipitur ex demonstratione predicta. Demonstratio igitur predicta aut predicte prestant mundum esse unicum et cum hoc causam propter quam mundus est unicus, et scire causam dat nobis scientiam huius rei cum sua causa, quod est perfectissimum modorum fidei, quapropter dissolvitur questio accidens in hoc.

---

39 Si igitur] et si *J* (*cf. supra, t. 95 [96], v. 2*) | composita] -tum *JY* 44 quantitatis] –tates *Y* | sit] sint *B* 46 erit] est *Z* 47 qualitate] quanti- *Z* 48 celum dicitur] *inv. BX* | aliud] + etc. idest est aliud *WZ* 49 dicitur] -catur *WY* 50 aliquo[1]] -qua *WXY* | aliquo[2]] -qua *Y* 51 quod] *om. YZ* | si] *om. B* 52 hoc etiam] *inv. BX* 53 dixit] *om. XY* | Remanet] + ergo *JZ* | ultimo] *om. BX*: alter *Y* 54 et] *om. J* | cum] *om. Y* 57 tota] + sua *J* 59 demonstratio] + est *JZ* | propter] *om. W* 62 hac demonstratione] *inv. W* 63 demonstratione predicta] *inv. Z* 64 prestant] -stat *XY*: probant *J*: est *Z* | unicum] unum *W* 67 questio] quomodo *Z*: quo *Y*

96. [97.] *Et remanet nobis declarare de celo cum sit compositum ex toto* 278b8 64va *corpore naturali sensibili;* | *sed antequam hoc faciamus, volumus hic perscrutari prius quid intendimus cum dicimus celum, et quot modis* 278b10 *dicitur celum, ut sit illud quod querimus de hoc magis declaratum est et magis manifestum ei qui utitur per nos per signum manifestum super viam nostram ad quam clamamus ignorantes ipsam, et vadit in rectum que intendit ipsam. Dicamus ergo quod celum dicitur secundum unum modum ultimum motuum totius, et corpus naturale quod est in ultimo motus totius; consueti igitur sumus appellare orizonta mundi et eius superius celum, et dicimus quod illic habitant spiritualia. Dicamus* 278b15 *etiam secundum alium modum celum corpus continuum cum orizonte motus totius mundi, ubi sunt sol et luna et quedam stelle; consuetudinem igitur habemus dicere quod sol et luna sunt fixi. Et etiam dicitur alio modo illud corpus contentum ex ultimo motus totius; et assueti* 278b20 *etiam sumus appellare perfectionem et totum celum; et manifestum est igitur quod celum dicitur tribus modis.* 278b21

Cum voluit declarare quod nihil est extra celum et quod materia celi est conclusa in eo, vult ante narrare secundum quot modos dicitur celum apud eos, ut sermo eius sit magis manifestus; dixit ergo *Celum dicitur uno modo* etc., et intendit per corpus totum neutrum quod movetur secundum motum motu diurno: hoc enim corpus est finis motus corporum naturalium motorum et mobilium recte. Et hoc intendebat cum dixit *Ultimum motus totius et corpus naturale* 64vb *quod est in ultimo motus totius*, idest et ipsum est corpus natu|rale in quo finitur motus corporum motorum mobilium ad superius, aut intendit et ipsum esse corpus naturale cuius motus est ultimus motus

---

96. [97.] 1 remanet] -nit *Z* 2 hic] *om. B*: hoc *WYZ* 3 prius] *om. XZ* | quid] quod *Y*: quo *Z* | dicimus] -ximus *WYZ* 5 manifestum[1]] + est *Z* | utitur] duci- *BX* 7 que] qui *WXY* | Dicamus] + etiam *Z* | ergo] igitur *BX* | celum] *om. BX* 8 motuum] motus *J* | totius] *om. YZ* 9 consueti] assueti *J (cf. infra, c. 96 [97], v. 29)* | eius] *om. WZ*: suum *J* 11 etiam] + quod *Z* | alium] alterum *Z* 12 totius] *om. X*: motus *Y* 13 fixi] + in celo JW_2G: + in elemento *Z* 14 illud] *om. WYZ* 15 sumus appellare] *inv. WY* | perfectionem] -fectum *JZ* | et[2]] *om. BX* 15-6 et[2]—modis] *om. J (sed cf. app. crit. ad t. 97 [98], v. 1)* 18 conclusa] + est *B* 21 motum] totum *B* 22 et[1]] *om. J* 26 et] *om. WZ* | corpus naturale] *inv. Z* totius, motus enim celi totius est ultimus omnium illorum mobilium que sunt apud nos, tamen vere est primus motuum.

Deinde dixit *Assueti sumus igitur dicere orizonta mundi et suum superius celum*: dixit quia hoc nomen celum dicebatur apud eos de toto mundo, et forte hoc erit magis famosum; ideoque dedit rationem et dixit *Dicimus quod illic habitant omnia spiritualia*, idest et signum quod celum dicitur grece de corpore quod circulariter movetur est quia consuetudo Grecorum est dicere quod in celo habitant T 77a spiritualia. Deinde dixit *Et celum alio modo* etc., idest orbis stellarum fixarum, quoniam hoc corpus est continuum cum fine motus omnium partium mundi que circulariter moventur, et propter hoc dixit *Continuum cum orizonte totius mundi ubi sunt sol et luna* etc., idest quod est continuum cum fine motus eorum que sunt sub eo ex corporibus celestibus in quibus sunt sol et planete. Deinde dixit *Habemus consuetum dicere quod sol et luna* etc., hec est significatio quod hoc nomen celum dicitur de orbe ultimo continente totum, quoniam stelle sunt in eo fixe vere; et quasi hoc nomen celum non dicebatur apud eos nisi de corpore in quo sunt stelle fixe.

Deinde dixit *Et etiam celum dicitur alio modo* etc., idest et dicitur celum de toto mundo; et ideo descripsit ipsum quod est corpus contentum ex motu totius, idest quod continetur sub ultimo motuum, et hoc dicitur apud nos mundos non celum; et hoc intendebat in hoc loco declarare esse unicum, quia extra ipsum nihil | *65ra* est ex sua materia, et quia de hac dicitur simpliciter et absolute totum et complementum ut in principio dictum est; et nomen celi dicitur etiam apud eos simpliciter, et complementum dedit rationem quod hoc nomen celum dicitur de toto mundo.

---

27 illorum] *om. BX* 28 sunt] + mobilium *B* | vere est] universalis est *Y*: est universalis *Z* 29 Deinde] *om. BWZ* 30 de] + eo quod est supremum in *JZ* 31 dedit rationem] *inv. W* 32 dixit] + et *BWX* | Dicimus] *om. YZ* 33 quod celum] *inv. B* | corpore] + ultimo *J*W_2*Z* 34 Grecorum est] *inv. Z*: Grecorum *Y* 35 celum] + etiam *B* | stellarum] + mobilium et *JZ* 37 que] qui *WY* 38 ubi] ut *B* | sunt] *om. JX* | sol] *om. W* 41 Habemus consuetum] consuetum igitur habemus *J*: Habemus consuetudinem *B (cf. supra, t. 96 [97], v. 12sq.)* | dicere] *om. BXY* 43 fixe] *ante* in *BX* | vere] *om. YZ* | quasi] quod *BWYZ* 45 Et etiam] *om. YZ* 50 et[1]] *om. Z* 51 complementum] -pletum *Y* 52 complementum] -pletum *W*

65rb2 97. [98.] *Dicamus ergo quod totum continens ex ultimo motus et unius* 278b21 T *est compositum ex toto corpore naturali sensibili necessario, quoniam extra celum nullum corpus est neque est possibile ut sit omnino. Volumus modo perscrutari de hoc etiam utrum extra celum est corpus, aut est possibile ut illic sit corpus. Dicamus igitur quod si extra celum sit corpus,* 278b25 *aut est ex corporibus simplicibus aut ex compositis, aut erit illic naturaliter aut accidentaliter, sed extra celum nullum corporum simplicium est omnino. Iam enim declaravimus superius et diximus quod impossibile est corpus rotundum mutare suum locum, et etiam impossibile est ut illud* 278b30 *corpus sit corporum que moventur a medio neque ad medium; quoniam si esset aliquod istorum, tunc ille locus non esset naturalis ei: loca enim istorum corporum simplicium sunt alia ab illo loco, et sunt cognita et manifesta. Si ergo ille locus qui est extra celum fuerit accidentalis illi corpori, erit naturalis alii corpori necessario; diximus enim multotiens quod cum aliquis locus fuerit accidentalis alicui corpori, necessario erit naturalis alii; sed non videmus aliud corpus ab istis corporibus simplicibus,* 65va *scilicet a corpore ro|tundo et a corpore quod vadit ad medium et a medio. Et cum ita sit, impossibile est quod aliquod corpus corporum simplicium sit extra celum; et cum hoc est impossibile, impossibile est etiam quod* 279a1 *illic sit aliquod corporum compositorum, quoniam cum corpus compositum fuerit, necessario erunt illic corpora simplicia.* 279a2

Cum declaravit quod individuum, cuius materia est in eo conclusa, est unicum numero, vult declarare quod hec est dispositio mundi, et dixit *Dicamus igitur quod totum continens* etc., idest et dicamus quod totus mundus, quod est totum continens omnia corpora naturalia, quod est ultimum rerum motarum quod est continens omnia que sunt in eo, est compositum ex omnibus corporibus T 78a naturalibus sensibilibus, que sunt materia eius.

---

97. [98.] 1 Dicamus] Cum igitur celum tribus modis dicatur *praem. J (cf. supra, t. 96 [97], v. 15sq., et app. crit. ad loc.)* | ergo] *om. J*: etiam *Z* | unius] totius *JW*$_2$*Z* 2 quoniam] *om. J* 3 extra] + enim *J* 4 celum ... corpus] *inv. B* | est^2] sit *Z* 5 Dicamus igitur] Et dicamus *J* 7 sed] si *BZ* | est] esse *Z* 9 corpus rotundum] *inv. Z* | suum locum] *inv. J* | locum] + non est possibile ut permutetur ex eo omnino *ZG* | etiam] dicamus etiam quod *J* 11 istorum] + corporum *BJ* 13 est] *om. B* 14 corpori] *om. X* | diximus—multotiens] multotiens enim diximus *Z* 15 necessario] *om. WZ*: non *Y* 16 aliud] + ut *B* 19 cum] *om. W* | hoc] *om. BX* | est^2] *om. BX* | etiam] *om. Z*: *ante* est^2 *W* | quod] + corpus *Z* 20 sit] fit *J* 23 conclusa] in- *XYZ* 24 igitur] *om. J*: modo *BX*: ergo *Y*

Deinde dedit causam et dixit *Extra enim celum corpus nullum est neque est possibile*, idest et hoc est necessarium, quia nos declarabimus quod nullum corpus est extra celum neque potest esse in futuro. Deinde incepit declarare quod nullum corpus modo extra celum sit: si extra celum est aliquod corpus, aut est compositum aut simplex; si simplex, aut in loco ei naturali aut accidentali. Deinde dixit *Et dicamus* etc., idest dicamus quod impossibile est inveniri corpus simplex extra mundum: corpus enim aut circulariter movetur aut recte; si circulariter, impossibile est ut sit in alio loco accidentaliter; declaratum est enim quod impossibile est, quod corpus quod circulariter movetur impossibile est transferi a suo loco, cum impossibile est ipsum *65vb* moveri violente. Et tacuit | ipsum esse possibile illic naturaliter, quoniam si extra mundum esset corpus celeste in loco naturali absque hoc, contingeret ut illic esset alius mundus extra hunc mundum, quoniam de necessitate esse corporis celestis est esse corporis quiescentis circa quod revolvitur, et ex necessitate esse corporis quiescentis est corpus mobile esse scilicet ignem, et ex necessitate essendi extrema est medium esse scilicet aerem et aquam.

Et cum declaravit quod impossibile est ut extra mundum sit corpus rotundum, incepit declarare quod impossibile est illic esse corpus quod recte movetur, et dixit *Et dicamus quod impossibile est quod aliud corpus* etc., idest dicamus etiam quod impossibile est ut extra mundum sit corpus simplex, scilicet quod recte movetur, scilicet aut grave aut leve, quoniam si illic esset aliquod istorum corporum, esset aut naturaliter aut accidentaliter: si naturaliter, impossibile est, locus enim totius idem est cum loco partis numero; neque etiam accidentaliter, quoniam contingeret ut alii esset naturalis, ut videtur de corporibus T 78b recte motis, verbi gratia quia superius est accidentalis terre est naturalis igni et econverso de inferiori; et si esset naturalis alicui,

30 est[1]] *om. Z* 32 modo] movetur *X* : *post* declarare *JZ* 33 si[1]] + enim *JZ* 33-4 si simplex] *om. W* 34 dicamus] -ximus *WZ* 35 inveniri corpus] *inv. B* : corpus *X* 37 alio] illo *BX* 38 impossibile—quod[2]] *om. J* 39 impossibile[1]] *om. Z* | est transferi] transferatur *WZ* 40 possibile] im- *JZ* 41 in] *om. BWXY* 42 contingeret] -get *YZ* : -getur *X* | illic] illi *Z* 43 est] + cum W_2Z 45 est] + principaliter *JZ* | mobile] -lis *XY* 46 et] aut *Z* 47 ut] quod *Z* : + quod *B* 48-9 illic—est] *om. homoiotel. YZ* 49 dicamus] + etiam *J* 50 aliud] illud *JWX* 51 scilicet[1]] *om. BX* 53 naturaliter ... accidentaliter] *inv. JX* | aut accidentaliter] *om. Y* 55 contingeret] –getur *X* | alii esset] *inv. Z* : alii *Y* : alii essent *B* 56 est] *om. B* 57 est] *om. W* 57-8 alicui contingeret] *inv. B*

contingeret duo loca numero inveniri uni corpori simplici, quod est impossibile, ut ante declaratum est. Et quia contentus est ex hoc quod apparet esse impossibile quod locus sit naturalis alii corpori, ad propalandum propria divisione istas duas partes oppositas sicut nos fecimus, scilicet ad dicendum scilicet quod si illic esset corpus simplex rectum aut illic esset naturale aut accidentale, quoniam posito illic esse ipsum accidentale, contingeret ut alteri esset naturale; et 66ra ideo excusatus fuit a propalatione | istius divisionis.

Et cum declaravit quod impossibile est illic esse corpus rotundum aut rectum, et ante non propalavit in suo sermone quod omne corpus simplex necesse est ei esse alterum istorum duorum modorum, propalavit hoc modo et dixit *Sed nos non videmus aliud corpus* etc., et hoc iam declaratum est in principio huius tractatus scilicet corpora simplicia esse tria; motus enim est triplex. Deinde dixit *Et cum ita sit, impossibile est* etc., idest si omne corpus simplex aut circulariter movetur aut a medio aut ad medium; et declaratum est modo ut aliquod istorum sit extra mundum esse impossibile, ergo impossibile est ut corpus simplex sit extra mundum omnino; et cum non simplex, ergo neque compositum; et quod dixit est intelligendum per se, quia compositum non fit nisi a simplici.

98. [99.] *Et etiam impossibile est quod extra celum sit corpus; aut enim,* 279a2 *si fuerit, erit corporum simplicium, sicut diximus superius, aut compo-* 66rb *|sitorum, aut illic naturaliter aut accidentaliter. Sermo ergo et ratio quod impossibile est illic esse corpus est sicut sermo noster superius de ea; ratio* 279a5 T *enim eadem est scilicet quod nullum corpus est extra celum, neque est possibile omnino. Manifestum est igitur ex hiis quod extra celum non est corpus, neque est possibile ut sit in eo aliqua magnitudinum corporum;*

---

59 ante] *post* declaratum *Z* : *post* est[1] *Y* | est[2]] + et quod *B* 60 alii] alteri *BX* 62 ad—scilicet[2]] *om. homoiotel. Y* | scilicet[2]] *om. JZ* 63 posito] ponendo *BX* 64 accidentale] naturale *BXY* 66 declaravit] -ratum (+ fuit B_3) *B* | illic esse] *inv. BJ* : esse *Y* : illic *X* 68 duorum] *om. Z* 71 est triplex] *inv. JX* 73 et] + cum *J*W_2*Z* 74 esse] est *WXY* | est] *om. B* 75 cum] est *Y* 76 neque] nec *W* | est intelligendum] *inv. Z* : intellectum est *J* : est intellectum *B* : intelligendum *W* 77 fit] sit *X*

98. [99.] 1 etiam] *om. YZ* 3 aut[1]] + est *Z* | ratio] ratiocinatio *JWY* : + non *Z* 4 impossibile] poss- *Z* | ea] eo *JXZ* 5 celum] + et quod *Z* 5-6 neque—celum] *om. homoiotel. X* 6 est[1]] *om. BJ* 7 neque] + enim *Z* *totus enim mundus est compositus ex tota sua materia convenienti; et iam diximus superius quod materia mundi est corpus naturale sensibile. Si igitur ita sit, manifestum est celos non esse plures modo, neque in pre-* *279a10* *terito neque in futuro; hec igitur celum unum est perfectum sine simili.* *279a11*

Cum declaravit quod extra mundum impossibile est reperire modo corpus in actu, neque simplex neque compositum, vult declarare quod non solummodo hoc declaratur ex hac demonstratione sed quod impossibile est etiam reperiri in hora alia. Et cum incepit declarationem istius quesiti, rediit ad demonstrationem predictam qua declaratur quod impossibile est reperiri in actu; et cum pervenit ex ea ad aliquantulum idest quia complevit contradictiones quas dixit in sillogismo hipothetico quo destruxit voces predictas, scilicet quod declaratio qua est necessarium accipere opposita omnium istarum contradictionum est eadem cum declarationibus predictis in demon- *66va* stratione quam induxit super primam questionem, | dixit *Erit ergo sermo et ratiocinatio* etc., idest et secundum hanc viam complebitur demonstratio super hoc scilicet secundum viam qua completur super prius quesitum; ratiocinatio enim quod extra mundum non est corpus, aut quod impossibile est ut numquam sit, eadem est; cum enim impossibile est corpus esse in actu, impossibile est esse, nam possibili posito non sequitur impossibile, sed hoc posito contingit impossibile. Et cum declaravit quod extra mundum non est corpus simplex aut compositum, necessarium est ut ista materia sit conclusa in eo; quapropter necesse est unum esse.

Et cum declaravit quod nihil est extra mundum, dedit causam et dixit *Totus enim mundus compositus est* etc., idest quia mundus compositus est ex tota sua materia. Deinde declaravit quod est hec materia, T79b et dixit *Et iam diximus superius quod materia mundi est corpus*

8 tota sua] *inv. BJX* : sua *W* | et] + quia *Z* 10 modo] *om. Z* 11 igitur] enim *Z* 12 reperire] -riri *Z* 13 corpus] + neque *Z* 15 hora alia] *inv. JZ* : hora aliqua *X* 16 rediit] reddit *B* : reducit *WYZ* | predictam] *om. WYZ* | qua] que *B* : quod *Z* 18 idest] ibi *J*W_2*YZ* 19 quo] quod *B* | destruxit] -straxit *B* | scilicet] secundum *B* : dixit *J* 20 est necessarium] *inv. Z* 20-1 istarum contradictionum] *inv. Z* 22 questionem] + et *JX* | ergo] *om. W* 23 et[2]] *om. JYZ* 25 ratiocinatio] ratio *BX* 26 ut] quod *B* | est[2]] *om. YZ* 29 corpus simplex] *inv. Z* 31 est] *om. Z* | unum esse] *inv. B* 33 mundus[2]] + est *Y* 34 est hec] *inv. YZ* 35 superius] *om. YZ* | quod] *om. B*

naturale, idest corpora simplicia, quoniam ista sunt eius materia; et cum declaratum est quod nihil ex eis fit extra ipsum, declarabitur quod eius materia est conclusa in eo. Deinde dixit *Manifestum est igitur* etc., et intendit per celos mundos, et vult declarare quod numquam sunt plures, quoniam, in quacumque hora ponuntur plures, contingit impossibile. Quia igitur entia sunt plura uno individuo propter diminutionem sui esse, quoniam intentio fuit in hoc ut remanerent in specie eterna, cum impossibile fuerit numero, necessarium est ut illud quod non invenitur in eo nisi tantum unum individuum ut sit perfectum in esse, ideoque dixit *Hoc igitur celum est unum perfectum sine simili.*

66vb14 **99.** [100.] *Et etiam nullus locus est extra celum neque plenus neque* *279a11* *vacuus, neque tempus; et hoc manifestum est; possibile est enim ponere in omni loco aliquod corporum. Vacuum autem descriptum est ab Antiquis dicentibus quod est illud in quo nullum corpus est omnino, et possibile est esse. Tempus autem est numerus motus, et motus non est sine corpore naturali.* *279a15*

Cum declaratum est quod extra mundum non est corpus, declarabitur quod illud neque est locus vacuus aut plenus: si enim illic esset locus, aut illic esset corpus aut possibile esset; sed iam declaratum est quod illic non est corpus, neque possibile est, ergo neque locus, neque possibile est etiam ut sit illic vacuum. Dicentes enim ipsum describunt ipsum quod est illud in quo possibile est esse corpus: si igitur impossibile est illic esse corpus sicut declaratum est, ergo et

36 corpora] + tria *B* | simplicia] + tria *J* | quoniam] quod *Z* 37 eis fit] hiis sit *BX* : eis *J* 38 eo] ipso *YZ* : ipso idest in eo ipso *J* 41 contingit] -get *Z* 43 remanerent] -neret *BY* | fuerit] fuit *BX* 45 ut] *om. JZ* | perfectum] perfecti tantum *Z* 45-6 in—perfectum] *om. homoiotel. Y* 45 ideoque] deinde *Z*

99. [100.] 1 etiam] *om. YZ* 2 est enim] enim *B* : est *Y* | ponere] + corpus *J* 3 aliquod corporum] *om. BJWXY sed suppl.* W_2 4 nullum] intellectu W_2*YZ* | corpus] + non *Z* 4-5 et—est[1]] impossibile *B* 5 est[3]] + nisi *B* 7 Cum] *om. Z* | declarabitur] + iam *JZ* 8 neque] non *B* | enim] igitur *Z* 9 locus … corpus] *inv.* Y | esset[1]] + possibile *Z* | esset[2]] + esse *BJ* | sed] *om. Z* 10 illic—est[1]] neque est ibi *Z* | neque[1]] + hoc *JZ* | possibile est] *inv. Z* | ergo] + quod *Z* | neque[2]] *om. B* 11 sit illic] *inv. JZ* 12 esse corpus] *inv. B* 13 illic] illuc *Z* | illic esse] *inv. W* | esse corpus] *inv. B* | sicut] sic *B* | ergo et] ergo *Z* : igitur *Y* vacuum. Et similiter de tempore: tempus enim est accidentium motus, et ubi non est corpus non est motus, et etiam ubi non est locus non est motus; omnis enim motus est in loco. T 80a

67rb4 **100.** [101.] *Et iam declaravimus superius quod extra celum non est* 279a16 *locus neque vacuum neque tempus; quapropter quod est illic non est in loco, neque tempus potest corrumpere ipsum; et nullum extrinsecum ab* 279a20 *ultimo motus potest alterari et transmutari omnino, sed semper est permanens non transmutabile neque passibile. Vita ergo semper est illic permanens, perpetua secundum perpetuitatem seculi seculorum, quod non finitur neque destruitur, et est vita nobilissima. Et bene dixerunt Antiqui in explanatione huius nominis seculum; dixerunt enim quod finis et orizon continens tempus vite cuiuslibet rei vive post quod et retro quod non est tempus aliud naturale, est seculum illius rei. Secundum igitur* 279a25 *hunc modum dicimus quod finis orizontis totius celi et finis continens totum tempus perpetuum est seculum celi; et hoc nomen derivatur a suo actu, quoniam semper est perpetuum; et in quibusdam rebus est semper manifestum, et in quibusdam parum. Et iam diximus in libris de Phi-* 279a30 *losophia Extranea, quam posuimus vulgo, quod oportet spirituale ut non* 67va *transmutetur neque corrumpatur, quoniam est | causa omnis esse, et post ipsam non est alia causa. Et est secundum hunc modum neque transmutabile neque alterabile, perfectum et completum et eternum per secula; supra ipsum enim non est alia causa nota ut moveat ipsum. Si igitur inveniatur alia causa, tunc erit etiam perfecta existens eterna, et post ipsam non erit alia nobilior. Et etiam diximus quod non recipit aliam occasionem illud spirituale, et motus eius est perpetuus sine abscisione. Et* 279b1 *dignum est habere hunc motum quoniam omnes res quiescunt per motus cum perveniunt ad suum locum; hoc autem corpus nobile non quiescit*

---

14 vacuum] motum W_2*YZ* | tempore] + proprius *B* 16 locus] + ibi *BJ* | est[2]] *om. Z*

100. [101.] 1 iam] ideo *YZ* 2 neque tempus] *om. Z* 3 nullum extrinsecum] -lus -cus *B* 4 ultimo] + et *B* 5 non] *om. Y* | ergo] est igitur *Y* : igitur *Z* | semper] *post* illic *JZ* | est illic] *om. Y* | illic permanens] *inv. W* 9 continens tempus] –netur corpus *Y* 10 illius] istius *Z* 10-1 igitur hunc] *inv. BXZ* 11 dicimus] -ximus *WY* 12 celi] seculi *WYZ* 12-3 suo actu] sua operatione *JXY (cf. infra, c. 100 [101], v. 85sq.)* 14 de] *om. Z* 15 spirituale] -lem *B* 16 quoniam est] est enim *J* | causa omnis] *inv. Z* | esse] cause B_2*W* 19-20 ut—causa] *om. Y* 21 alia] *om. WYZ* | aliam] aliquam *BJ* 22 occasionem] actio- *WZ* | et[1]] quod *Z* | est] *om. YZ* 23 quoniam] quando *YZ* | motus] -tum *BJ* 24 nobile] mo- *BJ* eius tunc erit debilior apud audientes propter paucitatem exercitantium in distinctione demonstrationum et propositionum dantium fidem; et hoc intendebat cum dixit *Propositiones signorum.*

Deinde dedit aliam proprietatem secundum quam debet homo stare cum suo adversario si voluerit scire veritatem, et est ut non habeat odio suum adversarium sed sit pacificus in suis sermonibus: maxima enim causa diversitatis est amicitia et odium. Deinde demonstravit quid sit pacificum, et est ut recipiat ex sermonibus sui adversarii sicut ex suis, scilicet ut recipiat ex suo adversario demonstrationes que sunt secundum conditiones secundum quas ipse recepit ipsas sibi; et hoc intendebat cum dixit *Et scientia apud terminos demonstrationum.*

69vb4 **102.** [103.] *Incipiamus modo et dicamus quod omnes Antiqui conve- 279b12*
niunt quod celum est generatum, sed quidam eorum dicunt quod non
habuerit principium neque finem in tempore quidam autem dicunt
ipsum habere finem in tempore et cadere sub corruptione sicut alia cor-
pora generata, quidam vero dicunt ipsum aliquando corrumpi ali- 279b15
quando generari, et quod ita est semper sine fine et sine abscisione, ut
Empedocles Fragitensis et Heraclitus Fastinensis. Dicamus ergo quod
sermo dicentis celum esse mundum maximum, et quod ipsum generat res **T83b**
et quod est perpetuum sine fine et sine principio, est impossibile; non
enim concedimus nec credimus aliquam dispositionem nisi viderimus
ipsam in omnibus rebus aut in pluribus; hec enim dispositio scilicet celi,
dicere quod est generatum et quod non corrumpitur neque cadit sub cor-
ruptione, nos videmus contrarium in rebus, quoniam omnia generata 279b20
corrumpuntur. 279b21

Cum declaravit quod vult perscrutari de sermonibus adversariorum de mundo, incepit numerare opiniones eorum et primo de communi: omnes enim conveniunt in hoc quod celum est generatum, et

---

42 distinctione] destructio- *Z* 43 fidem] finem *Y* 44 dedit] dat *JZ* 46 habeat] -bet *Z* 48 sui adversarii] *inv. JZ* 50 conditiones] contradictio- *JZ* | ipsas] + ipsum *B* 51 dixit] -xerit *B*

102. [103.] 2 eorum] *om. JZ* 3 habuerit] -buit *XZ* : -bere *J* | neque] aut *JZ* 4 sicut] ut *YZ* : et *BX* 7 Heraclitus] Eradicatus *Y* : Ervolutus *X* | Fastinensis] Fastrensis *X* : Fradidensis *Z* : Ephesius *J* | ergo] modo *JZ* 8 maximum] magnum *YZ* 9 est perpetuum] *inv. BJ* | sine^2] *om. J* 11 enim] autem *JXY* 15 declaravit] narra- *JYZ*

70ra nullus ante ipsum dixit quod celum non est generatum. | Deinde incepit narrare proprias opiniones, et dixit *Sed quidam dicunt ipsum non habere principium aut finem in tempore*, idest sed quidam dicunt quod numquam corrumpetur neque habebit finem ita quod tempus succedat ei in futuro; et cum dixit *Non habere principium aut finem in tempore*, intendit quod nullum tempus veniet in quo dissolvetur in suum principium ex quo generatum est, neque esse eius habet finem in tempore futuro, et hec est opinio Platonis de mundo: videt enim quod generatus est postquam non fuit, tamen est incorruptibilis.

Deinde dedit opinionem secundam et dixit *Quod quidam autem dicunt ipsum habere finem in tempore*, idest quod fuit inventus mundus in tempore terminato et quod corrumpetur; et hec est opinio Anaxagore, sicut narravit de eo in principio octavi Phisicorum: Anaxagoras enim videt mundum esse generatum postquam non fuit, et corruptibilem, ita quod nullum ens erit ex eo omnino. Deinde dixit *Sicut alia corpora generabilia*, idest quod hoc est magis secundum cursum nature, scilicet quod, cum ponetur generatum, oportet ut ponatur corruptibile, cum plures partes eius sint huiusmodi preter celum.

Deinde dedit tertiam opinionem famosam in eius tempore, et dixit *Quidam vero dicunt ipsum aliquando corrumpi et aliquando gene-* **T84a** *rari*, idest quidam dicunt quod aliquando corrumpitur et aliquando generatur, et semper est huiusmodi; et sustinentes hanc opinionem dicunt quod iam generabatur infinities et corrumpebatur infinities, et generabitur et corrumpetur infinities, et sic dixit Empedocles. *70rb* Opiniones au|tem de mundo sunt quatuor: aut credere ipsum esse non generabile aut corruptibile, et hec est opinio Aristotelis et ipse fuit primus dicens hoc Philosophorum Grecorum, et hoc tenebat lex Zabiorum antiqua, et hoc intelligetur ex verbis occultis Caldeorum;

18 nullus] + est *Z* 19 proprias opiniones] *inv. JX* 23 veniet] -nit *BX* 26-8 tamen—fuit] *om. homoiotel. X* 26 est incorruptibilis] *inv. Z* 27 dedit opinionem] dixit questionem *Y* | Quod] *om. B* | autem] *om. JY* : *exp. W* 28 habere] esse *Y* 29 quod] *om. B* | est opinio] *inv. J* : opinio *WY* 30 narravit] dixit *Z* | principio] primo *XYZ* | octavi] *om. WYZ* 32 corruptibilem] -bile *X* : in- *Y* : corporalem *Z* | omnino] numero *W* 33 hoc] *om. BX* 34 oportet … ponatur] apparet ut patiatur *W* 35 corruptibile] corporale *Z* | sint] sunt *B* | preter] primi *Y* 36 dedit] dixit *WXY* 38 et] *om. JW* 40-1 corrumpebatur—et[2]] *om. homoiarch. BX* 41 sic] sicut *WX* | Empedocles] + opinor aes de constitutione *B* 42 esse] *om. JZ* 43 non] *om. Z* 44 Philosophorum Grecorum] *inv. W* | tenebat] -bit *W* 45 Zabiorum] rabio- *W* : tabio- *X* : talio- *YZ* : chaldeo- *J*

aut credere ipsum esse generabile et corruptibile, et hec est opinio Anaxagore ex Philosophis, et hanc sustinent tres leges que sunt modo note scilicet Maurorum et Christianorum et Iudeorum; opinio vero tertia est Platonis et est quod sit generabile et non corruptibile; quarta vero credere non generabile sed corruptibile, sed hoc nullus dicebat inquantum scimus nisi loquentes nostre legis, quamvis sit hoc tantum extra naturam.

Et cum dixit opiniones de mundo, incepit dicere quod contingit cuilibet opinioni de impossibili, et dixit *Dicamus modo quod sermo dicentis celum* etc., idest sermo vero dicentis quod celum, quod est mundus maximus, est generatum postquam non fuit, et quod ipsum generavit res cum generavit ipsum creator, et quod est perpetuum sine fine in tempore futuro, et quod non invenitur in eo ultimum neque principium in quod dissolvitur, est sermo valde impossibilis quia contradicit sensui. Deinde dedit causam de hoc et dixit *Non enim concedimus et credimus* etc., idest et iste sermo non creditur, quia inductio in qua inducuntur omnia particularia contradicit huic opinioni: sensimus enim quod omne generabile est corruptibile, hoc **T84b** ergo iudicium, scilicet aliquid esse generatum et insimul non esse *70va* corruptibile, non sentitur neque | in omnibus neque in pluribus, sed contrarium, et est quod omne quod est generatum est corruptibile.

Et hoc quod dicit est sufficiens, scilicet quod non debemus credere aliquid habere aliquam dispositionem donec inveniatur in maiori parte illius rei aut in omnibus; iudicium autem huius affirmatur per inductionem per demonstrationem iam affirmatum est, quamvis non sentiatur neque in quibusdam neque in omnibus: sed tamen videtur quod intendebat dicere quod dispositio credenda impossibile est ut eius contrarium sit in quisbusdam illius rei aut in omnibus nobis credentibus, quasi ergo diceret cum fides dispositionem quarundam rerum est sentiendo illam dispositionem existentem in omnibus illius

46 esse] *om.* *X* : + non *Z* 47 hanc] hoc *BXYZ* 48 note] nate *W* : + in nostro tempore *B* 49 generabile] –bilis *J* | et^{2}] sed *BX* 50 generabile] -ratum *BJ* 52 tantum] unum *YZ* : + unum JW_2 53 quod] quid *JY* 55 vero] *om.* *JYZ* 56 fuit] fuerit *Z* 59 quod] quo *B* 63 sensimus] senti- *BJX* 66 est quod] *om.* *WYZ* | quod est] *om.* *BX* | est^{3}] videtur *BX* 67 dicit] -xit *BX* | credere] *om.* *BX* 68 aliquid] -quod *Z* 70 inductionem] + et *JZ* 72 dicere] *om.* *Z* | credenda] + est *JX* 73 nobis credentibus] *inv.* *BX* 74 quasi] idest *YZ* | diceret] -camus *Z* : -cetur *X* | dispositionem] -tionum *YZ* 75 est] sit *JZ*

rei aut in pluribus, impossibile est igitur habere fidem de re cuius contrarium sentiatur in omnibus illius rei.

70vb3 **103.** [104.] *Et etiam dicimus quod res in qua non est potentia qua alteratur ex sua dispositione, necessario sue alterationis est causa ut alteretur.* 279b21 *Dicamus ergo quod si mundus fuit compositus ex rebus quarum dispositiones* 279b25 *sunt primo aliter a dispositione mundi, et fuit impossibile illas dispositiones mutari, impossibile est ut mundus fiat ex eis omnino. Si ergo res possibiles sint alterari et transmutari necessario et non sint secundum unam dispositionem perpetuam, si ita fuerint cum alterabuntur ille res, possibiles erunt dissolvi et diminui etiam. Si ergo dissolvantur, componuntur etiam; erunt ergo secundum hunc modum dissolvi et componi in* 279b30 *infinitum. Si ergo hoc sit secundum hunc modum, mundus cadit sub corruptione, non sicut dixerunt ipsum esse extra corruptionem et consumptionem.* 279b31

Idest quod res carentes potentia transmutationis et alterationis impossibile est ut transmutentur in aliqua hora, quoniam si ita esset, tunc alteratio esset causa suiipsius, idest esset alteratio sine possibilitate; et intendit per alterationem transmutationem simplicem. Et cum posuit hanc propositionem dicentem quod illud quod impossibile est alterari non alterabitur nisi esset idem causa suiipsius idest esset causa et causatum idem et inveniretur aliquid sine causa, dixit T85a
71ra *Si mundus fuit com|positus* etc., idest manifestum est per se quod ea ex quibus compositus fuit mundus secundum quod fingunt habebant contrariam dispositionem illi quam post habebant, et quod ille dispositiones mutabantur in alias quasi dicens quod ante mundum carebant positione et post habebant compositionem. Si igitur antequam ex eis componebatur mundus fuit impossibile componi ex eis, non

---

103. [104.] 1 dicimus] -camus *Z* 3 ergo] *om. B* 4 sunt] fuerit *B* : fuerunt *J* : fuerint *X* 5 ergo] igitur *JZ* 6 res] *om. BWXY* | possibiles] -bile *XY* 7 dispositionem] + mediarum *W* 8 possibiles erunt] -bile erit *BX* | etiam] *om. B* | Si] et *praem. BJ* | dissolvantur] -vuntur *YZ* | componuntur] et ponuntur *BX* 9 erunt] erit *BJX* | dissolvi ... componi] *inv. Z* 10 ergo] igitur *BXY* | sit] fit *Z* : ita *Y* : ita sit *J* | cadit] -det *BJX* 13 transmutationis] transumptio- *Y* 17 propositionem] + displicem *W* 18 esset idem] *inv. J* : esset *W* | suiipsius] ipsius *Y* 19 causatum] -sarum *B* | et[2]] quod *Z* 20 compositus] + fuit *WY* 21 habebant] -bebunt *WZ* 22 post] primo *Z* | habebant] -bebunt *W* | ille] iste *JZ* 23 alias] + secundum *Y* : + scilicet *J* | quasi] quod *YZ* : + quod *X* 24 positione] com- *JX* : dis- *Y*

fuit possibile componi ex eis nisi esset possibile ut idem esset causa suiipsius.

Et cum posuit quod mundus non generatur nisi a rebus possibilibus recipere formam mundi antequam reciperent ipsam, dixit *Si igitur res sunt possibiles alterari* etc., idest si igitur ille res ex quibus fit mundus possibiles sunt transmutari in formam mundi, et non remanere secundum eandem dispositionem cum alterabuntur, continget cum alterabuntur in formam mundi ex forma habita quod sint possibiles alterari ex forma mundi in formam primam quam habebant. Et hoc intendebat cum dixit *Si igitur ita sit* etc., idest si igitur ille res sunt possibiles alterari in formam mundi, continget quod cum alterabuntur in formam mundi (secundum quod elementa alterantur ex privatione compositionis in compositionem) quod sint possibiles dissolvi ad primam compositionem.

Deinde dixit *Et si dissolvuntur componuntur*, idest et si possibiles sunt dissolvi, dissolventur, et cum dissolventur, cum fuerint possibiles componi componentur; non igitur cessabit mundus esse in generatione et corruptione perpetua. Deinde dixit *Si dixerint ipsum esse extra corruptionem*, idest contingit eis quod contrario sermoni eorum si dixerunt ipsum esse generatum et cum hoc immunem a
71rb corruptione; et decla|rabitur post quod impossibile est ut possibile non exeat ad esse in tempore finito.

104. [105.] *Ratio autem quam dixerunt, cum laborant in affirmando* *279b32*
mundum esse generatum sed non corruptibilem non est vera, et destruitur per veram demonstrationem. Dixerunt enim: Sicut Geometre faciunt in figuris compositis ex pluribus lineis, ita fecimus nos in mundo. Geo- *279b35*
metre enim, cum voluerint docere quomodo figure sint, dicunt quod ista

26 possibile[1]] im- *Z* 29 reciperent] -piant *YZ* : -pient *W* 31 remanere] + cum *Z* 32 dispositionem] + secundum *JZ* 33 sint] sunt *JYZ* 36 sunt] sint *BW* | formam] -ma *Z* | quod] ut *W* 38 compositionem] –tione *BW* 39 primam compositionem] *inv. Z* 40 Et[1]] *om. Z* | si[1]] + ergo *Z* | si[2]] *om. Z* 41 sunt] sint *YZ* | dissolventur] -vuntur *JX* | cum[2]] si *BJX* | fuerint] sint *W* 42 igitur] ergo *BX* 43 dixerint] -xerunt *JZ* 44 esse] *om. YZ* | eis quod] eis *X* : eis ex *Y* : ei ex *Z* | contrario] + secundum motum *B* | sermoni] -mone *Z* 45-6 si—corruptione] eat *Y* 45 dixerunt] -xerint *BX*

104. [105.] 1 quam] qua *B* 2 corruptibilem] -bile *XYZ* 3 veram] naturam *B* 5 cum voluerint] conveniunt *Y* | docere] dicere *YZ* | sint] sunt *WY* *figura componitur ex ista linea deinde ex ista, ad demonstrandum hoc* 280a1 *discipulis; et ita nos fecimus cuilibet qui dixit quod mundus sit, non quia tempus fuit et mundus non fuit, sed ut doctrina sit manifestior discipulis, quia ex hoc generabatur prius, deinde ex hoc, sicut fecimus in figuris.* 280a2

Cum contradixerit dicenti mundum esse generatum et non corruptibile incepit contradicere exemplo eorum, scilicet quomodo imaginatur generabile non in tempore: tempus enim apud Platonem,
71va dicentem hunc sermonem, se|quitur motum celi, ut apud Aristotelem; et cum ita sit, continget ei aliud impossibile scilicet ut aliquid generetur non in tempore.

Quapropter Plato dat rationem in hoc de generatione figurarum ex superficiebus et lineis, dicendo Non diximus quod mundus sit ex rebus simplicibus ex quibus componitur ita quod simplicia sint priora compositis tempore: videmus enim quod tempus sequitur motum celi, sed volumus docere quomodo sint priora mundo scilicet hac forma composita, quoniam sicut Geometre volunt docere quomodo linea et superficies sint priores figura, et quod figura componitur ex eis et nullus intelligit per hanc prioritatem prioritatem temporalem, ita non est intelligendum ex sermone nostro quod res ex quibus mundus fit sint priora mundo secundum tempus sed priora tantum, et cum attribuuntur tempori, inveniuntur in eo insimul sicut est dispositio de lineis cum figuris; quasi igitur voluit declarare ex hoc quod omne in quo dicitur esse prius alio non dicitur tantum secundum tempus omnino. **T86a**

Et hoc verum est, sed contingit ei, si hoc prius non fuit tempore, ut sit prius in esse tantum, et res priores in esse sunt simul in tempore et sunt eterne, sicut dispositio de superficiebus et lineis cum figuris est cum non sit tertius modus prioritatis; Plato autem ponit

---

7 nos fecimus] *inv. BJX* 8 tempus … mundus] *inv. Y* | non] *om. BXY* 9 generabatur] -bitur *XZ* | prius] *ante* generabatur *Z* | sicut] *ante* ex² *B* : sed *Z* : + illud *J* 11 contradixerit] -dixit *BX* : dixerit *Y* | non corruptibile] incorr- *B* 12 imaginatur] -nantur *B* 15 continget] -geret *Z* | aliud] –quod *J* 20 sequitur] con- *Z* 22 volunt] nol- *Z* | volunt docere] *inv. JW* 23 priores] priora *YZ* | figura²] priora *Z* 24 et] quod *Z* : + quod *J* | prioritatem] *haplogr. BYZ* 27 insimul] *ante* in *Z* 28 quasi] quod *BYZ* | igitur] ergo *Z* 29 ex hoc] *ante* declarare *(28) Z* 31 fuit] fuerit *BX* 32 prius] prioris *Z* | priores] priora *Z*

illas res, ex quibus mundus fit, priora mundo secundum seculum; et nos loquemur de hoc cum Aristoteles contradicet huic.

71vb4 **105.** [106.] *Et nos contradicemus eis, dicendo quod hoc exemplum non* 280a2 *convenit eis ex quibus sit mundus. In factura enim figurarum et ex compositione earum figura non fit completa perfecta nisi per omnes lineas ex quibus componitur; et figura non fit figura nisi per alterationem linearum ex quibus componitur, neque etiam fit figura nisi omnes linee sint insimul. Si ergo voluerint dictum earum affirmare et verificare ipsum, non manifestabuntur omnes linee semel, sed declarabit lineam post lineam donec perficiant omnes lineas totius figure.* 280a4

Dicit quod hoc exemplum de generatione mundi ex simplicibus ex quibus componitur non convenit compositioni figurarum ex superficiebus et lineis, idest quod altera duarum compositionum non assimiletur alteri tribus modis. Quorum unus est quod figura non invenitur in aliqua hora diminuta in aliqua linea linearum ex quibus componitur essendo figura in actu; et hoc intendebat cum dixit *Figura non erit perfecta nisi per omnes lineas*, idest quod figura non sit in aliqua hora in actu si fuerit diminuta in aliqua linea; et non intendebat per completa perfectionem ultimam, in hoc enim assimilatur figura et generatur. Secundus est quod figura non fit ex lineis per alterationem et transmutationem sicut generatum fit per alterationem 72ra | et transmutationem eius ex quo generatur. Tertius est quod linee non inveniuntur in aliqua hora absque figura, elementa autem generati inveniuntur ante generatum; et intendit per totum hoc quod generatum invenitur in omnibus istis dispositionibus econtrario

36 loquemur] –quamur *Y* : -quimur *X* | huic] hiis *W* : illi *Z*

105. [106.] 1 eis] ei *B* | hoc exemplum] *inv. BX* 2 sit] fit *BXY* | et] *om. WYZ* | ex] *om. X* 3 fit] sit *B* | completa perfecta] perfecta et completa *J* | per—lineas] cum omnibus lineis *Z* 4 et figura] figura autem *Z* | fit] sit *BY* : est *Z* | nisi] *om. BZ* 5 etiam] enim *Z* | fit] sit *BW* | figura] *om. W* | omnes] visiones *B* | sint] sui *Y* 6 Si ergo] et si *J* : si igitur *Z* | dictum] -ta *BX* : domini *Z* 7 semel] simul *JXZ* 10 compositioni] –tionem *Y* | figurarum] linearum *W* 11 assimiletur] -latur *BXY* 12 alteri] aliis *YZ* | tribus] *om. WY* | invenitur] -nietur *Z* 14 hoc] *om. B* 15 Figura[1]] + enim *WZ* | erit] fit *JZ* : est *Y* | perfecta] + et completa *J* | non[2]] *om. W* | sit] fit *YZ* 16 in[2]] *om. WYZ* | aliqua] alia *XY* 18 generatur] -ratum *BJ* | fit] sit *BY* 19 alterationem[1]] altitudi- *B* | generatum fit] *inv. W* 20 est] *om. Z* 21 generati] -rari *W* figure: fit enim per alterationem eius ex quo generatur, et invenitur generatum in actu ante complementum ultimum, et inveniuntur elementa eius ante ipsum tempore. T86b

Dixit *Et si voluerint dictum earum* etc., idest et si voluerint assimilare generationem mundi non compositioni figurarum ex lineis in se sed compositioni in doctrina; quoniam Geometre cum utuntur compositione ad docendum erit compositio temporalis, quoniam quod declaratur ex hoc est aliquid post aliquid; quasi ergo diceret quod hoc exemplum aut intelligitur per hanc compositionem scilicet compositionem in doctrina, aut compositionem in se. Si igitur intendebat compositionem in se, ergo contingit quod ipsa ex rebus ex quibus componitur sit insimul: generatio autem non est ita si in doctrina erit compositio temporalis; sed ipsi non concedunt generationem mundi esse in tempore; quia ergo posuerunt quod mundus cum eix ex quibus generatur non est simul, et posuerunt quod ante mundum non est tempus, continget eis ut sit prius et post sine tempore; et tunc non erit assimilatio eius compositioni figurarum ex lineis vera.

Defendentes autem Platonem dicunt quod assimilatio generationis mundi (ex suis elementis) generationi figurarum (ex lineis et superficiebus) a Platone significat quod mundus semper fuit apud Platonem et quod ipse non intendebat per hunc sermonem nisi *72rb* quod | partes eius sunt priores natura data, non quod ipse fuit generatus vere; et si hoc intendebat, nihil differt ab Aristotele; sed tamen male fecit in dicendo manifeste mundum esse generatum et non corruptibilem: hoc enim non intelligitur ex generatione nisi cum difficultate, et maxime cum dixit quod fuit quod efficitur ex non ordine in ordinem.

---

24 fit] sit *B* 25 ante] aut *WZ* 26 tempore] -pus *XYZ* 27 earum] + assimilare *J* 31 declaratur] -ravit *W* : -ratum est *Y* : -rant *B* | quasi] quod *BZ* 34 intendebat] -debant *BX* : -dat *J* 35 in] *om. BWXY* 37 quia] qui *Z* | posuerunt] -suerint *Z* : -suit *Y* 39 continget] -git *BJ* | prius] proprius *Z* 42 Defendentes] -dentem *YZ* 43 generationi figurarum] –tionem figurant *Y* | ex^2] + suis *Z* 44 significat] -navit *WY* : -nat *XY* 46 data] dat *X* : dante *B* : toto *J* | quod^2] quia *Z* | fuit] fuerit *BX* : fiat *Z* | generatus] *om. Z* 47 vere] nature *Y* | nihil] nullus *B* | ab Aristotele] *ante* differt *Z* 48 manifeste] *om. JYZ* | corruptibilem] -bile *WXY* 49 intelligitur] -ligi *B* : -ligit *Y* | generatione] + et corruptione vel sermone suo *Z* 50 fuit quod] *om. J*

106. [107.] *Compositio autem mundi non est huiusmodi; per ea autem* *280a5*
ex quibus componitur iuvabimur in rationibus nostris, dicendo quod
mundus primo generatur ex hoc, deinde post hoc ex hoc. Hec igitur duo
secundum suum sermonem differunt in se; res enim erant sine ordine,
deinde ordinabantur et fit ex eis mundus, et impossibile est ut ordinatum T87a
et non ordinatum sint insimul, quoniam inter ea est tempus et generatio;
linearum autem que sunt in figura nulla est ante alteram, et prima *280a10*
earum non est alterius dispositionis ab alia. Manifestum est igitur quod
impossibile est quod mundus sit generatus, et est perpetuus sine consump-
tione et corruptione. *280a11*

72va Cum narravit quomodo figure componuntur ex lineis et res appropriatas huic compositioni, incepit notificare quid contingit eis quod compositio mundi est contraria huic compositioni, et dixit *Compositio autem mundi* etc., idest quod compositio, quam fingit habere mundum qua sit generatum non est secundum modum compositionis figurarum. Deinde incepit narrare differentiam inter duas compositiones, et dixit *Per ea enim ex quibus mundus* etc., idest ut mihi videtur quod natura rerum, ex quibus mundus componitur secundum quod fingit, differt a natura linearum ex quibus componuntur figure; apparet enim nobis ex hoc error exempli eorum et surget nobis ratio contra eos; quoniam ea ex quibus mundus fit, quia sunt corpora existentia in actu, continget ut sint in loco et tempore; et cum ita sit, necesse est ut sint priora mundo tempore.

Et ideo nulla ratio est in sermone Platonis qui ponit ea priora per seculum mundo: quod igitur innatum est reperiri in tempore non invenitur sine tempore, et quod non est innatum inveniri in tempore

---

106. [107.] 2 iuvabimur] -bitur *BX* : -buntur *JZ* 3 primo generatur] *inv. YZ* | hoc[1]] *om. WY* | ex[2] hoc] *om. XZ* 4 suum sermonem] *inv. YZ* 5 fit] sit *BZ* 8 est[2]] + quidem *BWX* 9-10 consumptione—corruptione] *spat. vac., deinde* et conservatione *Z* : corruptione et consumptione *WY* 11 componuntur] *om. Z* 12 compositioni] -poni *BWZ* | notificare] notare *XY* : narrare *W* 13 quod] *om. W* 14 mundi] + et *Z* | fingit] figerit *W* : signat *Y* : significat *Z* 16 incepit] intendit *YZ* 17 enim] autem *Y* | ex—mundus] *om. J* 18 quod] quia *XY* 20 nobis] + hic *BX* | exempli] -plo *Z* 23 priora] propria *Z* 24 tempore] *om. WYZ* : in tempore *BX* 25 sermone Platonis] *inv. Z* | ea] eam *B* 26 quod] quia *Z* | reperiri] inveniri *W* 27 invenitur] -nietur *XY* | innatum] natum *JWZ* non habet prius et posterius nisi secundum esse, sicut est de lineis cum figura, et magis de rebus abstractis, quoniam si aliqua generatio fuerit ex rebus non temporabilibus, tunc ille res non erunt priores nisi in esse tantum. Si igitur intendebant, cum dicebant mundum esse generatum, quia est ex causis prioribus in esse, erit verum; sed verba eorum non patiuntur hoc, aut ergo peccant aut deviant a veritate aut male docent.

Et cum dixit quod ex natura illarum rerum apparet differentia inter utrumque, scilicet exemplum et exemplar, incepit demonstrare hoc et dixit *Quod mundus primo generatur ex hoc, deinde ex hoc*, idest **T87b** *72vb* ut mihi videtur quod contin|git eis ut ille res inquantum sunt entia sint alicuius qualitatis qua inducunt esse inordinata: componuntur igitur ex qualitate et quali et differenti et delato, et similis materie et forme; erunt ergo ille res primo generate ex aliquo quod est ex eis quasi materia; et erit iste mundus iste individualis primo generatus ex istis rebus. Deinde dixit *Hec igitur duo differunt ut dicunt*, idest contingit eis ut ea ex quibus mundus fit differant a mundo naturaliter sicut dixerunt.

Deinde dixit *Res enim erant sine ordine* etc., idest et similiter dixerunt ipsi in rebus ex quibus mundus generabatur cum mundo, dicunt enim quod ea ex quibus mundus fit erant inordinata; *Deinde ordinabantur, et factus est ex eis mundus*, idest quia erant non intelligibilia neque recipientia diffinitionem, deinde efficiebantur recipientia diffinitionem. Deinde dixit *Et impossibile est ut ordinatum et non ordinatum sint* etc., idest et impossibile est ut eadem res sint recipientia diffinitionem et non insimul, quoniam omnino erit privatio diffinitionis prius tempore habitu diffinitionis; et non est ita de lineis, que fiunt; et intendebat per hoc quod cum ponitur mundus

---

28 posterius] post *BX* 29 aliqua] qua *Z* 31 tantum] malum *B* | intendebant] -debat *Z* | dicebant] -cebat *Z* 32 erit] *om. B* 33 ergo] igitur *BX* | aut[2]] hic *praem. Z* : + hoc *J* 37 et] quod *Z* 39 sint] sunt *JXZ* | qua] quam *BJ* : quia *WY* | inducunt] dicunt *BJX* | inordinata] *om. YZ* : -nate *X* : -natam *J* | componuntur] -nitur *Z* 40 igitur] ergo *JZ* | et[2]] *om. JZ* | differenti] + decenti *Z* | similis] -milibus *J* : -mul *Z* 41 erunt ergo] *inv. W* 42 quasi materia] *om. Y* | iste[2]] *om. JYZ* 43 duo] due *Z* | ut] et *W* 44 contingit] contra *Z* | differant] -ferunt *W* : + et *B* | naturaliter] *ante* a mundo *BX* 47 ipsi] ipsum *Z* 48 mundus fit] *inv. W* | inordinata] in orizonta *W* | Deinde] + dixit *B* 50 recipientia[1]] -cuperantia *Z* 50-1 efficiebantur—deinde] *om. homoiotel. WYZ* 52 res] *om. YZ* : *exp. W* | recipientia] -cuperantia *Z* 53 privatio] principium *B* 54 diffinitionis[1]] + est *Z* 55 hoc] hos *Y* | quod] quia *BJY* | cum ponitur] componitur *BXZ*

esse factus, continget ut ea ex quibus fit sint priora eo tempore, et intendebat per hoc quod dixit quod generatio et tempus sint inter ea idest quod transmutatio eorum ex inordinatione ad ordinationem est generatio, et omnis generatio est in tempore. Deinde dixit *Linearum autem* etc., idest quia linee non sunt priores figura in tempore cum extra animam non inveniantur nisi per figuram: intellectus autem abstrahit eas a figura. *Et non est prima earum* etc., idest et non inveniuntur linee primo idest per se secundum dispositionem oppositam *73ra* dispositioni habite cum fuerint in figura; | contingit igitur ut transmutatio earum a prima dispositione in ultimum sit alteratio, et quod sint priores figura tempore.

Deinde dixit *Manifestum est igitur* etc., idest ex hoc quod primo **T88a** dictum est quod omne generabile oportet esse corruptibile: dicere autem mundum esse generatum a nihilo sicut loquentes dicunt contingit eis ut non sit generatum; generatio enim non est ex privatione essentialiter, quemadmodum calor non fit ex frigore essentialiter nisi contrarium recipiat suum contrarium. Cum igitur privatio non recipit esse, quia essentia privationis non transmutatur in esse, aut mundus erit modo privatus aut semper fuit; sed impossibile est ut sit privatus, ergo semper fuit; ideoque quidem nullus Antiquorum dicebat hoc, quoniam est impossibile et inopinabile.

73rb **107.** [108.] *Dicentes autem mundum generari aliquando et corrumpi,* *280a11* *volunt ponere mundum semper esse sine consumptione, sed forma eius alteratur et transmutatur successive de forma in formam, verbi gratia ex pueritia in virum, ex viro in pueritiam; aliquando igitur corrumpitur* *280a15* *puer et fit vir, et aliquando corrumpitur vir et fit puer. Dicamus igitur quod iste sermo est falsus: cum enim elementa veniunt adinvicem, non fiunt ex eis dispositiones diverse sed una ordinatio et una dispositio. Et*

56 eo] *om. Z* 57 quod[2]] *om. Z* : quia *W* | sint] sunt *JXY* 60 quia] quod *YZ* | figura] -rarum *B* 62 est] de *B* 63 idest] *om. Z* : + et *X* 64 fuerint] *post* figura *B* | contingit] contra *Z* 65 ultimum] + cum *Z* 66 figura] -gure *Z* : + in *B* 67 idest] *om. W* 72 igitur] enim *Z* 74 mundus] mundo *B* 75 ideoque] ideo quod *Y* : scilicet quod *W* | quidem] *om. JXZ* : quidam *Y*

107. [108.] 2 mundum semper] *inv. YZ* | consumptione] -sumatione *Z* 4 pueritiam] puellam *Z* 5 corrumpitur] *om. YZ* | igitur] ergo *Y* : nos *J* 7 fiunt] sunt *WYZ* : erant *J* | Et[2]] *om. Z* *ipsi concedunt hoc quia ponunt generationi et corruptioni causam, sed tamen causam corruptionis contrariam esse generationis. Dicamus ergo quod si corpus mundi est continuum totum, et aliquando fuerit secun-* 280a20 *dum hanc dispositionem et aliquando secundum aliam, et semper revolvitur secundum hunc modum, et fuerit compositio mundi et celi secundum quamcumque dispositionem sit, tunc mundus non erit generatus neque transmutabilis neque corruptibilis, sed dispositiones eius sunt que consumuntur et corrumpuntur.* 280a23

Cum complevit contradictionem opinionis dicentis mundum esse generatum sed non corruptibilem, et contradictionem exemplorum, nunc potest contradicere opinioni dicentis ipsum esse generabilem et corruptibilem revolutione, scilicet quod semper transfertur ex gene- **T88b**
73va ratione in | corruptionem et ex corruptione in generationem, et dixit *Dicentes autem mundum* etc., idest contingit eis qui hoc dixerunt scilicet quod mundus sit perpetuus secundum suam formam substantialem sine consumptione et generatione in substantia, quoniam transmutatio non est de forma in formam sed de accidente in accidens idest de dispositione in dispositionem. Et hoc intendebat cum dixit *Ex forma in formam*, et signum eius est quod post dixit *Verbi gratia ex pueritia*, idest quod hoc quod diximus de mundo est quemadmodum diximus in aliquo individuo ipsum esse eternum sed aliquando transmutatur ex pueritia in iuventutem et aliquando ex iuventute in pueritiam.

Deinde incepit narrare modum ex quo contingit eis hoc, et dixit *Dicamus nos quod iste sermo* etc., idest quoniam dicentes hoc fingunt quod elementa ex quibus fit iste mundus congregantur aliquando et aliquando separantur, ex quibus fit generatio et corruptio. Et cum ita sit, substantie non erit generatio et corruptio, cum diffinitio elemen-

8 concedunt] + et *Z* | quia] quod *YZ* | generationi … corruptioni] -tionem … –tionem *Z* 9 esse] cause *WY* | ergo] igitur *WX* 12 modum] mundum *Z* 13 sit] sed fit *YZ* : sic *J* 15 consumuntur et] *om. BJ* : sumuntur et *Z* 18 nunc] nec *Y* : non *Z* : vere *B* | et] sed non *Z* 20 in corruptionem] in -tione *X* : et –tione *Y* | in generationem] *om. JZ* 21 idest] + et *Z* | contingit] -get *Z* | hoc] *om. BWXY* 25 dispositionem] -tione *B* 26 dixit[2]] -ximus *W* 27 ex] quod *YZ* | quod[1]] *om. J* 28 diximus] -cimus *BX* : ut diximus *Z* 29 transmutatur] -tantur *BXY* | ex] in *Z* 30 pueritiam] -tia *B* 31 Deinde] + dixit *BWXY* | eis hoc] *inv. BXY* 32 nos] *om. Z* 33 iste mundus] *inv. BX* 34 et corruptio] *om. B* 35 substantie] -stanti *Z* | erit] est *Z*

torum et ordinatio sit eadem apud congregationem et separationem; videtur igitur quod intendebat dicere cum dixit *Non erant ex eis dispositiones diverse*, idest forme diverse; et signum est huius quod dixit quod ex eis erit una ordinatio.

Et potest intelligi quod dispositiones ex quibus transmutantur elementa non erunt infinite in specie, sed ex eis erunt dispositiones eedem in specie, et diffinitione que semper iterantur in elementis secundum reciprocationem; et intendebat per hoc quod si dispositiones essent infinite in specie, contingeret ut omnis dispositio earum esset corruptibilis simpliciter, cum numquam revertatur; quasi igitur *73vb* vult demonstrare quod contingit sustinentibus opinionem hanc | quod individualis mundus sit eternus in substantia et in accidentibus eternus in specie. Et dixit hoc quia si accidentia eius non essent non **T89a** eterna in specie, possibile esset dicere ipsum aliquo modo corruptibilem; et necesse fuit ut illa accidentia sint eterna in specie cum illa accidentia sint opposita, et iteranda super mundum ut congregatio et separatio quam dixit dicens istius sermonis. Et ipse innuebat cum dixit *Et ipsi concedunt hoc quia ponunt* etc., idest quia ipse Empedocles concedit istas dispositiones contrarias, et ponit dispositionem quam dicit generationem causam aliam a causa quam ponit dispositioni, quam dixit corruptionem, idest quia ponit causam generationis amicitiam et causam corruptionis litem.

Et cum narravit quod contingit eis quod elementa sint eterna individua in substantia eterna specie in dispositionibus, incepit demonstrare quod tale non debet dici absolute, neque esse generationem mundi neque corruptionem, et dixit *Dicamus etiam quod si corpus* etc., idest potest intelligi ex isto sermone quod si corpora ex quibus mundus componitur fuerint neque generabilia neque corruptibilia; et hoc intendebat cum dixit *Continuum totum*, et huiusmodi corpora aliquando inveniuntur secundum aliquam dispositio-

36 sit] fit *BX* | separationem] -parant *B* 37 erant] erunt *Z* 38 diverse[2]] *om. WZ* | est huius] *inv. JYZ* | quod] quid *BY* 41 infinite] -nita *W* 43 reciprocationem] replica- *WZ* 44 specie] -ciem *B* 45 revertatur] ever- *YZ* | quasi] quod *Z* 46 opinionem hanc] *inv. JZ* 48 in] *om. W* | quia] quod *YZ* | eius] *om. B* | non[2]] *om. J* : *exp. W* 50 illa[1]] alia *B* : nulla *Y* 50-1 eterna—sint] *om. homoiotel. X* 50 specie] -ciem *B* 52 ipse] ipsum *YZ* | innuebat] invenie- *B* 53 ponunt—quia] *om. homoiotel. W* | idest] *om. Z* | quia[2]] quod *BXY* | ipse] *om. J* : ipsi *B* 56 dixit] -cit *XYZ* 57 et] *om. WXZ* 58 sint] sunt *JZ* 62 etc.] *om. XZ*

nem et aliquando secundum aliam et semper reciprocentur, necesse est ut mundus sit eternus. Visum est igitur quod intendebat, cum dixit *Et erit compositio mundi et celi secundum quamcumque dispositionem sit*, idest ex hiis duabus dispositionibus reciprocantibus, scilicet ut intelligatur per dispositionem compositio que est in ipso mundo.

74ra Et possumus intelligere hic per disposi|tionem aliquam aliam intentionem, scilicet compositionem corporum ex quibus constat mundus, non compositionem mundi ex eis, idest dum non dissolvitur compositio illorum corporum simplicium ex quibus fit mundus secundum quemcumque modum sit compositio eorum, scilicet aut ex materia et forma, aut ex partibus indivisibilibus; hec autem corpora non transmutantur in dispositionibus nisi propter continuationem aliquando et separationem aliquando; contingit enim ex hoc quod mundus sit eternus, neque generabilis neque corruptibilis in substantia. Deinde dixit *Sed tamen dispositiones eius sunt que corrumpuntur*, et potest intelligi compositio corporum simplicium sicut diximus adinvicem et per compositionem apud congregationem et segregationem quas dixit Empedocles.

Et possumus intelligere per dispositiones congregationem et segregationem et per compositionem compositionem corporum simplicium in se, non cum adinvicem; hec enim compositio contingit eis **T 89b** ut sit eterna, cum non sint hic corpora simplicia in que fiet dissolutio istorum corporum simplicium ex quibus mundus componitur. Necesse est igitur ut compositio corporum ex quibus mundus componitur sit compositio necessaria: compositio autem mundi ex illis corporibus non est necesse apud eos ut sit necessaria; sed dissolutio huius compositionis non est generatio neque corruptio, cum generatio et corruptio apud omnes non est nisi in compositione corporum ex quibus mundus componitur; et propter hoc contingit eis secun-

68 erit] fuerit *J* : eat *Y* 73 compositionem corporum] *inv. Z* 74 compositionem] -tione *B* 76 modum] mundum *WY* | sit] fit *YZ* 81 tamen] *om. J* : cum *Y* | eius sunt] *om. Z* | corrumpuntur] -pitur *B* 83 et—compositionem] *om. JXY* 84 segregationem] separatio- *JW* 84-5 quas—segregationem] *om. homoiotel. BY* 85 segregationem] congre- *W* 87 cum] est *YZ* : tamen *J* | enim] igitur *W* 88 in] ut *Z* 90 corporum] + simplicium *J* 92 necesse] + est *Z* 94 apud] + eos *J* | est] sit *J*

dum hanc opinionem ut mundus non sit generabilis et corruptibilis in substantia, sed in accidentibus.

74rb12 **108.** [109.] *Dicentes vero quod mundus sit corruptibilis et non revertitur* 280a23 *omnino ad generationem, respondebimus eis quod si mundus sit unus, impossibile est ut sermo eorum sit verus omnino, quia antequam mundus fuit, fuit illud ex quo componebatur; et nisi illud esset ante* 280a25 *generationem, non diceremus quod illud alteraretur in ipsum. Si igitur habuit potentiam ut alteretur, possibile est ut mundus ex eo revertatur; si autem mundus non sit unus sed plures, tunc forte verum dicunt de mundo. Et nos dicemus post utrum est possibile aut non.* 280a28

Modo vult contradicere opinioni dicentium mundum esse generabilem et corruptibilem corruptione eterna sine generatione, et dixit *Dicentes vero quod mundus* etc., idest dicentes autem quod mundus corrumpetur post generationem et quod numquam revertetur ad generationem mentiuntur; et sic destruemus sermonem eorum, scilicet quod mundus non corrumpitur nisi in ea ex quibus generatur. 74va | Et hoc erit cum fuerit unus individuo: necesse est ut revertatur ad generationem cum hanc dispositionem habebat ante generationem. Ea enim possibilia sunt ut ex eis sit mundus post corruptionem sicut post generationem vel ante, et nisi essent possibilia ut ex eis fieret mundus, non esset mundus; cum ergo posuerimus hoc esse possibile post corruptionem quod ante est possibile, necessario exibit in actu, **T90a** et si non, possibile reverteretur in impossibile; necessarium est ut generatio mundi revertatur; et hoc intendebat cum dixit *Et nisi illud esset ante generationem* etc., idest nisi illud ex quo generatus fuit esset possibile ut mundus sit ex eo ante generationem mundi, non diceremus in eo quod alteraretur et transmutaretur in mundum.

---

108. [109.] 1 revertitur] -tetur *BJ* 3 quia] *om. BXY* 6 habuit] -buerit *W* : -buimus *X* 7 mundus] *om. W* 8 post utrum] *inv. Y* : utrum *Z* | est possibile] *inv. Z* 10 sine] sive *Z* 11 Dicentes] -cente *B* 12 corrumpetur] -pitur *J* : -pere *X* | numquam] ante- YZ_3 15 hoc] + etiam *J* | unus individuo] individui unus *B* : unus individuus *Z* 16 hanc] habeat *Z* | habebat] -bebant *B* | ante] autem *Z* 18 ex] *om. B* 19 mundus^2] mundum *B* 20 necessario] -cesse *B* 21 revertere-tur] -tetur *BWY* | necessarium est] *inv. Z* 23 esset^1] esse *B* 25 alteraretur] alteretur *XYZ* | transmutaretur] -mutetur *XYZ*

Deinde dixit *Si igitur habuit potentiam ut alteretur* etc., idest si igitur illud aut illa ex quibus mundus generatur habebant potentiam alterationis in mundum, necessario alterabantur in ipsum post corruptionem sicut ante generationem, mundus ergo revertetur post corruptionem multotiens. Deinde dixit *Si mundus non sit unus* etc., idest si autem mundus corrumpatur secundum mutationem ex hac forma in formam illi contrariam diffinitione, ista autem forma secunda remaneat eterna, aut transmutetur in formam ei et prime contrariam scilicet tertiam, et sic in infinitum numquam revertetur illa forma. Deinde dixit *Et nos dicemus post* etc., hoc quidem iam dixit superius scilicet quod impossibile est ut mundi sint plures uno et quod impossibile est transmutari in formam contrariam sue forme; et forte rectum est reverti ad sermonem de hoc et complere contra-
74vb dictionem huius po|sitionis ex propriis illius, scilicet eius qui ponit plures mundos diversos in forma; et si declaratum est hoc esse impossibile, cum sit impossibile esse corpora simplicia alia ab hiis corporibus, et impossibile est ut ex eis componatur mundus compositione alia ab hac compositione.

109. [110.] *Sunt autem alii qui dicunt Possibile est ut aliquid non* *280a28*
generetur, tamen corrumpitur, et possibile est ut aliquid generetur et non
corrumpatur, ut dictum est in Timeo quod celum est generatum et est *280a30*
perpetuum ceteris temporibus sine corruptione et consumptione. Nos vero
dicimus quod isti non celum disponunt nisi tantum sermone naturali. Et
nos perscrutabimur de huiusmodi verificatione perscrutatione continente
omnia. Si enim hoc fecerimus, manifestabitur tunc nobis utrum celum **T90b**
sit generatum sine corruptione. *280a34*

Cum demonstravit mundum non esse generabilem aut corruptibilem, vult perscrutari utrum cum hoc sit in eo potentia ad corruptio-

26-7 igitur illud] *inv.* *B* 29 sicut] si *W* 30 Si] + autem *JZ* | non] *om.* *B* 31 secundum] + hanc *Z* 33 remaneat] -net *BX* | ei] eius *B* 35 illa] ista *BJ* 36 dixit] -ximus *W* 38 rectum est] *post* hoc *Z* | complere] -pleri *BX* 39 illius] illis *YZ* : illi *X* 40 esse] est *BZ* 42 et] *om.* *YZ* : non *X* 43 alia] illa *Y*

109. [110.] 2 est] *om.* *Y* 3 Timeo] Taneho *Y* : Tymeo *X* | est generatum] *inv.* *BJ* 4 perpetuum] + in *BJX* 5 naturali] + sive particulari *Z* | Et] *om.* *JZ* 6 huiusmodi verificatione] *inv.* *Z* 7 hoc] hec *XZ* 9 demonstravit] declara- *W* | non] *om.* *Z* | generabilem] demonstrabi- *B* 10-1 ad—potentia] *om. homoiotel.* *Y*

75ra nem; quoniam | qui concedit quod in eterno sit potentia ut corrumpatur, possibile est apud eum ut aliquid generetur et remanet post eternum vel perpetuum, ut Plato in Timeo; tunc enim habebit potentiam ad corruptionem absque corruptionem, quoniam antequam generabatur erat corruptibile in actu. Cum igitur generatum fuit, corruptibile est in potentia, et similiter possibile est secundum hanc opinionem ut aliquid quod in preterito semper fuit, in futuro tamen corrumpatur; potentia enim corruptionis in eo tempore infinito fuit scilicet preterito, deinde exibit in actum post tempus infinitum; et dixit, in sermone dicentis quod celum est generatum et non corruptibile, quod non disponebant celum dispositione naturali, quoniam manifestum est per inductionem universalem quod omne generatum est corruptibile.

Et dixit *Nos autem perscrutabimur de hoc perscrutatione continente omnia*, quia qui voluerit perscrutari de hoc in celo, et dimiserit perscrutari de hoc in eo quod est universalius scilicet in omnibus entibus, est ille qui perscrutaretur de tribus angulis quod deberent esse equales duobus rectis in triangulo equilatero, non de triangulo simpliciter inquantum triangulo; declaratum est enim quod tales propositiones non sunt prime neque essentiales; ideoque dixit *Cum igitur hoc fecerimus*, idest cum perscrutabimur de hoc per propositiones essentiales.

75rb22 **110.** [111.] *Sed antequam hoc faciamus, dividamus prius intentiones* *280b1* *generationis et corruptionis, et declaremus quid est dicere generabile aut non generabile, aut non corruptibile aut corruptibile, quoniam cum idem fuerit recipiens diversas intentiones que non distinguuntur, et aliquis consideraverit illud sine distinctione et sine divisione, stupefiet in intellectu, et dicet de eo sermonem sine divisione et distinctione necessa-* **T91a** *rio, nesciet enim de cuiusmodi natura dicitur illud, et cuius intentione.* *280b5*

---

12 remanet] -neat *JYZ* 13 enim] + non *B* 15 generabatur] -bitur *W* | igitur] enim *Z* 16 est[1]] *om. WXY* 17 ut] quod *J* | in[1]] *om. B* 20 dicentis] -centium *BJ* : -centi *Y* 25 dimiserit] -misit *YZ* 26 eo] hoc *YZ* | universalius] unus *YZ* 27 qui] quod *YZ* : quasi *X* | deberent] -bent *JW* 28 equilatero] -latere *Z* : -angulo *Y* 29 triangulo] –gulus *J*

110. [111.] 2 et[2]] *om. W* | quid] quod *Y* : quia *W* 3 non[2]] *post* aut[2] *Z* 4 distinguuntur] -guantur *B* 5 consideraverit] -ravit *W* | et—divisione] *om. B* | sine] *om. Z* 6 distinctione] + et *B* 7 illud] *om. Z*

Quia vult perscrutari utrum aliquid sit eternum et deinde post
corrumpatur, aut aliquid sit generatum deinde post remaneat eter-
75va num, ge|nerabile autem et corruptibile et non generabile et non cor-
ruptibile dicuntur multis modis, vult primo determinare secundum
quot modos dicuntur hec nomina, deinde ostendere illud quod
intendit, quoniam si hoc non faceret, proveniret error sicut dixit. Et
dixit *Cum enim idem fuerit recipiens plures intentiones*, potest intelligi
per idem dictio, et quasi diceret quoniam cum aliqua dictio signifi-
cet plures intentiones; et potest intelligi per idem subiectum istius
questionis, quoniam cum idem subiectum fuerit diversis intentioni-
bus, et non distinguantur ille intentiones et accipiantur quasi
eedem, necessario apparebit illud subiectum habere diversas disposi-
tiones quoniam tunc erunt propositiones non eedem, ut dictum est
in Peri Hermenias, et hec explanatio est quasi apertior. Deinde dixit
Nescit enim etc., idest quia nescit de cuiusmodi natura dicitur una-
queque earum intentionum cum ille intentiones dicantur de naturis
diversis.

75vb **111.** [112.] *Dicamus modo quod aliquid dicitur non generabile multis* *280b6*
modis. Quorum unus est cum primo fuerit in non esse et deinde post fuit
in esse sine generatione et sine alteratione, ut quidam dicunt de tactu et
motione: dicunt enim quod tactus et motio non sunt generabiles. Dicitur
etiam alio modo cum antequam fiant est possibile ut fiant, sed non fiunt *280b10*
nisi per generationem. Et dicitur etiam non generabile alio modo cum
fuerit impossibile generari et sine potentia in eo ad generationem, et non
sit aliquando, et aliquando non sit. *280b12*

---

8 Quia] qui *Z* 9 remaneat] -net *B* 10 autem] *om.* Y | et[1]] + non *Z* 11 dicuntur] -centur *B* 12 quot] hoc *B* | ostendere] -det *BX* : -deret *Y* 13 faceret] -cerit *B* | proveniret] per- *JWY* | dixit] -cit *B* | Et] *om. YZ* 14 dixit] *om. WYZ* | idem] *om. Z* | fuerit] fuit illud *Z* 15 dictio[1]] dictum *Z* : ductio *B* 15-6 dictio[1]—idem] *om. homoiotel. X* 18 ille] illius W_2Z | et[2]] ut *YZ* | quasi] quod *Z* 19 diversas dispositiones] *inv. J* : dispositiones *B* 21 Peri Hermenias] perãr̃ *B* : periar̃ *X* : pervatm̃ *Y* | hec] huic *YZ* | explanatio est] *inv. B* | quasi] quod *Z* 22 natura] non *B* 23 intentionum] + sunt *Z* : + et *X*

111. [112.] 1 modo] *om. Z* | dicitur non] *inv. B* 2 est] *om. X* : + unus est *Z* | post] *om. BJ* 5 cum] quod *W* | fiunt] fuerint *Y* : fit *J* 8 sit[1]] fit *XY* : fuit *B* | non] *om. JYZ*

Intendit per esse formam, et per non esse privationem forme, quasi igitur diceret Dicitur non generabile cum aliquid fuerit prius privatum deinde post fuit ens, cum exitus eius ex privatione ad esse fuerit sine transmutatione et sine alteratione, scilicet quod generatio T91b eius fit sine tempore. Et hoc intendebat cum dixit *Sine generatione et sine alteratione, sicut dicunt quidam de tactu et motu illos non esse generabiles*, quia caruit eo quod necesse est in generatione, scilicet quod exitus eorum sit a non esse ad esse per transmutationem et in tempore, et illud quod plus invenitur huiusmodi est in rebus relationis. Et manifestum est quod hoc non dicitur non generabile nisi per similitudinem, quia omne quod post non esse est necessario est generatum; et apparet ex hoc quod non dicimus ipsum non generatum proprie nisi apud suum esse; Themistius autem dicit quod dicitur in ambobus scilicet apud esse et non esse.

76ra Deinde dixit *Dicitur etiam alio modo* etc., idest et dicitur non generatum de aliquo non ente sed quod possibile est esse per transmutationem et in tempore; et dixit hoc quia illud quod non est movens apud suum non esse potest dici non generabile, sed tamen conditionatur in eo ut sit possibile esse ad differentiam eius ad impossibile; impossibile enim dicitur semper non generabile, hoc autem non dicitur non generabile nisi in aliquo tempore. Sed tamen hoc non generabile, quod dictum est hic, non opponitur generabili, quapropter visum est quod non numeravit hanc intentionem de intentionibus non generationis et dimisit intentionem de qua dicitur non generabile apud suum non esse cum fuerit possibile, sed suum esse est sine causa generationis, quoniam cum fuerit ens non dicetur generabile sed etiam dicetur non generabile, quasi igitur vult hic numerare de modis non generabili illud quod in sua compositione dicitur generabile, cum illud de quo vult hic perscrutari sit huius-

9 per^{2}] *om. W* 10 Dicitur] + enim *Z* 11 fuit] fuerit *BX* | ex] a *JY* 12 transmutatione] -sitione *Z* | sine^{2}] *om. B* 13 fit] sit *JX* | Sine^{2}] *om. YZ* 14 dicunt quidam] *inv. BX* 15 caruit] -ret *JZ* : -rent *XY* 16 sit] est *J* : fiat *B* | a] *om. BWXY* 17 relationis] -solutionis *B* 19 post] potest *WY* | est] *om. WYZ* 21 suum] + non *J* | dicit] -xit *BJ* 22 et] *om. B* 25 et^{1}] *om. Z* | non est] *inv. BX* : est *WY* 26 movens] ens *JX* 28 enim dicitur] *inv. B* | semper] *om. BX* 29 tamen] cum *Z* : in *Y* 31 non] affirmative *J* : cum *Z* 32 dimisit] divi- *JZ* 34 dicetur] -ceretur *Z* : -cere *B* 35 quasi] + diceret *WY* | igitur] + diceret *Z* 36 numerare] innum- *B* | modis] modo *X* : mundo *B* | generabili] -bilis *J* : -bilibus *Y*

modi generis. Themistius autem dicit quod non dixit possibile in eo nisi ad distinguendum inter ipsum et impossibile; et intendit per possibile grave possibilitatem.

Deinde dixit *Et etiam dicitur non generabile alio modo cum aliquid fuerit impossibile ut generetur*, idest et dicitur etiam de aliquo ente esse non generabile cum fuerit impossibile generari, quia caret potentia ad generationem, et hoc est quia aliquando invenitur ens et aliquando non ens. Et hoc intendebat cum dixit *Sine potentia que sit in eo ad generationem et non sit aliquando et aliquando non sit*, et addidit T92a
hanc conditionem quia aliquod ens non est impossibile generari nisi cum in eo non fuerit potentia ad generationem; et hoc est de quo
76rb vere dicitur | non generabile, scilicet quod numquam fuit non ens neque habet potentiam ad non esse. Et secundum hoc non intelligitur ex non generabili impossibile sicut intellexit Themistius, sed dignius ens, ut videmus quod Themistius intelligit ex hac tertia intentione impossibile.

Hec enim intentio tertia dividitur in duo, quorum alterum est quod in eo non sit aliqua potentia omnino, et hoc est impossibile, alterum autem est ut non sit in eo potentia nisi difficilis, sed ipse Themistius dicit quod hanc intentionem secundam earum duarum divisionum oportet intelligere ex secunda intentione prime in qua scilicet dixit *Et dicitur etiam non generabile alio modo*, et fingit quod Alexander explanabat hunc sermonem secundum intentionem qua nos explanavimus idest res generabilis antequam generetur.

Et cum explanatio fuit secundum Themistium, tunc non generabile dicetur tribus modis: dicitur enim de eo quod fit absque causis generationis, et de difficili generationis tamen possibilis, et de impossibili generationis, et secundum hoc dividitur ex eis de quibus dicitur non generabile illud quod est magis famosum dici, et est illud quod

38 dicit] -xit *JWY* 40 grave] -vem *XYZ* : + per *WY* 41 etiam] *om. Z* 42 ut] *om. JYZ* | generetur] -rari *JY* 42-3 idest—generari] *om. YZ* 44 quia] quod *BZ* 45 sit] fit *YZ* 46 non[2]] *om. JYZ* 47-8 generari—generationem] *om. WYZ* 48 non] *om. BX* 49 vere dicitur] *inv. Z* 50 Et] + hoc *W* 51 ex] de *W* | intellexit] vult lex *BX* 52 ens] est *Z* | intelligit] -lexit *JW* 55 sit] fit *W* 55-6 et—potentia] *om. Z* 56 potentia] + omnino *W* : + aliqua omnino *Y* 57 secundam] -dum *XYZ* 58 divisionum] -sionem *YZ* | secunda] sua *Z* 62 cum] *om. WZ* | fuit] fuerit *BJX* | non] *om. Z* 63 dicitur] -cetur *WXY* 64 de[1]] *om. BX* : dici *Z* | de[2]] dum *Z* : deinde *B* : cum *Y* 65 secundum] *om. B* | dividitur] -mititur *J* : -minuitur *BX* | de] ex *W*

intenditur scilicet eternum quod numquam fuit generatum, et etiam diminuitur ex hoc illud quod non est generatum tamen possibile est facile generari. Illud enim de quo dicitur non generabile aut vere dicitur aut non vere dicetur: si non vere dicetur de generato absque causis generationis, et de eo quod est difficilis generationis, dicitur enim non vere, quia collocatur sub diffinitione generati, quod est ens post quam ens non fuit; de hiis igitur duobus dicitur non generabile non vere. Vere autem similiter dicitur de aliis duobus scilicet de non gene-
76va rato in actu idest in aliquo tempore, et de non gene|rato simpliciter sed neque in aliqua hora; et hoc quidem dicitur de eo quod semper est et dicitur de eo quod semper non est: triangulus enim non est generatum simpliciter, et habere plus quam tres angulos est non generatum etiam simpliciter in rei veritate; ergo sunt quinque modi.

Videtur ergo quod forte Aristoteles aggregavit ens simpliciter, et non ens simpliciter sub tertio modo, et aggregavit non ens in aliquo tempore quod est difficilis generationis, et facilis generationis sub secundo modo, et secundum hoc erunt intentiones quas dixit tres, tamen vere sunt quinque; et hoc fecit propter breviloquium.

112. [113.] *Similiter dicitur generabile multis modis. Quorum unus est,* *280b14* T
cum aliquid fuerit impossibile generari, deinde fuit in postremo non ens, aut per generationem aut sine generatione, et est aliquando ens et aliquando non ens. Et dicitur etiam generabile alio modo, cum aliquid quod non fuit est possibile generari. Et etiam dicitur generabile alio modo, cum aliquid non fuit et deinde fuit per generationem. *280b20*

76vb Cum finivit de numeratione intentionum de quibus dicitur non generabile, vult dicere de quibus dicitur generabile. Et quia una intentionum de quibus dicitur non generabile est illud quod est pos-

70 non[1]—si] numere sed *Y* 72 quia] quod *Z* | est] *om. WYZ* 73 ens—fuit] fuit non ens *BX* | de] ex *W* | igitur] ergo *BX* 74 Vere[2]] *om. YZ* | aliis duobus] *inv. Z* | de[2]] *om. Z* 76 sed] scilicet *BX* 77 non[2] est] *inv. JZ* 79 sunt quinque] *inv. YZ* 80 aggregavit] congre- *JZ* 80-1 ens—aggregavit] *om. homoiotel. B* 81 aggregavit] congre- *J*

112. [113.] 1 Similiter] simplic- *Y* : Et *praem. J* | dicitur] + non *Y* 2 in] *om. Z* | non] *om. Z* : *exp. BW* 5 est possibile] impossibile est *Z* | dicitur generabile] *inv. YZ* 6 fuit[1]] fuerit *XY* 7 finivit] finium *B* : fiunt *Z sed corr.* Z_2 : + dictum *J* 8 vult—generabile] *om. homoiotel. BZ* 8-9 vult—generabile] *om. homoiotel. Y* 8 dicitur generabile] *inv. X* | Et] *om. B* 9 non—est[2]] *om. B* sibilis generationis quod adhuc non est; hoc autem est duobus modis, aut fiendum per causas generationis aut sine causis eius; et quia hoc est, oportet ut hoc nomen generabile dicatur etiam de oppositis harum duarum intentionum scilicet de eo quod est in actu post non esse absque causis generationis, et de eo quod est in actu cum causis generationis.

Et ideo incepit et collocavit has duas intentiones in uno sermone, et dixit *Et similiter dicitur generabile multis modis, quorum unus est cum aliquid fuerit impossibile generari, deinde fuit in postremo non ens, aut per generationem aut sine generatione*, idest quia generabile dicitur de eo quod fuit in actu postquam non fuit, deinde post non est; et hoc est duobus modis quorum unus est ut fiat sine causis generationis, alius cum causis. Et dixit *Cum aliquid fuerit impossibile generari*, idest et non dicitur de istis esse non generabile nisi secundum statum in quo non est generandum scilicet cum fuerit in actu, quoniam cum aliquid fuerit completum in generatione, impossibile est generari. Et ille due intentiones in quas prius divisum fuit non generabile sunt contente hic in hac prima divisione; et utraque istarum divisionum dicentium sunt opposita duabus intentionibus de quibus dicitur non generabile scilicet prime sue explanationis et secunde. Et possibile est generabile et non generabile ex eis dici de una intentione sed diversis modis: generatum enim absque causis
77ra generationis possibile est dici apud suum esse | generabile et non generabile insimul: non generabile vero quia est sine causis generationis, generabile autem quia est postquam non fuit; sed tamen melius est ut non dicatur non generabile nisi apud non esse et generabile apud esse; quapropter dixit *Cum aliquid fuerit impossibile generari* etc.

Et cum dixit hanc intentionem huius nominis generabile que opponitur duabus primis intentionibus huius nominis non generabile, T93a dixit etiam secundam intentionem generabilis et dixit *Et etiam*

12 etiam] *om. YZ* : et *BX* 13-4 de—generationis] *om. Y* 18 aliquid] *post* impossibile *YZ* : aliud *X* 20 deinde] dictum *YZ* | post] primo *Y* 21 est[1]] esse *WYZ* 22 aliquid] –quod *Y* 23 de] in *XY* 24 scilicet] sed *YZ* | fuerit] fuerint *B* 26 prius divisum] *inv. Z* 28 divisionum] divisio *XY* 29 non generabile] ingenerabile *B* 30 Et[1]] *om. B* 30-2 de—dici] *om. homoiotel. Y* 34 quia] *om. X* : quod *B* 35 non[2]] *om. BY* | generabile[1]] *om. B* : signabile *Y* 39 opponitur] compo- *B* 40 generabilis] -bile *Z*

dicitur generabile alio modo etc., idest dicitur generabile etiam de eo quod est generatum potentia, non generatum actu; et Themistius intelligit ex hac intentione illud quod est difficilis generationis, et dixit quod hec intentio secunda quam dixit ex hoc nomine generabile eadem est cum opposita secunde intentionis quam dixit ex eis de quibus dicitur non generabile scilicet ei quod est difficilis generationis secundum suam sententiam non secundum sententiam Alexandri.

Deinde dixit tertiam intentionem, *Et dicitur generabile alio modo*, etc., idest et dicitur generabile de eo quod est generabile in actu vero, et est illud quod prius fuit non ens deinde fuit ens via generationis. Et estimatur de hoc esse collocatum sub primo dividente; et quod dixit in eo cernit aut per generationem aut sine generatione; tamen videtur quod non vocavit ipsum hic generabile nisi inquantum ens est post non ens secundum viam generationis, non inquantum est ens post non ens, quod intendebat in primo modo; et Themistius dixit quod intendebat per hoc illud quod necessario generatur; sed tamen visum est quod illud quod necessario generatur non fit secundum viam generationis sed per alterationem.

77rb14 **113.** [114.] *Et similiter corruptibile dicitur multis modis. Quorum unus* 280b20
est, cum aliquid fuerit primo ens et postremo non ens sive per alterationem sive non. Et forte dicitur corruptibile alio modo, cum aliquid fuerit possibile ut non sit non ens per corruptionem; et dicitur etiam corruptibile alio modo, cum aliquid fuerit velocis corruptionis, ut illud quod est facilis corruptionis. 280b25

41 etiam] *post* eo *W* : et *J* 42 est generatum] *inv. J* : generatum *B* : generatu *X* 43 intelligit] -lexit *BJ* 44 generabile] + verum *Y* 46 ei] *om. WYZ* 48 intentionem] + et dixit *J* 48-9 alio—generabile] *om. homoiotel. B* 49 vero] vere *J* : retro *B* : natura *Y* 50 prius] primum *W* | ens[1]] + et *B* 51 esse] omne *B* | primo] -ma *Z* 52 cernit] *om. BX* : crevit *Y* | aut[1]] *om. YZ* 53 ipsum] + esse *Y* | hic] hoc *JZ* 54 est[1]—ens] *om. homoiotel. WXYZ* 54-5 secundum—ens[2]] et *B* 54 non[2]] *om. Y* 55 quod] + non *W* | modo] *om. W* 56 necessario generatur] *inv. W* 56-7 sed—generatur] *om. homoiotel. YZ* 57 tamen] *om. X*

113. [114.] 2 primo] post *BZ* 3 dicitur] -cetur *Y* : vocabimus *J* 4 etiam corruptibile] *inv. B* 5 illud] quo *B* 6 facilis] fallacis *B* : facit *Y*

Idest quorum unus est cum aliquid fuerit prius ens deinde post fuit non ens; dicitur enim in eo apud non esse suum consequens esse T93b
quod est corruptibile et quod iam corruptum est, sive suum non esse fuerit per alterationem, scilicet corruptione vera aut sine alteratione; et hoc est corruptum in actu.

Deinde induxit secundam intentionem et dixit *Et forte vocabimus aliquid corruptibile alio modo* etc., idest et forte dicemus corruptibile cum aliquid fuerit possibile non esse per aliquam causam causarum corruptionis, ad demonstrandum quod cum hoc nomen corruptibile
77va dicitur de eo quod est in potentia | huiusmodi, scilicet quod intrat ipsum non esse sine causis corruptionis, dignius est dici de eo quod est in potentia non ens cum fuerit non ens per causas corruptionis; et Themistius, secundum quod reputo, intelligit ex pronuntiatione hac corruptionis cum corrumpitur sine alteratione et sine corruptione sicut corruptione fluvii magni cum transmutatur de loco in locum.

Deinde dixit tertiam intentionem, et est illud quod non est corruptum in actu sed est facilis corruptionis (de hoc enim dicitur etiam corruptibile), et dixit *Et dicitur etiam corruptibile alio modo* etc., intendit dicere quod illud quod est in potentia dicitur duobus modis de difficili (et hoc raro et in similitudine remota) et de facile corruptibili (et est hoc magis in usu). Et abbreviatio huius est quod corruptibile dicitur de corrupto in actu et in potentia: in actu dicitur de eo quod corrumpitur per alterationem sue substantie et transmutationem, et de eo quod corrumpitur per transmutationem alterius et eius transmutationem; in potentia dicitur de eo quod est difficilis corruptionis, et de eo quod est facilis corruptionis.

---

7 est] + etc. W_2Z | aliquid] *om. Z* 8 suum consequens] *inv. B* 10 corruptione vera] –nem -ram *Z* 11 corruptum] -ruptio *Y* : corpus *Z* 12 Et^{2}] quod *Z* 13 dicemus] -cimus *XY* : + quasi *W* 16 quod^{1}] + non *B* 18 est] *post* ens^{1} *B* 20 hac] huius *J* 21 corruptione] -ruptio *JY* | in] ad *JZ* 23 intentionem] mutatio- *Z* 24 enim] etiam *B* | etiam] *om. B* : et *Z* 25 etiam] *om. YZ* 28 est hoc] *inv. BJ* | quod corruptibile] *inv. B* 31 transmutationem] –sumptionem *Z* : alterationem *J* 32 transmutationem] –sumptionem *Z* 33 et—corruptionis] *om. homoiotel. WYZ*

77vb8 **114.** [115.] *Et similiter dicitur illud quod non cadit sub corruptione* *280b25*
multis modis. Quorum unus est, cum aliquid fuerit aliquando ens et aliquando non ens, ut tactus: aliquando enim est ens et aliquando non ens absque esse sub corruptione. Et dicitur etiam non corruptibile alio modo illud quod est ens et habet potentiam ad non esse, vel illud quod in
futuro erit non ens modo autem est ens, et est secundum hoc cadens sub *280b30*
corruptione, in futuro enim erit non ens. Modus vero verus, scilicet quod non cadit sub corruptione, est ut illud quod modo est ens et non habet potentiam ad corruptionem ita quod non sit modo ens et post non ens. Et etiam dicitur non corruptibile alio modo, cum non corrumpitur et possibile est in postremo ut sit non ens sine causa corruptionis. Et etiam dici-
tur non corruptibile alio modo, cum aliquid non cadit velociter sub cor- T94a
ruptione, sicut illud quod non est facilis corruptionis. *281a1*

Idest dicitur non corruptibile illud cuius corruptio est sine causis corruptionis, sicut dicitur non generabile illud cuius generatio est
78ra sine causis gene|rationis. Iste autem modus potest dici corruptibile et non corruptibile duobus modis diversis: corruptibile vero inquantum non est post esse, non corruptibile autem quia suum non esse non est per causas corruptionis. Ista autem sunt, sicut dixit, que aliquando sunt et aliquando non sunt, quorum esse non est per causas generationis et quorum non esse neque per causas corruptionis verbi gratia sensus tactus: sensus enim tactus non est transmutatio ut fiat aut corrumpatur in tempore, sed est consequens transmutationem, unde est non in tempore; et similiter omnia elementa.

Deinde dixit *Et dicitur etiam non corruptibile alio modo*, idest et dicitur non corruptibile de eo quod est corruptibile in potentia et est

---

114. [115.] 1 illud] *om. J* 2 et] *om. JZ* 3 ut—ens[3]] *om. homoiotel. X* 4 etiam] + illud *Z* | non] *om. WY* | corruptibile] in- *WYZ* : + illud *VWXY* 6 erit—ens[1]] non est *J* | modo] modus *Y* 6-7 modo—ens] *om. homoiotel. X* 6 ens[2]] *om. JY* 7 erit] *om. J* | vero] autem *JZ* | quod] quia *XY* 8 et] + tamen *B* 9 ens[2]] est *Y* : est ens *Z* 10 etiam dicitur] *inv. JZ* | cum] + res *J* : + aliquid adhuc *Z* 10-2 non[2]—cum] *om. homoiotel. BWXY sed suppl.* W_2 10 corrumpitur] -patur adhuc *J* 11 est] *om. W suppl.* W_2 | Et] Et—corruptionis *(v. 9-11) iter.* W_3Z 12 aliquid] -quis *B* 14 est sine] fuit *Z* 15 dicitur] *post* generabile *B* 16 Iste] ita *YZ* | modus] *om. WYZ* 18 autem] aut *Y* | quia] quod *Z* | non[3]] *om. BX* 19 Ista] isti *B* : ita *Z* | que] quia *B* 20 et] *om. Z* 21 neque] non est *J* 22 enim] *om. X* : autem *Z* 23 consequens] -tingens *Z* 24 elementa] relativa *JX* : relatam *B* 26 dicitur] + et *B* ens in actu et est possibile ut non sit post. Deinde dixit *Aut illud quod in futuro non est modo autem est*, et hec est altera pronuntiatio eiusdem intentionis scilicet corruptibilis in potentia.

Deinde dixit *Et modus autem verus* etc., idest modus autem de quo dicitur non corruptibile veriori nomine est illud quod est ens in actu, in quo non est potentia ad non esse ita quod aliquando sit ens et aliquando non ens. Deinde dixit *Et dicitur etiam incorruptibile alio modo cum aliquid adhuc non corrumpitur et possibile est in postremo ut sit non ens*, et post hoc esse illud quod in postremo est non ens sine causis corruptionis, et differt per hoc ab eo in quo primo dixit cum aliquid fuerit ens et habet potentiam ut sit non ens.

Deinde dixit *Et dicitur etiam non corruptibile alio modo cum ali-*
quid non cadit velociter sub corruptione, et hoc etiam manifestum est
esse quod est aliud condividens a predictis; et abbreviatio huius ser- T 94b
monis est quod incorruptibile dicitur aut quia corruptio eius sit sine
78rb causis corruptionis aut | quia est non corruptum in actu; et possibile
est corrumpi in futuro, et hoc est duobus modis, aut sine causis corruptionis aut cum causis; dicitur itaque de hoc non corruptibile duabus intentionibus, prima scilicet et secunda; tertius autem est verus, et est illud quod est ens in actu et non habet potentiam ad corruptionem neque est possibile ut in aliqua hora sit corruptum; quartus vero est illud quod est difficilis corruptionis.

115. [116.] *Et cum de generatione et corruptione et suis intentionibus* *281a1*
iam diximus aperte, perscrutemur modo de illo de quo dicimus ipsum habere potentiam aut non habere potentiam. Incorruptibile enim verum est illud in quo non est potentia ad recipiendum corruptionem, neque
habet potentiam ut aliquando sit ens et aliquando non ens. Illud autem *281a5*
quod non generatur est illud quod non habet potentiam ut sit tempera-
tum naturaliter. *281a7*

---

27 Aut illud] *inv. B* 28 autem] aut *XY* | hec] *om. B* | altera] alia *BX* 29 scilicet] *om. Z* | corruptibilis] corporalis *Z* 30 Et] *om. JZ* | autem^1] aut *Y* : alter *BX* 31 veriori] verior *Z* : veriorum *Y* | est^2] *om. B* 32 et] *om. YZ* 35 post hoc] potest hic *J* | illud] aliud *J* 37 fuerit] fuit *Z* 39 etiam] *om. JYZ* 40 esse] *om. JZ* 42 quia] quod *Z* 44 itaque] autem *W* 47 quartus] quamvis *Y* 48 illud] *om. BX* : id *J*

115. [116.] 1 et^3] *om. B* 3 aut—potentiam] *om. homoiotel. WYZG* 4 neque] non *BY* 5 sit] + non *Z* 6 temperatum] -ratus *B* : equale *JG*

78va Idest perscrutandum est ante introitum in eis que intendimus quid est dicere Hoc habet potentiam aut non habet; illud enim de quo dicitur vere non corruptibile est quod non habet potentiam ad corruptionem neque etiam est possibile ut aliquando sit ens et aliquando non ens. Deinde dixit *Illud autem quod non generatur* etc., idest illud autem quod vere non generatur est quod non est in potentia in aliquo tempore; et cum ita sit, est igitur illud in quo non est potentia ut sit ens, in quo enim est potentia ut generetur, cum generabitur non habebit potentiam iam ad generationem; illud ergo quod non generatur non habet potentiam ad generationem. Et hoc intendebat cum dixit *Non habet potentiam ut sit equale naturaliter*, idest quia semper fuit equale per naturam; et intendebat per equale naturaliter illud quod est in sua perfectione; et innuit per hoc, secundum quod reputo, opinionem Platonis dicentis quod mundus ante hoc T95a fuit equalis per naturam idest equaliter in potentia, deinde post hoc fuit equalis per naturam. Et est hec ordinatio quam intendebat cum dixit quod mundus prius fuit non ordinatus deinde post venit ad ordinationem; et voluit hic notificare quod mundus si semper fuit ordinatus non est in eo potentia ad ordinationem, et si venit ex inordinatione ad ordinationem, potentiam habet ad ordinationem et non ordinationem.

78vb15 **116.** [117.] *Dicamus modo quod si potentia alicuius fuerit moveri per* 281a8 *centum miliaria et ferre centum harenas, poterit etiam in partibus que* 281a10 *sunt inter ea; oportet ergo nos diffinire potentiam. Dicamus ergo quod diffinitio potentie est ultimum potentie rei: quod enim potest ambulare per centum leucas potest ambulare per duas leucas, et quod potest ferre centum libras potest ferre duas libras. Manifestum est igitur quod poten-*

8 Idest] *om. Y* | ante] autem *Y* | introitum] intextum *B* : innotum *Y* 9 enim] vero *Z* 10 dicitur vere] *inv. JYZ* 13 non[1]] *om. Y* 15 in quo] *om. Z* : in aliquo *Y* | enim] *om. WY* : non W_2*Z* | generabitur] *om. Z* 16 potentiam iam] *inv. JZ* 16-7 illud—generationem] *om. homoiotel. W* 18 naturaliter idest] qualiter enim *Y* 19 equale[1]] -lis *X* : -liter *WY* | equale[2]] -liter *YZ* 20 per] ex *JZ* 21 hoc] + non *JY* 22 equaliter] -lis *WXY* 23 est hec] *inv. JZ* 24 prius fuit] *inv. B* : fuit *X* 25 ordinationem] ordinem *JYZ* | si] non *WYZ* 26 inordinatione] ordine *Z* 27 potentiam habet] *inv. BX* | et non] in *B*

116. [117.] 1 modo] *om. X* : + etiam *J* 2 ferre] forte *W* | harenas] aurenas *BJ* : leucas *J* 5 centum—per] *om. homoiotel. Y* | potest[1]] et post *X* *tia diffinienda est in fine eius quod potest; debile autem diffinitur per* 281a15 *minus quam potest, quoniam si aliquid non potest ferre aliquod pondus, non poterit ferre maius, verbi gratia si non poterit ferre centum libras, non potest ferre centum et unam, et quod non potest ambulare per mille leucas non poterit ambulare per mille et unam.* 281a17

79ra Cum narravit quod oportet perscrutatorem de potentia scire ante quid est dicere Hoc habet potentiam super hoc aut non habet, et utrum (cum de aliquo dixerimus ipsum habere potentiam ad esse) est in toto tempore sui esse aut est in ultimo istius temporis, et similiter utrum cum dixerimus quod aliquid non habet potentiam ad esse, utrum sit in toto tempore sui non esse aut in suo eterno vel suo principio, et dixit *Dicamus modo etiam quod cum potentia alicuius fuerit in moveri per centum leucas* etc, idest et cum manifestum est per se quod si aliquid potest aliquem actum quod potest illud quod est sub eo, manifestum est quod potentia eius non est diffinienda nisi per ultimum illius actionis; et cum ita sit, manifestum est quod potentie rerum non terminantur nisi per suos fines et per eos distinguitur que- T95b libet potentiarum rerum habentium potentias diversas, verbi gratia quod cum voluerimus diffinire potentiam illius quod potest moveri per centum leucas, non dicemus quod habet potentiam moveri per minus quam per centum; et cum ita sit, potentia igitur non diffinitur nisi per finem sui actus, non per illud quod est ante finem.

Et cum demonstravit quid est dicere Hoc habet potentiam, et quod ipse intendit per hoc finem potentie non quemcumque, incepit dicere modum ex quo diffinitur debilitas potentie et suum defectum, et dixit *Debile autem terminatur per minus quam potest*, idest defectus autem potentie terminatur per minimum in posse; cum enim non poterit minus, necessarium est ut non possit maius, econtrario deter-

---

8 potest[2]] poterit *WXY* 9 poterit[2]] potest *BJ* | centum] duas W_2*YZ* 10 potest[1]] poterit *WXY* | et[1]] *om. YZ* | unam] libras *YZ* | mille] tres *Z* 11 mille] tres *Z* 12 perscrutatorem] -scrutari *XZ* | scire ante] ante scire autem *Z* 14 cum] *om. WY* : inde *Z* | dixerimus] -ximus *X* : -camus *Z* 15 toto tempore] *inv. Z* | est] *om. XY* 17 suo[1]] *om. Z* : sui *J* | eterno] *post* suo[2] *WZ* : termino *J* | vel] *om. Z* : + in *BX* 18 dixit] -ximus *Z* | etiam] *om. YZ* 19 et cum] *om. B* | est] + quod potentia *B* 23 et—eos] per eas *B* 28 illud quod] aliquid *B* 29 quid] quod *WZ* 30 quecumque] quamque *Z* : quantum quia *B* 31 modum] -dus *B* | suum defectum] suus –tus *J* 33 autem] *om. JYZ* | posse] possibile *Z* 34 est] *om. B* | maius] minus *WXY*

minatione actionis potentie, scilicet quia potentia diffinitur ex fine sui actus, defectus autem eius diffinitur et primo in posse; cum enim non poterit ferre quinque libras quod est plus, dicemus quod est
79rb impotens | ferre quinque libras et non plus, quoniam si quinque non potest, necessario plus non potest; et cum ita sit, manifestum est cum dicimus Hoc habet potentiam, intendimus terminum potentie idest finem; et similiter cum diximus Hoc non habet potentiam, non intendimus nisi terminum.

117. [118.] *Sic ergo terminus potentie finis eius potentie est ultimum* *281a18*
eius potentie secundum magis. Non igitur estimetis si aliquis dixerit
nobis quoniam forte aliquis contradicet sermoni nostro, quod diffinitio *281a20*
per quam diffinimus potentiam non est necessaria neque recta: qui enim videt aliquid per distantiam unius leuce, potest videre minus ex distantia minori una leuca. Sed contradicit huic sermoni, quia qui potest
videre punctum aut audire minimam vocem potest videre plus quam **T96a**
punctum et audire magnam vocem. Dicamus ergo quod hoc non destruit
79va *nostrum | sermonem de diffinitione potentie, quia diffinitur aut in fine*
rei aut in ultimo sue potentie. *281a25*

Idest sit ergo terminus iste potentie secundum magis quod est finis sue actionis; et quia in hoc quod dixit de potentia est questio, incepit munire homines ab illa questione, demonstrando eius contradictoriam, et dixit *Non igitur estimetis* etc., idest audientes nostrum sermonem non estimetis ipsum esse falsum propter sermonem alicuius contradicentis, quia qui diffinit potentiam per maius non recte facit,

35 quia] que *WXYZ* 36 diffinitur] + vel *Z* : + per minus *JZ* | posse] possibile *XZ* 37 quinque libras] *inv. Z* | quod[1]—plus] *om. BJ* 38 libras] *om. Z* | et] *om. XY* | non[1]] *om. JZ* 41 similiter cum] *inv. W* | diximus] –xit *JZ* 42 intendimus] -tendit *JZ* | terminum] + Sic—magis *(t. 117. [118.], v. 1sq.) WYZ*

117. [118.] 1-2 Sic—magis] *iter. WYZ (cf. app. crit. ad c. 116. [117.], v. 42)* 1 eius] sue *J*Y_2 : suus *W sed corr.* W_2 1-2 est—potentie] *om. homoiotel. X* 2 igitur] enim *Z* | aliquis] -quid *Z* 3 quoniam] quod *XY* | sermoni nostro] *inv. Z* 4 diffinimus] -nivimus *X* : destruimus *Y* 5 per—leuce] leuca *J* | distantia] *om. J* 6 una] *om. J* 7 aut] vel *JZ* | minimam] minorem *B* 11 Idest] *om. XY* | sit] sint *BZ* | terminus] *om. BZ* : + iste *J* | magis] maius *WX* 12-3 quia—illa] *om. X* 14-5 etc.—estimetis] *om. homoiotel. JYZ* 14 idest] + quasi *W* : + et *X* : + o'*B* | nostrum] hunc *X* 15 falsum] factum *Y* : finem *WZ* 16 contradicentis] dicentis *Z* | qui] *om. BZ*

qui enim videt aliquod quantum terminatum ex spatio terminato potest videre minus ex minori spatio, et non potest videre illud minus ex spatio ex quo vidit maius; non ergo omnis potentia que maius comprehendit potest minorem comprehendere, nec etiam omnis potentia habet finem terminatum in sua actione, nam qui aliquid ex una leuca comprehendit comprehendit maius ex illa leuca et maius minori et sic in infinitum. Et hoc intendebat cum dixit *Qui enim vidit aliquid ex leuca potest videre minus ex minori leuca*, idest quod non potest videre illud minus ex leuca et videt maius ex illa leuca; *Sed huic est contradictio sermoni* etc., idest sed hoc quod apparet de visu contrarium est dicto iam enim dictum est quod illud cui data est aliqua potentia non potest agere plus illa potentia, et hic apparet de potentia visus esse econtrario, quoniam si minus comprehendit maius comprehendet, et similiter de voce, nam qui aliquam vocem ex aliquo spatio audierit, audit maiorem ex illo eodem spatio.

Deinde incepit dissolvere hanc questionem dicendo *Dicamus igitur quod hoc non destruit nostrum sermonem* etc., idest et hoc exem-
79vb plum non destruit uni|versalitatem sermonis predicti: omnis enim potentia aut est activa aut passiva; si activa, terminatur per ultimum T96b
finem eius quod agit, si passiva per minimum eius a quo patitur; nam cum a minori patiatur, patietur a maiori, et potentia que est fortioris passionis patitur a minori; hec quidem est dispositio potentie visus. Cum enim patiatur a grano milii ab aliquo spatio, patietur a maiori ex illo eodem spatio; sensus igitur fortes sunt qui comprehendunt minimum eorum que sunt possibilia ut comprehendantur de spatiis ex quibus desperatur visio; et cum ita sit, potentia non terminatur nisi ex suo fine, sive fuerit activa sive passiva.

---

17 qui] quod *WY* 18 videre[1]] -deri *W* | minori] -nore *Z* 19-21 que—potentia] *om. homoiotel. X* 20 maius] magis *WY* 22 una leuca] sua *Z* | comprehendit] *haplogr. Z* | maius] magis *BYZ* 23 minori] maiori *BJX* 24 potest—leuca] *om. homoiotel. X* | videre] -deri *W* | minori] -nore *Z* 26 huic] *post* contradictio *BX* 27 contrarium—dicto] dicto est contrarium *JZ* 28 agere] *om. B* 29 econtrario] -verso *Z* | comprehendit] -det *JZ* 30 comprehendet] -dit *B* | et] *om. JYZ* | voce] + si *YZ* | nam] autem *YZ* 31 audierit] *om. Z* 33 nostrum sermonem] *inv. BXZ* 35-6 si—passiva] *om. homoiotel. Y* 35 terminatur] -nanda est *BX* 40 maiori] minori *Z* 41 comprehendantur] -datur *BX* 42 desperatur] -pertur *B* : separatur *Z*

118. [119.] *Volumus modo, postquam hoc distinximus, scilicet poten-* [281a28] *tiam et debilitatem, determinare residua per ea que sunt similia. Dicamus igitur quod si aliquis habuerit potentiam ad esse et non esse, omnino* [80ra] *est tempus distinguens inter suum | esse et suam destructionem, scilicet* [281a30] *tempus in quo illud habet potentiam ad esse et non esse. Et hic quidem sermo est in omnibus predicamentis, ut homo et album et tricubitum et aliis predicamentis; si igitur hec potentia non fuerit in aliqua hora temporis et tempus eius fuerit continuum, et non inveniatur ei tempus neque modicum neque multum, illud tunc habebit potentiam ad esse et non* [281b1] *esse in tempore infinito; quod est impossibile.* [281b2]

Cum determinavit potentiam ex ultimo sue actionis, et defectum potentie ex prima re que non potest aliquid facere (et hoc in rebus activis et passivis), vult ostendere quod similiter contingit in rebus generabilibus et corruptibilibus: entia enim per potentiam que est in eis ad generationem generantur, et per potentiam que est in eis ad esse sunt, et corrumpuntur per defectum istius potentie. Et hoc intendebat cum dixit *Terminare residua per ipsum quod est simile*, idest per mutationem iudicii in hoc ad generabile et corruptibile.

Deinde dixit *Et iste sermo est in omnibus predicamentis*, idest et hec intentio potentie defectus est eadem in omnibus predicamentis, in substantia scilicet ut homo: in homine diximus ipsum habere potentiam ad esse centum annorum et non habere potentiam esse plus quam centum annorum; et similiter dicimus in quantitate et qualitate et aliis predicamentis cum attribuimus eis potentiam, scilicet quia terminamus omnes has potentias et ultimus suarum actionum [80rb] in tempore scilicet ex ultimo tempore in quo possibile est ut | istud remaneat ens: corruptionem autem terminamus ex primo tempore in quo possibile est corrumpi, et similiter terminamus potentiam gene-

---

118. [119.] 1 distinximus] destruxi- *Z* : dixeri- *X* | scilicet] + et *Z* 3 aliquis] -quid *BJXY* | habuerit] -buit *BXZ* | et non esse] *om. B* 5 et[1]] + ad *WY* 6 tricubitum] attributum *WXY* 9 modicum] medio *Y* 11 determinavit] declara- *J* : termina- *X* : demonstra- *Z* 12 aliquid facere] *inv. BY* : facere *Z* 17 Terminare] de- *JX* 19 Et[1]] *om. B* 21 homine] + enim W_2 : + etenim *J* : + et *Z* 23 annorum] *om. BXY* | dicimus] *om. JYZ* 24 et] + in *BX* | attribuimus] -buerimus *YZ* | eis] ei *W* 25 quia] quod *JXYZ* 27 primo] quo *Y* | in] *om. YZ* : ex in *W* 28 quo] quod *Z* 28-9 et—est] *om. Y* 28 generandi] -rati *B* randi ex primo tempore in quo possibile est generari cum hoc generari in hoc sit hoc corrumpi.

Et cum demonstravit hoc de potentia, dixit *Si igitur hec potentia non fuerit in aliqua hora temporis* etc., idest et cum apertum est de potentiis quod tempora earum sunt terminata: si igitur aliquis finxerit potentiam esse non terminatam in aliquo tempore, scilicet non habere principium neque finem, erit utraque potentia que est ad aliquid et ad suum oppositum infinita; quod est impossibile.

80va10 119. [120.] *Incipiamus modo et declaremus quomodo est impossibile; et* *281b2*
dicamus quod falsum et impossibile non sunt idem: sunt enim quatuor, impossibile et possibile et verum et falsum. Possibile quidem est ut trian- *281b5*
gulum esse isochelem; impossibile vero ut diametrum esse equalem coste. Falsum enim et verum non sunt idem, sicut diximus superius: qui ergo dicit aliquid surgere cum non surgit falsum dicit sed non impossibile; et *281b10*
qui de cantore dicit ipsum cantare dum non cantat falsum dicit sed non impossibile; dicere autem Sortem esse surgentem et sedentem simul non tantum falsum sed impossibile. Impossibile enim est hominem habere *281b15*
potentiam qua sit surgens et sedens insimul in eodem tempore, quia in tempore in quo est ei unum est in illo tempore potentia alterum etiam: quia in eodem tempore habet duo, tamen non in actu, sed surrectio in uno tempore et sessio in alio. *281b18*

Cum declaravit quod contingit dicenti quod in aliquo quod est
80vb ens per tem|pus infinitum est potentia ad non esse, ut due potentie insimul sint infinite, et quod tempus actionis unius potentie non sit aliud a tempore secunde potentie; et narravit quod ab hac positione sequitur impossibile, incepit dicere quomodo sequitur, et proponit

---

29 possibile est] *inv. Z* 30 hoc[1]] quo *W* 31 cum] *om. WYZ* | demonstravit] -straverit *B* 33 sunt] sint *BX* 34 in] *om. WYZ* 35 utraque] ita- *B*

119. [120.] 2 et] *om. WX* 3 verum ... falsum] *inv. WX* | quidem est] *inv. B* : est *J* 4 isochelem] -cheles *BX* | equalem] equale *B* 6 dicit[1]] -xit *BX* 8 Sortem] *om. B* : idem *J* 9 tantum] enim *J* | falsum] + dicit *W* | enim est] *inv. WX* | hominem] *om. WZ* 10-12 in[1]—duo] *om. G* | quia—tempore] *om. homoiotel. XY* 11-2 in[1]—tempore] *om. homoiotel. BW sed suppl.* W_2 11 ei] ea *Z* | unum] + utrorumque W_2Z 12 quia—tempore] *om. Z* | tamen non] *inv. W* 13 uno tempore] *inv. W* 14 est] *om. B* 15 ad non] et non ad *Z* 18 proponit] -posuit *JZ* : opponit huic *B* : preparavit *Y* : + huic *X*

hanc propositionem scilicet quod falsum est duobus modis impossibile scilicet et non impossibile, et quod falsum quod est non impossibile est illud quod est possibile ut sit quando ponitur ens in actu antequam sit, verbi gratia quod si posuerimus Sortem qui est surgens esse sedentem, est falsum possibile non impossibile; et universaliter, quemadmodum dixit quod sunt quatuor intentiones, falsum, possibile, impossibile, verum. Deinde dixit exemplum de possibili ut triangulum esse isochelem, quod est esse possibile; et ideo laborant Geometre declarare quomodo possit inveniri; impossibile autem est ut diametrum quadrati esse communicantem coste, et hoc intendebat cum dixit equalem.

Deinde dixit *Falsum autem et verum non sunt idem*: ita fuit in libro primo; sed debet esse falsum autem et impossibile non sunt idem sicut diximus superius: impossibile enim est falsum sed non convertitur, sicut dixit in surgentem qui non est surgens, quia falsum dicit non tamen impossibile cum possit quidem surgere: qui autem dixerit Sortem esse surgentem et sedentem simul falsum dicit et impossibile; falsum ergo est duobus modis, falsum possibile et falsum impossibile.

Deinde dixit *Quia in tempore in quo* etc., idest quia in tempore in quo habet unum oppositorum, non habet in illo tempore nisi potentiam ad secundum oppositum, non potentiam ad oppositum habitum in actu; quoniam quod habet in actu non habet potentiam ad
81ra ipsum; cum ergo caruerit illo oppo|sito et habuerit secundum oppositum, convertetur hoc, ergo non invenientur in eo due potentie insimul sed cum non inveniuntur in eo due actiones insimul in eodem tempore; et voluit ostendere per hoc quod sic est de potentia ad opposita sicut de ipsis oppositis, scilicet quia sicut duo opposita non T97a

20 scilicet] *om. JZ* 21 illud] aliud *BX* 22 Sortem] Socratem *J* : Sor *X* : secundo *BY* : secunde *W* : secundum quatuor *Z* 23 falsum] tantum *B* | non] *om. X* : et non *JYZ* 23-5 et—verum] *om. Y* 24 quod] *om. J* : que *Z* | falsum] + enim *W* | possibile] + et *B* 25 dixit] + in *WXY* | possibili] -bile *B* 26 isochelem] -cheles *BX* | esse possibile] *inv. JZ* : esse im- *X* 28 communicantem] –nicationem *Y* 29 equalem] equale *Y* 30 fuit] sive *B* 31 autem et] et *J* : aut *Y* | impossibile] verum *Y* 32 sicut] ut *Z* | diximus] + sicut *B* | enim est] *inv. B* : enim *WXY* 33 surgentem] -gente *YZ* | quia] quod *B* 35 Sortem esse] secundum esse *B* : secundo esse *W* : idem *JZ* : secundo *Y* : sor. *X* | simul] in- *WX* 37 quia[2]] *om. JZ* 38 unum] + duorum *BX* 41 caruerit] + in *Z* 43 non] *om. JZ* 45 opposita[1]] -situm *BJ* | sicut[2]] *om. WYZ* : sint *X*

congregantur insimul, sic et earum potentie non simul congregantur in eadem materia.

120. [121.] *Si igitur aliquis dixerit aliquid habere potentiam ad plura* 281b18
in tempore infinito, significavit quod illud non facit illas actiones in
tempore sed omnes in eodem tempore insimul. Et si ita est sicut dixit, erit 281b20
aliquid eorum cadentium sub corruptione semper in tempore infinito
habet potentiam ut non sit sub corruptione semper ad infinitum, quo-
81rb *niam illud quod habet potentiam | ut non sit, erit etiam. Et si ita sit,*
erit illud generatum et non generatum insimul in actu; et hoc est falsum
non possibile: est quidem falsum, quia eius propositio fuit falsa; et nisi
esset falsa et impossibilis insimul, conclusio non esset impossibilis etiam. 281b25
Manifestum est igitur quod omne eternum in esse non cadit sub corrup-
tione neque sub generatione; quoniam si caderet sub corruptione et esset
generatum, haberet potentiam ut non esset in tempore. 281b27

Cum declaravit quod potentie que inveniuntur in eadem re ad esse et non esse sint finitorum temporum et diverse, dixit quod si invenirentur in aliquo potentie ad oppositas res non terminatorum temporum, sed tempus cuiuslibet potentie est infinitum, contingit ut ille potentie agant suas actiones in eodem tempore insimul, non quod unum eorum agat in tempore aliquo alio a tempore in quo aget secunda. Et hoc intendebat cum dixit *Si aliquis dixerit aliquid habere* etc., idest si aliquis finxerit aliquid eternum habere potentiam ad corruptionem, finget quod aliquid habebit potentias ad res oppositas in tempore infinito: potentia enim ad esse et non esse invenitur in eo per tempus infinitum; et qui finxerit quod in eodem invenientur potentie ad res oppositas in tempore infinito ponet quod illud non
facit illas actiones oppositas in temporibus diversis sed in eodem tem- T97b

120. [121.] 1 aliquis] -quid *Z* 2 significavit] -cat *BXY* | facit] -cet *B* : habet *Z* 3 sicut] sed *Z* : illud *XY* 5 habet] –bens *J* 8 non] nam *WYZ* | quia] quod *B* : quod quia *Y* | fuit falsa] falsa fuerit *B* 9 non esset] *inv. Z* | etiam] *om. JYZ* 11 neque—corruptione] *om. homoiotel. Y* 12 in] de *WXY* 14 dixit] *om. BW* : et *praem. J* 18 tempore aliquo] *inv. J* : tempore *BXY* | alio] *om. WZ* | tempore^{2}—quo] tempore quo *J* : quo tempore *Z* : tempore a quo *WY* 19 Si—dixerit] *om. Y* 20 aliquis] -quid *Z* : quis *Y* | aliquid] -quod *BX* 21 finget] -git *B* | potentias] -tiam *B* 22 non] + in *B* 23 per] quod *Z* 25 sed] et *W* pore insimul, quod est impossibile scilicet actiones contrarias exire in actu.

Et quia dicens aliquid eternum habere potentiam ad corruptionem 81va non ponit ipsum corruptum in actu, incepit narrare quomodo | sequitur impossibile hanc positionem, et est quod sequitur dicentem hoc ut positio duarum actionum insimul sit falsa possibilis non falsa impossibilis, et dixit *Et si sit sicut dixit, erit aliquod cadentium sub corruptione semper ad infinitum, quoniam illud quod habet potentiam ut non sit erit*, idest et cum declarate sunt hee propositiones, si ita fuerit sicut fingit hic dicens scilicet esse aliquid in quo est potentia ad corruptionem et illud esse semper in infinitum, necessarium est ut habeat potentiam ut non sit in tempore in quo existit in actu; positio ergo eius quod sit non consistens in actu est falsa possibilis, sed si ita fuerit, continget ut sit ens et non ens, quod est falsum impossibile; sed falsum impossibile non sequitur falsum possibile, ut dictum est in Analeticis; illud ergo quod positum fuerit possibile scilicet eternum corrumpi, impossibile est. Et hoc intendebat cum dixit *Et si ita sit, illud erit generatum et non generatum in actu, quod est falsum impossibile*, idest quod si in eterno esset potentia ad corruptionem, esset tempus actionis potentie cum tempore sui esse idem; quapropter necessarium est quod positio eius scilicet non esse ipsum in tempore sui esse sit falsa possibilis; sed est falsum impossibile, et falsum impossibile non sequitur possibile, eternum igitur non est possibile ut corrumpatur.

Deinde dixit *Et nisi eius positio esset falsa et impossibilis insimul, conclusio non esset impossibilis*, idest et nisi propositio dicens quod eternum est corruptibile esset falsa et impossibilis, non sequeretur eam impossibilem. Sillogismus autem ita componitur: mundus est

---

26-7 exire—actu] in actu exire *JY* : in actu sit exire *Z* 28 potentiam] -tias *W* 29 non ponit] *om. X* : *inv. JZ* : non *Y* | actu] -tum *B* 31-2 possibilis ... impossibilis] *inv. J* 32 et[1]] *om. JYZ* 33 quoniam] quod *B* 34 sit] fit *B* | erit] erit etiam *W* : est etiam *J* : est *Z* : ens *B* | declarate] -terminate *JZ* | hee propositiones] *inv. JZ* 35 fingit] finxit *B* : fixerit *X* | esse aliquid] *inv. JY* : aliquod esse *Z* 36 semper] *post* infinitum *Z* : *om. W* | in] *om. X* 37 habeat] -beant *Z* 38 ergo eius] *inv. B* | consistens] exi- *Y* | actu] + in tempore in quo est existens in actu *BX* | falsa] -sum *Z* 39 continget] -gens *B* 40 sequitur] + ad *J*W_2*Z* 41 fuerit] fuerat *J* : fuit *BX* 43 illud erit] *inv. BX* | actu] -tum *B* 46 scilicet] *om. Y* 47 sit] ut *B* 50 eius] *om. BX* | positio esset] *inv. W* 51 idest et] nisi et idest *B* 53 eam] eum *W* | Sillogismus autem] simul dum *YZ* eternus et corruptibilis, ergo aliquid eternum est corruptibile, quod *81vb* est impossibile; ergo illud ex quo sequitur est im|possibile scilicet mundum esse corruptibilem; impossibile enim non sequitur falsum **T98a** possibile.

Deinde dixit *Manifestum est igitur quod omne eternum* etc., idest manifestum est igitur quod omne eternum caret potentia ad corruptionem ex hoc sermone; quoniam autem impossibile est ut aliquid sit eternum postquam generabatur apparet ex eo quod addidit hic, quoniam si fuerit non ens in actu in aliquo tempore, erit non ens in potentia cum fuerit ens in actu, necessario igitur erit in eo potentia sicut dixit ut non sit, et quia in futuro invenitur eternum, habebit potentiam ut non sit per tempus infinitum et ut sit insimul; et sequitur impossibile predictum. Et non debet aliquis dicere Forte cum fuerit ens in actu non habebit potentiam ad non esse; quoniam dicemus quia quod caret potentia ad aliquid, impossibile est omnino ut sit in eo; si igitur caret aliquid potentia ad non esse, impossibile est ut in aliquo tempore sit non ens; sed iam positum fuit possibile ut in aliquo tempore non sit ens: quod ergo impossibile est non esse possibile est non esse, quod est impossibile. Et etiam dicimus quod cum sit non ens in actu et non ens in potentia, utrumque (tempus sui esse in actu et non esse in potentia) est idem tempus; si igitur dicatur ipsum esse idem, erit non impossibile ut ponatur non ens et ens insimul; et hoc est falsum impossibile, non possibile; et si tempus non esse fuerit aliud a tempore actionis potentie ad esse, erit tempus duarum potentiarum necessario finitum.

Et debes scire quod potentia ad esse aliquid non est infinita in preterito tempore cum fuit accepta in loco propinquo generationi scilicet finita ex duobus extremis; si autem potentia ad aliquid fuerit *82ra* accepta in prima materia, tunc eterna fuit in preterito, | sed tamen

58 quod—eternum] *om. Z* 58-9 etc.—eternum] *om. homoiotel. Y* 59 igitur] *om. X* 60 impossibile est] *inv. JZ* 61 addidit] addit *BJ* 65 et[2]] + post *Y* 68 quia] *om. JZ* | omnino] in eo *Y* 69 in eo] omnino *Y* 70 in[1]] *om. Z* | fuit possibile] est esse possibile *J* : possibile est *Z* : fuit *BWXY* 70-2 possibile—non] *om. BWXY sed suppl.* B_2 72 esse] omne *Y* 73 non[1]] *om. JWY* | non[2]] *om. BWXY* | utrumque] utrum *XY* | sui] + non *BXY* 74 in[1]] *om. Z* | non] *om. BWXY sed suppl.* W_2 | in[2]] *om. WZ* 77 esse[2]] rem *Y* : + tunc *J* 79 debes] -bet *Z* | esse aliquid] *inv. J* : esse ad *B* | est] esse *Z* | infinita] finita *WY* : finita nisi *B* : infinita nisi *X* 80 fuit] fuerit *B* | scilicet] sed *JW* : s. *BYZ* 81 finita] + nisi *B*

ista potentia que semper fuerit in prima materia est individui entis per accidens vel accidentalis. Et cum quidam homines acceperant hanc potentiam que est per accidens quasi essentialiter, tenuerunt esse possibile ut aliquid non sit per tempus infinitum in preterito deinde post erit, quapropter est potentia ad suum esse infinita; et cum putaverint quod potentia ad esse possibile est ut sit infinita, estimaverunt autem quod possibile est ut potentia ad corruptionem in aliquo sit infinita in futuro. Error igitur horum est quia acceperunt essentialiter quod est accidentaliter deinde assimulaverunt essentialem potentiam ad corruptionem accidentali potentie ad generationem: **T98b** impossibile est enim inveniri in aliquo potentiam ad corruptionem infinitam neque accidentaliter neque essentialiter potentiam autem ad generationem possibile est infinitam reperiri accidentaliter non essentialiter reperiri; secundum igitur hunc modum fuit error istorum hominum in hoc loco.

82rb10 **121.** [122.] *Dicamus ergo quod illud quod cadit sub corruptione est* *281b27* *illud quod primo fuit ens, deinde modo non ens, aut possibile ut non sit ens in futuro. Generatum autem est illud quod primo fuit non ens, deinde post est ens. Et quod est eterne quiditatis non habet potentiam ut* *281b30* *sit non ens in tempore finito neque in tempore infinito; impossibile est igitur ut idem habeat potentiam qua semper sit ens et semper non ens. Et impossibile est aleconeis ut sit verum, ut dicimus quod impossibile est idem esse ens semper et non ens semper. Et cum ita sit, impossibile est ut aliquid sit semper ens et cum hoc cadat sub corruptione. Sunt igitur duo* *282a1* *termini, quorum unus est ante alterum, et impossibile est ut alter terminus sit absque essentia primi, quoniam si primus non fuerit, non erit alter omnino. Et cum ita sit, et res eterna perpetua que impossibile est ut*

---

83 ista] ita *Z* : in *B* | fuerit] fuit *BJX* 84 accidentalis] -taliter *WXY* | acceperant] -pant *W* 85 accidens] + vel *Z* 87 est] erit *BXY* 89 autem] etiam *BJX* 90 aliquo] alio *Y* | quia] quod *Z* 91 essentialiter] *om. BWXY sed suppl.* W_2 | accidentaliter] + esse essentialiter *BX* 93 est enim] *inv. JYZ* | inveniri] *post* aliquo *JYZ* 94 neque[1]] *om. B* 94-6 potentiam—reperiri] *om. homoiotel. Y* 95 infinitam] –nita *J* 95-6 accidentaliter—reperiri] *om. homoiotel. X* 96 reperiri] *om. JZ*

121. [122.] 1 est] + possibile *B* 2 non sit] *inv. JYZ* 3 quod] quo *Z* 4 est ens] *inv. X* : esse *Z* | quod—eterne] res eterna *J* 5 finito ... infinito] *inv. YZ* 7 impossibile[1]] poss- *BX* | aleconeis] -conem *J* : aloconem *Z* | est[2]] *om. BXY* 9 sit] *om. XY* | cadat] -dit *Z* 11 primi] + sic *W* 12 que] *om. B* : quod *XY* *sit in aliquo tempore fuerit non ens impossibile est ut sit generata omnino.* *282a4*

82va Vult declarare quod ex diffinitione generabilis et corruptibilis apparet quod impossibile est ut aliquod generabile non sit non corruptibile neque corruptibile non generabile, et dixit *Dicamus ergo* etc., idest dicamus modo quod corruptibile dicitur duobus modis, in actu scilicet et in potentia: in actu est illud quod fuit in tempore preterito ens et modo non ens, in potentia autem est illud quod modo est ens et in futuro non ens possibiliter. Et cum diffinivit corruptibile, diffinivit generabile, et dixit *Generabile autem* etc., idest generabile autem est quod in tempore preterito est non ens, modo autem est ens.

Deinde dixit *Et res eterne quiditatis non habet potentiam* etc., idest **T99a** et cum declaratum est quod generabile et corruptibile conveniunt in hoc quod in aliquo tempore non sunt, et quod est huiusmodi habet potentiam finitam ut corrumpatur scilicet in tempore finito, eternum igitur non habet potentiam ut corrumpatur neque in tempore finito neque in tempore infinito; et intendebat per hoc quod si in eo inveniatur potentia in tempore infinito, continget predictum impossibile. Videtur igitur quod in primo sermone posuit quod si in eterno inveniatur potentia ad corruptionem, continget ut sit infinita; in hoc autem posuit quod si in eterno inveniatur potentia ad corruptionem, aut est finita aut infinita: si finita, continget ut eternum sit non eternum, si infinita consequetur ex possibili impossibile.

Deinde dixit *Impossibile est ut aluchanez sit verum* etc., idest impossibile est ut contraria inveniantur insimul, ut diximus quod impossibile est ut aliquid sit non ens semper et ens semper. Et videtur quod induxit hoc ad demonstrandum quoniam sicut est impossi- *82vb* bile quod contraria invenian|tur insimul in actu, ita est impossibile ut due potentie inveniantur insimul ad duo contraria scilicet imme-

---

13 aliquo] alio *WYZ* | tempore] + vel *Z* : + et W_2 : + etiam *J* | fuerit] *om. B* | generata] -rati *Z* 14 omnino] opinio *Y* 16 aliquod] -quid *WY* | non[1]] *om. XY* 19 in[1]] *om. B* 21 possibiliter] *ante* non *J* 22 dixit] + quod *Z* | autem] *om. Z* 23 est non] *inv. X* : ens non *Y* 28 finito] in- *Z* 29 igitur] autem *J* 35 est] erit *WXY* 36 si] sed *Z* 37 aluchanez] alechonez *WXYZ* : aleconem *BJ* | verum] unum *YZ* 39 ens[1]] eius *Z* 41 inveniantur insimul] *inv. BX*

diata; sed potentie succedunt in subiecto sicut contraria. Et quod dixit hic videtur propter hanc intentionem et propter illud quod predixit scilicet quod in eterno non est potentia infinita neque finita ad corruptionem, quasi ergo dicat Et signum quod in eterno non est potentia ad corruptionem neque in tempore finito neque infinito est quod impossibile est ut congregentur contraria insimul.

Deinde dixit *Et cum ita sit, impossibile est* etc., idest et si impossibile est ut contraria congregentur, impossibile est ut aliquid sit semper ens et sit corruptibile in aliqua hora aut antequam generatur, ut dicunt facientes eternum generari, aut post ut qui ponunt eternum esse in preterito corruptibile in futuro. Deinde dixit *Sunt igitur* etc., et intendit per terminos duos generabile et corruptibile: res enim primo est generata deinde post corrupta; et qui posuerit eternum esse corruptum in postremo, ponet quod postremus terminus invenietur absque primo, quod est impossibile. Deinde dixit *Et cum ita sit, et res* T99b
eterna etc., idest et cum contingit quod corruptum non fuerit ante generatum: et generatum est illud quod in aliqua hora non fuit, eternum autem est illud quod impossibile est ut aliquo tempore sit non ens; manifestum est quod eternum numquam fuit generatum, et cum non generatum, non corruptibile est.

83ra16 **122.** [123.] *Et etiam dicimus quod aluchanez eius quod habet poten-* *282a4*
tiam ut semper sit ens, est illud quod non habet potentiam ut semper sit ens; contrarium autem eius est illud quod habet potentiam ut semper sit non ens, aluchanez etiam istius est illud quod non habet potentiam ut semper sit non ens. Dicamus igitur quod possibile est ut inter ista aluchanez sit aliquod medium quod possibile est ut sit ens et non ens in tem-

43 quod] *om. WXY* 44 propter[1]] per *B* | predixit] dixit *B* 45 infinita neque] *om. BWXY* | infinita … finita] *inv.* W_2 46-7 quasi—corruptionem] *om. homoiotel. X* 47 finito … infinito] *inv. Z* | est] *om. JZ* 48 impossibile est] *inv. JZ* 49 est] *om. Z* 50 congregentur] -guntur *WX* | est[2]] *om. BX* | sit semper] *inv. BX* 51 ut] et *B* 52 ut] aut *B* 53 etc.] + idest *BX* 55 deinde] *om. JZ* | post] *om. Y* : postea *JZ* 56 postremus] posterior *JZ* | invenietur] -nitur *JZ* 58 non] *om. XY* | ante] + quam *XY* 60 autem] *om. WY* : quod *Z* | impossibile est] *inv. JZ* | ut] + in *W* 62 est] erit *BXY*

122. [123.] 1 dicimus] -camus *J* : -cemus *Z* | aluchanez] -chonem *B* : -chenez *Y* : -chonez *X* : elochanez *Z* : aleconem *J* 3 ens—sit] *om. homoiotel. BY* | eius] istius *JZ* 4 aluchanez] -chanem *B* | istius] eius *W* 5 sit] *om. BWX* | aluchanez] -chanem *B* 6 sit[1]] *om. B* | sit[2]] *om. XY*

pore et tempore, quoniam aluchanez utriusque eorum erit non ens in 282a10
tempore ens et in tempore ens erit non ens, et hoc erit illud quod non habet potentiam ut semper sit non ens, sed in aliquo tempore ens et in alio tempore non ens, et quod sit in aliquo tempore ens et in aliquo
83rb *tempore non | ens. Erit igitur in eodem potentia ut sit ens et non ens in*
tempore et tempore, et est medium aluchanez, sicut diximus superius. 282a14

Hec est demonstratio tertia, sed prosequitur ipsam per primam demonstrationem: omnes enim demonstrationes quas inducit in hac questione prosequitur per demonstrationem ducentem ad impossibile dicentem quod potentia eterni quod sit non ens est falsa non impossibilis, et dixit *Dicamus etiam quod aluchanez quod habet potentiam ut semper sit ens* etc., idest et dicamus quod contradictorium sermonis dicentis quod aliquid habet potentiam qua sit semper ens, est quod non habet potentiam ut semper sit ens: istorum enim duorum
si alterum fuerit falsum, alterum erit verum. Deinde dixit *Contra-* T100a
rium autem istius est illud quod habet potentiam ut semper sit non ens, idest contrarium autem sermonis qui est quod hoc habet potentiam ut semper sit ens est hoc quod habet potentiam ut semper sit non ens; et intendebat per hoc quod contraria habent medium, contradictoria autem non. Deinde dixit *Aluchanez autem istius est* etc., idest et sicut contradictorium eius quod habet potentiam ut semper sit ens est quod non habet potentiam ut semper sit ens similiter est contradictorium eius quod habet potentiam ut semper sit non ens, est quod

7 quoniam] quando *B* | aluchanez] -chanem *B* : allu- *WZ* | erit] est *Z* 7-8 in tempore] *om. BY* 8 ens^{1}] *om. BJY* : erit ens *Z* | et^{1}] *om. BY* | erit^{1}—ens] *om. J* 9 sit] + ens vel semper *J* | aliquo] alio *Y* | tempore] + non *WZ (cf. infra, c. 122. [123.], v. 49sq.)* 9-11 ens^{2}—tempore] *om. homoiotel. B* 10 tempore^{1}] *om. X* | ens^{1}] + habebit *(*habebit *J* : erit *XZ)* igitur idem *(*idem *J* : in aliqua *XZ)* potentia ut sit semper *(*sit semper *inv. XZ)* non ens *JXZ* 10-1 et^{1}—ens^{1}] *om. homoiotel. Y* 10 quod] ut *JZ* 11 tempore] *om. X* | ens^{1}] + et hoc est impossibile *JZ* | Erit] habebit *J* | in—ens^{3}] idem potentiam ut non sit ens et ut sit ens *J* | ens^{2}—ens^{3}] non ens et ut sit ens *Z* | et—ens^{3}] *om. B* 12 diximus] -xit *W* 13 Hec] Et *praem. BX* | demonstratio] questio demonstrativa *BX* | primam] ipsam *Z* 14 inducit] -xit *W* 15 per] *om. B* | ducentem ad impossibile] *om. homoiarch. B* 16 falsa non] *om. Z* 17 aluchanez] -canem *B* : -canez *W* : aleconem eius *J*W_2 : allichanez *Z* 18 semper sit] sit semper permanens *W* 19 qua] ut *J* 22 autem] *om. Z* 23 quod hoc] *inv. BW* 24 hoc quod] *inv. Y* : hoc *WX* 25 quod] quia *BY* 26 Aluchanez] –canez *BW* : aleconem *J* | autem^{2}] *om. Z* : etiam *J* | istius est] *inv. B* : istius *Z* 28 ut—ens] *om. B* 29-31 est—ens] *om. homoiotel. Y*

non habet potentiam ut semper sit non ens, est quod non habet potentiam ut semper sit non ens.

Et cum declaravit contradictorium utriusque harum duarum propositionum contrarium scilicet hoc quod habet potentiam ut semper sit ens et habet potentiam ut semper sit non ens, ex quo declaravit
83va eas esse contrarias non contradictorias, | dixit *Dicamus igitur quod possibile est ut ex hiis aluchanez sit aliquod medium*, idest et cum declaratum est quod dicere Hoc habet potentiam ut semper sit et hoc habet potentiam ut non sit semper, dicere hoc scilicet contraria sunt; et quia in pluribus contrariis existit medium, possibile est ut in istis inveniatur medium, et est ens et non ens scilicet habens potentiam ut sit et non sit non in eodem tempore; et hoc intendebat cum dixit *Possibile est ut sit ens in tempore et tempore*. Deinde reddit causam huius possibilitatis et dixit *Quia aluchanez utriusque eorum erit non ens*, idest quia contradictorium sermonis quod hoc est ens in aliquo tempore est quod non est ens in illo tempore; et voluit per hoc demonstrare quod non est impossibile dicere de eodem esse in aliquo tempore et non esse in illo tempore.

Et cum demonstravit hoc non esse impossibile, dixit quid est talis nature et dixit *Et hoc erit illud quod non habet potentiam ut semper sit ens sed in aliquo tempore ens et in aliquo tempore non ens*, idest et hoc medium est illud quod non habet potentiam ut semper sit, sed in quibusdam horis sit et in quibusdam non sit. Deinde dixit *Habebit igitur idem potentiam ut sit semper non ens et ut sit in aliquo tempore non ens et in aliquo tempore non ens*, idest secundum quod reputo quod si aliquis posuerit iudicium medii inveniri in uno duorum extremorum contrariorum verbi gratia in eo quod est non ens sem-

---

33 contrarium] -riarum *X* : -dictoriarum *Y* | quod] *om. WXY* : + semper *BX*   34 et] + hoc *J*   36 hiis] eis *JYZ* | aluchanez] aleconem *J* : albucaliz *WYZ* : albucanem *B* | idest] *om. YZ*   40 habens] -bet *B*   42 ens] *om. WYZ* | et] + ut non sit ens in *JZ* | tempore[2]] *om. WX* | Deinde] *om. W*   43 Quia] quoniam *J* : quod *Z* | aluchanez] aleconem *J* : albu- *BWXYZ*   44 ens[1]] + in tempore ens et in tempore ens *(*et—ens *om. B)* erit non ens *BX (cf. surpa, t. 122. [123.], v. 7sq.)* | quia] quod *W*   45 est[1]—tempore] *om. homoiotel. Y* | voluit] -lui *Z*   47 illo] alio *J* : aliquo W_2*Z*   48 quid] quod *JYZ* : quidem *BW* | est] *om. B*   50 ens[1]] + vel semper non ens *J*W_2*Z* | aliquo[1]] alio *XY* | aliquo[2]] alio *X* | tempore[2]] *om. W*   54 non[1]] *om. JBX* | non[1]—tempore] *om. Y* | aliquo] alio *Z* | non[2]] *om. Z* | secundum] + hoc *B* | quod] + ego *BX*   55 posuerit] + quod ponunt qui ponunt eternum generabile corruptibile aut utrumque W_2*Z*   56 est—ens] non est *JZ*

per, continget ut idem habeat potentiam ut semper sit non ens et T100b potentiam ut in aliquo tempore sit non ens et in aliquo ens; quod est 83vb impossibile. Deinde dixit *Habebit igitur idem | potentiam ut non sit ens et ut sit ens in tempore et tempore*, idest et est medium aluchanez sicut diximus, idest et cum impossibile est ut habens potentiam ut in aliquo tempore sit ens et in aliquo non ens, sit illud quod habet potentiam ut semper sit ens, aut illud quod habet potentiam ut semper sit non ens, oportet quod sit medium inter hec duo extrema contraria, sicut superius diximus; et intendebat per hoc quod potentia que dat alicui non esse in aliquo tempore non reperitur in eo quod semper est non ens.

123. [124.] *Sit ergo A et B contraria non cadentia in eodem insimul, et* 282a14 *sit C contrarium ad A et D contrarium ad B. Et dicamus quod illud quod non est in A neque in B erit in toto C et D. Sit ergo H, quod est inter A et B: illud enim quod non est alterum duorum contrariorum, medium est inter illa; erit ergo H sicut C et D necessario.* 282a19

84ra
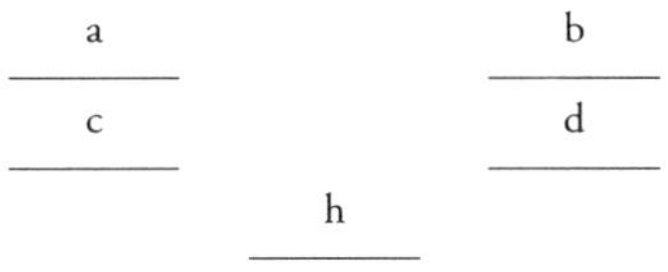

Cum declaravit ratione quod illud quod non habet potentiam ut sit in alia hora et ut non sit in alia hora non est illud quod habet potentiam ut semper sit, neque illud quod habet potentiam ut semper non sit, sed medium vult declarare hec litteris, et dixit *Sint ergo A et B* etc., idest sit illud quod habet potentiam ut semper sit A, et eius

57 continget] -git *BW* 58 aliquo^{2}] + tempore *Z* 59 non sit] *inv. BXY* 60 et^{1}—ens] *om. B* | et est] *om. WXY* | aluchanez] aleconem *J* : albucanez *BWY* : albuchanez *Z* 61 et] *om. JZ* | aliquo] alio *W* 62 et] *om. B* | aliquo] alio *WX* 64 quod] ut *W* | inter] *om. B* 65 superius diximus] *inv. BXY* 66 esse] + est *XY*

123. [124.] 2 dicamus] + igitur *JZ* | illud] *om. X* : id *Z* 3 erit] est *JZ* | est^{2}] *om. WXY* 4 illud] et *praem. B* 5 C et D] A et C *Y* 7 alia^{1}] aliqua *B*W_2 | et—hora] *om. homoiotel. B* | ut] *om. X* | alia^{2}] aliqua *W* 9 sed] neque *B* | hec] hoc *WZ* | Sint] si *B* : sunt *Z* : sit *WXY* contrarium quod semper non sit B, et sunt que impossibile est ut congregentur in eodem insimul. Deinde dixit *Et sit C contrarium ad A et D contrarium ad B*, idest et ponamus contrarietatem que est in hoc sermone verbi gratia illud quod est ens in aliqua hora et non ens **T101a** in aliqua hora C et D, et sit C contrarium ad A quod posuimus sem- *84rb* per ens: ex C ergo intelligendum est illud quod non est | in aliquo tempore, et erit D, quod est ens in aliquo tempore, contrarium ad B quod est semper non ens: erunt ergo tria, semper ens scilicet A, et semper non ens scilicet B, et esse et non esse scilicet B D.

Et cum posuit hoc, quesivit de qua re non dicitur esse et non esse, ad demonstrandum quia, cum non dicatur de extremis, necesse est ut dicantur de natura media inter duo extrema, et dixit *Dicamus ergo quod illud quod non est in A neque in B erit in toto C et D*, idest et cum ita sit, si posuerimus aliquod ens de quo non dicitur A quod semper est, neque B quod semper non est, necesse est ut dicatur de eo D et C scilicet ens in aliquo tempore et non ens in aliquo; quo- *84va* niam inquantum non est ens semper, oportet | ut sit non ens in alia hora; et hoc intendebat cum dixit *Si igitur H* etc., idest et contingit in ista natura que est H ut sit media inter A et B. Deinde dixit *Erit ergo H sicut C et D necessario*, idest invenientur in A hec duo scilicet entia in aliquo tempore et non entia in aliquo tempore.

124. [125.] *Revertamus modo ad nostrum sermonem et dicamus quod* *282a21* *illud quod est semper ens neque est generabile neque corruptibile. Et cum ita sit, manifestum est quod illud quod est generabile et corruptibile est non ens perpetuum eternum; si enim esset perpetuum eternum, habebit potentiam ut semper esset ens et semper non ens; quod est impossibile omnino, sicut diximus superius in eis que declaravimus.* *282a25*

---

11 semper non] *inv. W* | sit] + contrarium *W* | que] quod *BJ* 13 contrarietatem] *om. WYZ* | que] quod *Z* 14 quod] *om. YZ* | ens[2]] est *Z* 15 aliqua] alia *JXY* 16 non est] + ens *Y* : est non ens *BX* 17 erit] est *Z* 18 erunt] sunt *W* 19 B D] C D *J* 20 quesivit] que sunt *B* : cum finit *Z* | re] *om. BWXY suppl.* W_2 21 non] *om. B* | extremis] + et *B* 22 dicantur] -catur *JXYZ* 23 illud—est] *om. J* | non est] *inv. B* | erit] est *JZ* 24 dicitur] -catur *JY* 26 aliquo[1]] alio *XY* | aliquo[2]] alio *BXZ* 27 non[2]] *om. WY* 29 in] *om. W* | que] quod *W* 30 A] H *J* 31 aliquo[2]] alio *XY* | tempore[2]] *om. WX*

124. [125.] 1 sermonem] *om. BWXY* 1-2 quod illud] *inv. Z* 2 est semper] *inv. J* 4 perpetuum[2] eternum] *inv. B* : eternum *J* 5 et—ens] *om. homoiotel. Y* 6 diximus superius] *inv. JZ*

Cum declaravit quod illud quod dicitur non ens in aliquo tempore T101b oportet dici ens in alio tempore, quoniam illud quod dicitur non ens in aliquo tempore impossibile est dici semper non ens aut semper ens; dicitur igitur in aliquo tempore ens et in aliquo tempore non ens. Et quia potest aliquis dicere quod hoc non est nisi ex sermonibus insufficientibus (quoniam possibile est ut aliquid sit generatum deinde numquam corrumpatur, et quod hoc manifestabitur ex hoc *84vb* | sermone) est quod illud quod est ens semper in preterito et in futuro impossibile est ut non sit in aliquo tempore, et similiter illud quod semper est non ens in preterito et in futuro, impossibile est ut sit in aliquo tempore: quod autem est non ens semper in preterito non est impossibile ut sit in futuro, quemadmodum non est impossibile ut illud quod est ens semper in preterito sit in futuro; et propter hanc questionem contingentem sue demonstrationi confirmavit illam per primam demonstrationem, cum illa demonstratio sit firmior et illa est sustentamentum huius quesiti. Et dixit *Et dicamus ergo quod illud quod semper est ens* etc., idest et dicamus igitur quod si aliquid positum fuerit semper ens, continget ut semper sit in toto tempore, et quod illud in cuius natura est receptio eternitatis non recipit eternitatem in preterito absque futuro neque in futuro absque preterito.

Deinde fecit memorationem predicte demonstrationis et dixit *Si enim esset eternum, haberet potentiam* etc., idest quoniam si ita esset, si poneretur corruptibile, esset falsum possibile; sed est falsum impossibile, illud ergo quod positum fuit possibile est impossibile. Et debes scire quod impossibile est imaginari ens habere potentiam ad esse et non esse infinitum in tempore preterito absque futuro neque in tempore futuro absque preterito, nisi infinitum adderet super infinitum, quod est impossibile: quamobrem contingit, si aliquid inve-

---

8 oportet—quod] impossibile est quod semper *Z*   9 dici semper] *inv. Y* : semper dicitur *Z* : ut semper dicatur *J*   10 igitur] ergo *JZ* | in[1]] *om. BXY* | aliquo[2]] alio *BXYZ* | tempore[2]] *om. JZ*   12 insufficientibus] suffic- *BXY*   13 corrumpatur] -petur *BX* | hoc[1]] *om. BXY*   13-4 hoc sermone] *inv. B*   15 est] *om. BW* | aliquo] alio *XY*   16 in[2]] *om. Z*   17 est[1]] sit *JZ* | non[2]] *om. B*   20 confirmavit] + demonstrationem *JZ*   22 Et[2]] *om. WX*   22-3 quod—quod[1]] *om. Z*   23 est] *om. Z* | si] *om. Y* : cum *BX*   25 quod] *om. B* | illud] + quod *JZ* | cuius] eius *JWZ*   27 memorationem] rememo- *WXY*   28 enim] igitur *Z* | esset[1]] esse *B*   29 possibile—falsum] *om. homoiotel. Y*   30 illud ergo] *inv. JZ* | possibile est] ponere non *Y*   31 debes] -bet *Z*   32 infinitum] -tam *BY*   34 quamobrem] quoniam *Y*

niatur habens potentiam infinitam, ut actio illius potentie sit in tempore; si igitur potentia fuerit infinita, continget ut sit non ens in 85ra omni tempore, et si | potentia ad esse fuerit infinita, contingit ut sit ens in toto tempore; et si aliquis posuerit aliquid esse ens in toto T102a tempore, sed in eo est potentia ad corruptionem absque eo quod corrumpitur, continget ut due potentie inveniantur insimul infinite scilicet potentia ad esse et potentia ad non esse; erit igitur altera earum otiosa, cum nulla actio perveniat ex ea. Et in hoc non errat nisi qui estimat hoc non esse impossibile propter hoc quod res non exit ex potentia in actum nisi propter agens, quod si igitur agens non agat, non est remotum ut remaneat et non exeat in actum, et hec quidem imaginatio inducit Platonem in hunc sermonem.

Et tunc non est remotum dicere in responsionem huius sermonis scilicet sermonis Aristotelis, quod si ita esset, oporteret ut aliquod possibile, si poneretur in actu ens, esset falsum impossibile non possibile falsum: impossibilitas enim non hic sequitur naturam possibilem sed sequitur defectum actionis agentis, quoniam corruptio mundi ut dicunt possibilis est ex parte materie, impossibilis quoad agens, aut quia agens aliquid non est innatum corrumpere ipsum, aut quia agens sapiens non est innatum corrumpere suum factum, aut quia corrumpens est contrarium generati aut quia non corrumpetur omnino.

Et cum intuebimur istas omnes questiones, apparebit destructio eorum ex demonstratione Aristotelis hic per quam incepit declarare hoc quesitum: possibile enim aut est possibile ex se et ex rebus extrinsecis, ut notum est de possibili simpliciter, aut quia aliquis dicet aliquid esse possibile in se et impossibile ex extrinseco, secundum opinionem dicentis mundum esse possibilis corruptionis in se et 85rb impossibilis quoad extrinsecum. | Manifestum est igitur quod Aristo-

---

35 in] + toto *J*W_2*Z* 36 potentia] + ad non esse *J*W_2*Z* | fuerit] sit *JYZ* | continget] -git *BJ* 37 infinita] finita *Z* | contingit] -get *WXY* 38 posuerit] -suit *Z* 39 corrumpitur] -patur *JZ* 40 inveniantur] -niuntur *B* : -nientur *WXY* 42 in] *om. B* 43 exit] exeat *JY* 45 remaneat] -moveat *B* 46 inducit] -duxit *BX* 47 tunc] cum *BXY* | in] *om. BZ* | responsionem] -sione *JXY* : -sio *Z* 49 ens] eius *Y* | esset] esse *B*W_2 : est *X* 50 non hic] *inv. J* : hic *Z* 53 aut[1] quia] *inv. W* : aut quod *B* 54 agens] + aliquid *W* 55 generati] + mundo autem genito non est contrarium *J*W_2 57 intuebimur] -bitur *X* : -batur *BZ* | destructio] –stinctio *Y* 58 hic] *om. Z* 59 quesitum] que sint *Z* : que sunt *B* 60 dicet] –ceret *JZ* 63 quoad] quod ad *B*

telis quod potentia alicuius possibilis ex se et impossibilis ex extrinseco est potentia contradicens diffinitioni possibili; dictum est enim quod cum possibile poneretur esse ens, non sequitur impossibile scilicet quia erit falsum possibile non impossibile falsum et cum concessum fuerit quod diffinitio cuiuslibet possibilis est huiusmodi, si aliquis posuerit aliquod possibile esse et impossibile ab extrinseco, continget ut eius potentia in actu sit falsa impossibilis non falsa possibilis.

Et intendit Aristoteles ad declarandum hanc intentionem propter hoc possibile, quia notum est per se quod illud quod est possibile ab extrinseco est possibile ex se, sed econtrario latet scilicet quod illud quod est possibile ex se est possibile ab extrinseco: estimatum est **T102b** enim quod non est impossibile quod aliquid sit possibile in se impossibile ab extrinseco. Et dispositio in hoc de possibili est quasi contraria ei que est de impossibili, quia concessum est quod omne impossibile in se est impossibile secundum extrinsecum, et non est concessum quod non est impossibile ex extrinseco impossibile ex se, ideoque intendit Aristoteles istam declarationem quam diximus; et etiam possibile non est possibile nisi in respectu agentis et patientis; et impossibile est imaginari aliquid esse possibile in respectu patientis et impossibile in respectu agentis, quoniam contingeret ut sit patiens impossibile idest possibile impossibile.

Et possibile est uti in hac demonstratione accepta ex loco a contrarietate, et est quia impossibile ex se est contrarium possibili ex se et impossibile ex se est impossibile quoad extra, ergo possibile in se est possibile quoad extra scilicet quoad omnes causas extrahentes ipsum;
85va sed demonstratio Aristotelis est firmior quia est sumpta a diffi|nitione, hec autem a contrariis, quamvis iste locus intret scientias aliquo modo scilicet quando predicatum fuerit proprium contrario et

---

66 scilicet] sed *JW* 67 erit] est *JZ* : esse *W* 69 esse] + scilicet ex se *JZ* 70 ut] quod *JZ* 73 est] et cum *B* | econtrario] –verso *W* : conversio *J* 74-5 est enim] *inv. Z* 75 possibile … impossibile[2]] *inv. Z* 76 quasi] *om. Z* : quod *Y* 77 que] + est *JW* | impossibili] poss- *Z* 79 est] *post* extrinseco *BX* | ex] *om. B* | impossibile[2]] + est *JZ* 81 nisi] *om. Z* 83 in] *om. B* 84 impossibile[1]] poss- *BWXY sed corr.* W_2 | possibile] + et *W* 85 possibile est] *inv. W* | hac] hoc *BJ* | demonstratione] + et *B* | a] ex *B* : et *WY* 86 est[1]] *om. B* | possibili] im- *B* 87 ex—impossibile] *om. homoiotel. YZ* | est[2]] *om. BWY* : + predicto *Z* 88 omnes] *om. WXY* 91 predicatum fuerit] *inv. W* | proprium contrario] proprium *X* : contrarium proprio *YZ* : contrarium predicato proprio *J*

subiectum habuerit aliquod contrarium, tunc necesse est ut contrarium predicati inveniatur in contrario subiecti.

125. [126.] *Dicamus etiam quod illud quod non cadit sub generatione,* 282a25 *scilicet illud quod non est generabile, est eternum sine principio et sine fine necessario; et similiter illud quod non cadit sub corruptione est eternum sine fine. Et dico non generabile et non cadens sub corruptione vera intentione mei sermonis, quod illud quod est vere ens est non generabile,* 85vb *et illud quod est vere non ens est non cadens sub* | *corruptione. Modo igitur volumus diffinire ea, et dicamus quod illud quod est non generabile est modo ens, et non dicitur quod non fuit primo non ens, neque dicitur quod postremo erit non ens. Et cum ita sit, erit alterum horum duorum* 282a30 *consequens alteri: illud enim quod est non generabile non cadit sub corruptione, et omne illud quod non cadit sub corruptione est non generabile etiam. Et utrumque eorum necessario est eternum, quoniam si aliquid fuerit non generabile erit semper eternum, et si aliquid non cadit* 282b1 *sub corruptione erit semper eternum; et hoc manifestum est et existens, et* **T 103a** *sunt ea que superius diximus.* 282b2

Cum declaravit quod impossibile est aliquod eternum corrumpi, manifestabitur ei quod illud quod non est generabile est eternum; et cum declaravit quod impossibile est ut aliquid sit generabile non corruptibile, manifestabitur etiam ei quod non corruptibile est etiam eternum; et cum declaratum fuit ei quod non corruptibile est eternum et quod non generabile est eternum, declaravit per hoc quod

---

92 habuerit] –buit JZ | contrarium[1]] + et *YZ* : + aut *X* | tunc] cum *Y* : enim *B* : cui *X* : + enim *W*

125. [126.] 1 generatione] corruptio- *Z* 2 scilicet] est *W* | non] *om. Z* | generabile] -bilis *W* 3-4 est—corruptione] *om. homoiotel. Y* 4 sine fine] *om. JVWYZ* | generabile] -bilem *B* | corruptione] + et *B* | vera] natura *Z* 5 mei] sui *BX* | vere ens] *inv. JZ* 6 est[1]] *om. Z* | est[2]] *om. WYZ* | non[2]] *om. BWXY* 7 volumus] vult *WY* 8 primo] post *W* | non ens] movens *Y* 9 erit[2]] est *Z* | horum duorum] *inv. B* 10 est non] *inv. BJX* 11 et—corruptione] *om. homoiotel. X* | est] *om. BWXY suppl.* W_2 | non[2]] *om. BYX* | generabile] *om. BWXY suppl.* W_2 12 etiam Et] erit *B* : etiam *YZ* | est eternum] eternum esse *B* | quoniam] quod *B* 13 fuerit non] *inv. J* | erit] *om. Z* 13 et si aliquid] *post* est *(14) Z* 16 corrumpi] + et *B* 17 quod[2]] *om. Z* 19 etiam ei] *inv. WY* : ei *JZ* | non] + illud quod *Z* | etiam[2]] *om. JZ* 21 et—eternum] *om. homoiotel. WYZ*

non corruptibile est non generabile et econverso; et hec est sua intentio in hoc capitulo. Et dixit *Dicamus etiam quod illud* etc., idest et hoc est quod ante declaravit; et est una duarum propositionum quas posuit ad declarandum conversionem; deinde dixit secundam propositionem et dixit *Et similiter illud quod non cadit* etc., idest et similiter declaratum est ex nostro sermone quod illud quod est non corruptibile est eternum, cum demonstratum est quod impossibile est
86ra aliquid | esse generabile non corruptibile.

Et cum posuit has duas propositiones, scilicet quod non generabile est eternum et quod non corruptibile est etiam eternum que multis modis dicuntur, dixit quod non intendit in hoc loco nisi veram intentionem inter alias intentiones, et dixit *Et dico non generabile* etc., idest et intelligendum est ex nostro sermone non generabile et non corruptibile vera intentione. Deinde explanavit illam intentionem et dixit *Quod est vere ens est non generabile*, idest et est illud quod ante diximus, quod illud quod vere est ens scilicet existens non potest non esse, dicitur non generabile vere, et premunivit nos etiam in hoc quia non generabile etiam dicitur de generabili quod est sine causis generationis aut difficilis generationis; et similiter dicitur non generabile illud cuius generatio est finita in hora in qua finiebatur: intendit igitur per non generabile illud cuius esse non precessit non esse, non **T103b**
illud quod est impossibile de quo etiam dicitur non generabile.

Deinde etiam explanavit intentionem non corruptibilis, et dixit *Et illud quod vere est non ens est cadens sub corruptione*, idest et debes intelligere hoc per non corruptibile illud quod est oppositum ei quod vere est corruptibile, et est illud quod deficit post esse; quasi ergo diceret Debes intelligere per non generabile illud quod opponitur vere generabili scilicet enti post non esse, et per non corruptibile illud quod opponitur vere corruptibili scilicet non enti post esse.

22 non^{2}] *om. B* | est^{2}] *om. B* 24 ante] non *YZ* : iam *W sed corr. W_2* 25 dixit secundam] dedit causam *BX* 26 Et similiter] *om. Z* 28-9 est aliquid] *inv. W* : aliquid *BXY* 29 generabile] + et *JZ* 31 est etiam] *inv. B* 33 Et^{2}] *om. YZ* 34 generabile] -bilem *B* 35 vera intentione] –ram -nem *B* 36 dixit] + quod illud *BX* 37 vere est] *inv. WYZ* | ens] + quod *JZ* | non^{2}] nunc *Z* 38 premunivit] -munit *BJ* | quia] quod *BJX* 39 etiam dicitur] *inv. WXY* : dicitur *B* | de] *om. B* : + non *Z* | generabili] *om. B* 40 dicitur non] *inv. Z* 41 in^{2}] *om. JZ* 42 per] quod *BX* 43 etiam] *om. Z* : et *WY* 44 etiam] *om. JYZ* 45 non] *om. B* | est^{2}] + non *JW_2* | debes] -bet *JZ* 46 quod^{1}] + non *Z* 47 est illud] *inv. JZ* 48 Debes] oportet *JZ* 49-50 et—esse] *om. BW sed suppl. W_2* 50 non enti] *om. X*

Et cum posuit has duas propositiones, dixit *Modo igitur volumus diffinire* etc., idest manifestum est ex nostro sermone quod diffinitio non generabilis est illud quod modo est ens et ante non fuit non ens 86rb neque in futuro erit non ens: hec | enim diffinitio est non generabilis quia convertuntur. Deinde dixit *Et cum ita sit* etc., idest quod non generabile et non corruptibile convertuntur adinvicem; deinde incepit declarare quod hec conversio apparet ex duabus propositionibus quas predixit, scilicet quod non generabile est eternum et quod non corruptibile est etiam eternum, et dixit *Et utrumque eorum est necessario eternum*, idest et signum conversionis eorum est quia eternum predicatur de utroque illorum predicatione conversionis; et cum idem predicatur de duobus predicatione conversionis, necessario illa duo convertuntur adinvicem. Et hoc intendebat cum dixit *Quoniam si aliquid fuerit non generabile erit eternum* etc., idest et cum de non generabili predicatur eternum predicatione conversionis, similiter etiam de non corruptibili, manifestum est quod omne non generabile est non T 104a corruptibile et quod omne non corruptibile est non generabile.

Et quia hoc declaratum est ex Analecticis, dixit quod manifestum est, declaratio quidem sic existit: omne non generabile est eternum, et omne eternum est non corruptibile, ergo omne non generabile est non corruptibile. Deinde diximus omne non corruptibile est eternum, et omne eternum est non generabile, ergo omne non corruptibile est non generabile, que est conversa prime; et hoc declaratum est in Analecticis scilicet quod cum tres termini convertuntur adinvicem, necesse est ut conclusio convertatur.

86va8 **126.** [127.] *Dicamus igitur quod si aliquid fuerit corruptibile, necessa-* 282b2
rio erit generabile; et si fuerit non generabile, non erit corruptibile; et si

---

52 est] + igitur *BX* 53 non[1]] *om. B* 54 futuro] + non *YZ* | non[1]] *om. B* | diffinitio est] *inv. WY* : diffinitio *BX* | non[2]] *om. BWXY* 55 quia] quod *JZ* | ita] *om. B* | idest] + erit *Z* 56 incepit] cepit *B* 59 Et[2]] + etiam *JZ* 61 illorum] eo- *B* : duobus *Y* 63 convertuntur] -tentur *Z* 64 erit] est *Z* 67 corruptibile[2] … generabile] *inv. Y* 68 ex] in *WXY* 68-9 manifestum est] *inv. WXY* 69 omne] *om. WYZ* 71 Deinde diximus] dicamus *J* 72 est] *om. WYZ* 73 conversa] diversa *Z*

126. [127.] 1 fuerit corruptibile] *inv. JYZ* 2 erit[1,2]] est *Z* *aliquid fuerit generabile, necessario erit corruptitible; et si non fuerit corruptibile, erit igitur non generabile, sicut diximus superius.* 282b5

Vult declarare in hoc capitulo quod generabile et corruptibile convertuntur cum declaratum est quod non corruptibile et non generabile convertuntur, et dixit *Dicamus igitur quod si aliquid* etc., idest et cum declaratum est quod non generabile et non corruptibile convertuntur, manifestum est quod omne generabile est corruptibile. Demonstratio eius: quia si non fuerit corruptibile, erit non corruptibile; omne enim aut est corruptibile aut incorruptibile; si igitur non corruptibile, ergo non generabile; sed iam posuimus ipsum esse generabile, sed hoc est impossibile. Et cum omne corruptibile est generabile, demonstratio eius quod si non fuerit generabilis, erit ingenerabilis, quilibet enim aut est generabile aut non generabile, et si fuerit non generabile, ergo non corruptibile, sed iam fuit positum corruptibile.

Deinde dixit *Si aliquid fuerit generabile, necessario erit corruptibile*, idest et etiam diximus quod omne generabile est corruptibile, 86vb econverso prime. Deinde induxit demonstratio|nem et dixit *Si igitur non fuerit corruptibile non erit generabile*, idest et demonstratio huius est quod si non fuerit corruptibile, est non corruptibile, et si fuerit non corruptibile erit non generabile; quod est inconveniens. **T104b** Et hoc iam declaratum est in Analecticis esse locum generalem scilicet quia cum fuerint duo termini oppositi dividentes verum et falsum, et fuerint alii duo termini dividentes verum et falsum, et fuerit alterum unius divisionis convertibile cum uno alterius divisionis et reliquum reliquo.

---

3 erit] est *Z* | si] + igitur *JZ* 7 igitur] ergo *W* 8 et[1]] *om. Y* | cum] *om. YZ* 10 quia] quod *JW* | erit non] est non *JZ* : *inv. X* 13 est[1]] *om. BWY* | Et] *om. Z* | est[2]] *om. WXY* 14 generabilis] -bile *W* | erit] est *JZ* 15 ingenerabilis] -bile *WX* | quilibet] quod- *JYZ* | est] *om. WXY* | generabile[2]] in- *J* : corruptibile *Z* 16 sed] et *JW* 18 erit] est *Z* 19 diximus] -cimus *BX* 20 Deinde] *om. Z* | induxit demonstrationem] *om. YZ* | et] *om.* Y 21 non erit] *inv. Z* : est non *J* 22 huius] eius *YZ* | quod] quia *BX* | est[2]] erit *WXY* 23 erit] est *JZ* 24 iam] *om. Z* | generalem] -rabilem *YZ* 25 quia—fuerint] quod cum fuit *B* 26 et[1]] + si *B* | et[1]—falsum] *om. homoiotel. WYZ* | fuerint] + etiam *BX* | termini] + oppositi *J*

127. [128.] *Si igitur non generabile et non corruptibile non sequantur* 282b5 *adinvicem, ergo non sunt eterna. Et verificatio huius est quod alterum illorum consequitur alterum, quemadmodum ego sum dicens quod generabile est etiam corruptibile et omne corruptibile est necessario genera-* 87ra *bile. Et verificatio huius est illud quod | diximus superius, quod ex illo* 282b10 *quod est semper ens et ex illo quod semper est non ens sit aliquod medium diversum ab utroque illorum, et est illud quod est generabile et quod corruptibile; habet enim potentiam ut sit ens et ut sit non ens, sed tamen in tempore et tempore, scilicet ut in tempore ens et in alio non ens.* 282b14

Cum declaravit quod non corruptibile et non generabile non convertuntur nisi ex eo quod declaratum est quod utrumque eorum est eternum, fecit de hoc rememorationem et dixit *Si igitur non generabile* etc., et quia declaratum est etiam, cum posuerimus quod omne generabile est corruptibile et quod hoc convertitur scilicet quod omne non generabile est non corruptibile et quod hoc convertitur (et hec est conversio prime demonstrationis), voluit convertere secundam scilicet ut prima declaret quod omne generabile est corruptibile et econverso. Deinde ex hoc declarabitur quod omne non generabile est non corruptibile et econverso, et dixit *Et verificatio huius est quod alterum illorum sequitur alterum quemadmodum modo sum dicens*, idest et iam declaratum est quod non corruptibile est non generabile, et quod hoc convertitur ex eo quod generabile et corruptibile conver- T105a tuntur. Et hoc est conversum prime demonstrationis, et non est petitio principii, quoniam propositiones quibus utitur in una non utitur in altera.

Et quia in hac declaratione vult ponere conversionem generabilis et corruptibilis esse fundamentum ad conversionem non generabilis et non corruptibilis, incepit declarare quod generabile et corruptibile 87rb convertuntur, et dixit *Quod verificatio huius est illud quod | superius*

---

127. [128.] 1 sequantur] -quuntur *JX* : sequitur *B* 3 illorum consequitur] eorum sequitur *Z* 4 etiam] *om. JZ* 6 est semper] *inv. WX* : + non *Y* 7 medium diversum] *inv. W* | est[1]] *om. Z* : cum *Y* 9 ut] *om. W* | et[2]] + non *Z* 10 non[2]] *om. Z* 12 eternum] + et *B* 13 etiam] et *YZ* 15 hoc] hec *BZ* 15-22 et[2]— convertitur] *om. homoiotel. X* 16 est] *om. WYZ* | conversio] *om. Z* | demonstrationis] + est *W* 17 declaret] -raret *WY* 18 declarabitur] -bit *B* 20 alterum[2]] *om. Z* | modo] ego *JYZ* 23 Et[1]] *om. Z* 27 esse] est *Z* 28 non] *om. Z* 29 illud] *om. BXY* *diximus quod ex eo* etc., idest et verificatur quod generabile et corruptibile convertuntur ex eo quod diximus superius, scilicet quod impossibile est ut generabile sit non corruptibile aut corruptibile sit non generabile: hec enim duo sunt medium inter illud quod semper est ens et inter illud quod semper est non ens, et non est alterum eorum sed habet potentiam ut sit ens et ut sit non ens. Et cum dixit quod hec duo sunt medium inter illud quod semper est non ens et inter illud quod semper est scilicet ens in aliqua hora et non ens in alia, incepit declarare hec literis.

128. [129.] *Dicamus igitur etiam quod omne corruptibile necessario est* 282b14 *medium inter duo, sicut diximus superius. Sit igitur A illud quod semper est ens, et B illud quod semper est non ens, et C generabile et D corruptibile: manifestum est igitur quod C est inter A et B, A enim et B* 87va *non | habent finem neque ultimum in tempore. Generabile enim habet finem et ultimitatem in tempore aut in actu aut in potentia; A vero et* 282b20 *B non habent finem in tempore neque in actu neque in potentia, quoniam non sunt in aliquo tempore ens et in alio non ens, sed alterum* 87vb *eorum est | ens in tempore infinito, et reliquum non ens in tempore infinito. Et similiter D habet finem et ultimum ut C, aut in actu aut in potentia. Omne ergo generabile est corruptibile, et omne corruptibile generabile; declaratum est igitur quod generabile et corruptibile se* **T 105b** *sequuntur.* 282b23

32 aut] + ut JW_2 | corruptibile^{2}] + ut *Z* 34 est^{3}] + ens *Z* 36 quod^{1}] et *Z* | medium] -dia *JZ* 37 aliqua] alia *J* 38 alia] aliqua *JW* | hec] hoc *BZ*

128. [129.] 1 igitur] *om. JZ* | etiam] *om. YZ* 2 Sit] idest *B* 2-3 igitur—est^{1}] istud semper *Z* 5 enim] aut *JZ* 8 alio] aliquo *BJ* : + tempore *W* 9 et—infinito] *om. homoiotel. BWXY sed suppl.* W_2 11 ergo] *om. YZ* : igitur *J* 12-3 se sequuntur] *inv. Z* : sequuntur *B* : se secantur *X*

87va10 Dixit *Dicamus etiam quod omne corruptibile necessario est medium inter duo*, et intendit per duo illud quod est ens semper et illud quod est non ens semper; deinde incepit demonstrare hoc literis, et dixit *Sit ergo* etc.; deinde dixit *Manifestum est igitur quod C est inter A et B*, idest et cum ita posuerimus, manifestum est quod C quod est gene-
87vb10 rabile cadit inter A quod est semper ens et in|ter B quod est semper non ens. Deinde dedit demonstrationem super hoc quod cadit inter duo extrema, et dixit *A enim et B non habent finem* etc., idest et signum quod generabile est medium inter illud quod est semper ens et inter illud quod est semper non ens est quia esse eius quod est ens semper et non esse eius quod est non ens semper est infinitum; gene-
88ra rabile autem habet finem in tempore idest quod est | ens in aliquo tempore et non ens in alio. Et iste sermo componitur sic in secunda figura; sequitur enim ex hoc quod illud quod habet generationem non est semper ens neque semper non ens, ergo necessario est medium, communicationem enim habet cum non esse quia invenitur post non esse, et habet communicationem cum esse quia exit a non esse ad esse. Et dixit *Generabile autem habet finem aut in actu aut in potentia*, quia generabile, quamvis ponatur perpetuum, continget ut sit corruptibile in potentia quamvis non corrumpatur in actu; et potest intelligi quod finis quem habet est per alteram istarum dispositionum, aut aliquando in actu aut aliquando in potentia, et est dignius, non quia intendit per hunc sermonem declarare quod generabile in actu est corruptibile in actu.

Et cum posuit quod generabile habet finem in tempore aut in actu aut in potentia, dedit propositionem secundam ad eam declarandam, T106a
et est quam primo posuit, et dixit *A autem et B non habent finem in tempore*, idest illud quod est semper ens et quod est non ens semper.

---

14 Dicamus] –cimus W 16 hoc] *om.* *YZ* | literis] -teras *Z* 17 ergo] igitur *BX* 18 cum] + hoc *B* | est[1]] + igitur *B* 22 generabile est] *inv.* *Y* : generabile *Z* 23 est semper] *inv.* *JXYZ* | est[2]] *om.* *WYZ* 24 et—semper[2]] *om. homoiotel.* *B* | non[1]] *om.* *JZ* | esse] ens *Y* | est[2]] in *Z* 25 quod] quia *BJX* | est] non *WZ* 26 tempore] *om.* *YZ* | et[1]—alio] *om.* *Y* | non] *om.* *W* : + est *J* | alio] aliquo *JZ* 27 sequitur] semper *B* 28 neque—ens[2]] *om. homoiotel.* *B* 29 enim] + neque semper non ens nisi *B* 30 quia] quod *WY* | a] de *BX* 31 autem] *om.* *BX* 33 corrumpatur] -pat *Z* 34 quod] quia *B* | dispositionum] duarum propositionum *BX* 35 aut[1]] ut *Y* | aut[2]] *om.* *Z* 36 non quia] *inv.* *J* 39 eam declarandam] *inv.* *BY* : declarandam *X* : + eam *W* 40 et[1]] ut *Z* | primo posuit] posuit post *Z* : post posuit *J* 41 semper ens] *inv.* *BX* | est non ens] ens non *B*

Deinde dedit declarationem super hoc et dixit *Quia non sunt in aliquo tempore ens et in alio tempore non ens*; et cum declaratum fuit ei quod necessarium est ut generabile sit medium inter ens semper et non ens semper, declaravit etiam quod corruptibile necesse est ut sit medium per illam eandem declarationem, et dixit *Et similiter erit D* etc., idest et similiter erit D, quod est corruptibile, habens finem in tempore aut in potentia aut in actu, econtrario illi quod est semper ens et quod est semper non ens.

Et cum declaravit quod generabile est medium inter hec duo et similiter corruptibile, conclusit quod corruptibile est idem cum generabili, scilicet quod omne generabile corruptibile et econverso; et hoc *88rb* non verificatur nisi cum | concessum fuerit quod inter ens semper et non ens semper est tantum unum medium. Dicentes autem ens semper in preterito sed in postremo corruptum et generatum, sed perpetuum in futuro, ponunt tria media inter hec duo, scilicet generabile et corruptibile et generatum non corruptibile et corruptum non generabile; sed videtur quod Aristoteles non inducit tales declarationes nisi ad iuvandum, et fortitudo eorum non est nisi fortitudo inductionis in qua perficiuntur omnia particularia, et est primus ordinum certificationis. Et ideo confirmat ipsam per predictam declarationem que est sustentamentum huius capituli, et visum est confirmare hoc dicendo quod illud in quo est potentia ad esse aut non esse infinita, impossibile est ut sit in preterito absque futuro neque in futuro absque preterito: infinitum autem comprehendit duo tempora necessario, nisi esset infinitum maius infinito; ponens autem eternum corruptibile aut generatum perpetuum, ponit potentiam infinitam non comprehendentem duo tempora, quod est impossibile; et cum ita sit, generabile igitur est ens in tempore finito **T 106b** et non ens in tempore finito; et cum ens in tempore finito et non ens in tempore finito convertantur, necesse est ut convertantur inter se.

43 alio] aliquo *BJZ* 46 illam] aliam *B* | Et[2]] *om. Z* | erit] *om. J* : est *Z* 47 idest] *om. Y* | idest—D] *om. Z* | et] *om. W* 49 quod] *om. Z* 50 hec duo] *inv. W* 52 generabile] + est *J* : + et *B* | et[1]] *om. B* 53 inter] totum *Z* 54 semper[1]] *om. BWXY* | est] *om. B* 55 perpetuum] *post* futuro *(56) W* 57 corruptibile[1]] non generabile *Z* 59 eorum] ea- *JZ* 61 Et ideo] *om. Z* : alias *(obsc.)* non *Y* | confirmat] -mavit *J* : non firmat *Z* 63 confirmare hoc] *inv. JZ* | aut] + ad *YZ* 65 neque] aut *JZ* 67 autem] aut *Z* 69 finito] in- *BZ* : + et similiter corruptibile *J* 71 finito] in- *B* | convertantur[1]] + cum generabili et corruptibili *J*

88va14 **129.** [130.] *Sit etiam non generabile A et generabile C, et non corrup-* *282b24*
tibile B et corruptibile H: dico quod C et H consequuntur se cum fue-
88vb *rint | secundum quod narrabo ex literis hiis quas dicam; quia C et H*
consequuntur se, A autem et C non cadunt super idem, sed hoc super rem
et hoc super aliam, et similiter H et B. Si igitur ita sit, erunt H et C
consequentia se; sit igitur H non consequens C, tunc erit igitur C conse- *282b30*
quens A; H enim et C non cadunt simul super idem, B autem et C cadunt,
H ergo consequetur ad B; quod est impossibile propter hoc quod superius
diximus. Et similiter erit actio de A et B, propter hoc quod A sequitur B. *282b32*

a	b
c	h

89ra Vult in hoc capitulo declarare quod cum fuerint duo termini
oppositi dividentes verum et falsum in omnibus entibus, et duo alii
similiter et unus unius divisionis cum uno alterius, tunc necessario
reliquus cum reliquo. Et vult hoc declarare generaliter sicut fecit in
Analeticis, ideo quod accepit literas loco oppositorum, et cum hoc
declaravit declaratione universali, dicet quod generabile et non gene-
rabile et corruptibile et non corruptibile sunt ex huiusmodi opposi-
tis, quapropter concludetur quod cum unus duorum terminorum ex
una divisione horum quatuor scilicet generabile et corruptibile et T107a
non generabile et non corruptibile, convertetur cum alio termino
secunde divisionis, necessarium est ut reliquus cum reliquo.

Incepit igitur prius, et posuit quesitum quod declarare intendit, et dixit *Sit igitur non generabile A* etc., idest et manifestabitur modo quod generabile et corruptibile et non generabile et non corruptibile, si unum consequitur unum, et reliquum reliquum ex eo quod dicam. Deinde dixit *Dico quod C et H consequuntur* etc., idest et dico quod

---

129. [130.] 3 quia] modo *JZ* 5 ita sit] *inv. Z* | erunt] *om. BJ* | C] + sunt *BJ* 6 sit] si *B* | erit] est *WZ* : *post* igitur *JX* | C^{2}] *om. B* : H *W* 7 B] A *JZ* | C^{2}] B *J* : A *W* 8 ergo] igitur *BJ* | ad] *om. WXY* 9 erit] est *WZ* | et^{2}] + de *W* 10 quod] quia *BY* 12 et] + convertitur *J* : + convertat *W* | necessario] -cesse *B* 13 reliquus] -quum *BZ* : -quo *X* 14 ideo quod] et ideo quod *Z* : ideo *JX (leg.* ideoque *?)* 17 duorum] eo- *Z* 19 termino] vero *(obsc.) Z* 20 secunde divisionis] -dum divisionis *B* : divisionis secundum *W* 22 igitur] *om. X* | non] *om. BXYZ* 24 unum^{2}] verum *Z* | reliquum^{2}] falsum *Z* 25 Deinde] *om. YZ* | C] + etiam *Z* generabile scilicet C et corruptibile scilicet H necessarium est ut convertantur cum posuerimus quod A scilicet non generabile et B scilicet non corruptibile convertuntur: declaratio huiusmodi est secundum quod narrabo cum acceperimus loco literarum opposita et cum declaraverimus hoc in eis. Et mutavit declarationem ad opposita in eo quod sunt opposita, quia cum hec intentio verificabitur in oppositis, verificabitur etiam in eis que sunt sub oppositis; et fecit hoc ut demonstratio esset essentialis et prima, ut declaratum est in Posterioribus Analeticis, quoniam si hoc declarasset in generabili et non generabili et corruptibili et non corruptibili, esset demonstratio *89rb* | accidentalis; et quasi declararet quod triangulus isocheles haberet tres angulos duobus rectis equales: hoc igitur intendebat cum dixit **T107b** *Cum fuerint secundum quod narrabo ex literis istis quas dicam modo*, idest cum acceperimus has literas loco oppositorum simpliciter; et est illud quod narravit.

Deinde dixit *Et etiam C et H consequuntur se*, idest dico quod cum acceperimus has literas quas accepimus loco generabilis et non generabilis et corruptibilis et non corruptibilis, loco oppositorum simpliciter, ita quod A et B convertantur, et tunc C et H convertuntur; conditiones autem in istis duabus divisionibus sunt due, scilicet ut utraque divisio dividat verum et falsum in omnibus entibus, et ut unus unius divisionis convertatur cum uno alterius. Et cum posuit quod cum A et B convertuntur, necessarium est ut H et C convertantur, quamvis ipse non propalavit illud quod est quasi precedens sed illud quo est quasi consequens, induxit demonstrationem super hoc et dixit *A autem et C non cadunt super idem sed hoc super rem et hoc super aliam rem*, idest et demonstratio huius est ut proponamus

---

26 scilicet[1]] *om. Z* | C] *om. YZ* | scilicet[2]] *om. Y* | necessarium] *om. B* 28 declaratio huiusmodi] *inv. JZ* 29-31 et—opposita] *om. homoiotel. Y* 31 sunt] fuerit *B* 33 hoc ut] *inv. B* | esset essentialis] *inv. Z* 34-6 in—declararet] *om. homoiotel. Y* 35 non[1]] corruptibili *Z* 36 quasi] quod *BZ* | declararet] -claret *B* 37 rectis] *om. B* | equales] *ante* duobus *BX* 38 literis istis] *inv. BX* | quas] quasi *W* 41 Et[1]] *om. BY* | etiam] quod *BWY* | se] *om. W* | idest] *om. YZ* | quod] *om. JW* : autem *Z* 42 acceperimus] -pimus modo *B* 43 et[1]] *om. Z* 45 divisionibus] *om. WYZ* 46 entibus] essentiali- *W* 47 Et] + quod *Z* | cum posuit] composuit *B* 48 quod] quia *B* | cum] *om. Z* | convertuntur] -tentur *B* : -tantur *J* | necessarium] -cesse *BY* : -cessario *Z* | H et C] Z et H *BX* 49 quasi] *om. WZ* 50 quasi] *om. WYZ sed suppl.* W_2 51 C] + idest *JWZ* | super[1]] sub *B* 52 et] *om. WY*

huic quod proponendum est ex eis que proposuimus, et est quod A scilicet unum oppositorum et C scilicet alterum oppositorum non dicantur insimul de eodem, et similiter alia duo opposita scilicet B et H. Et cum posuit has duas propositiones, dixit *Si igitur ita sit erit, H et C consequentia se* cum posuerimus A et B consequentia se.

Deinde dedit demonstrationem huius ducentem ad inconveniens, et dixit *Sit igitur C non consequens H* etc., idest quia cum C non sequebatur H, idest cum C non dicatur de H, necesse est ut dicatur T108a
89va de hoc oppositum C quod est A, cum C et A non | dicantur insimul de eodem neque denudetur ab altero eorum ens. Et cum posuit quod si C non dicatur de H, necesse est ut dicatur A de ipso, contingit huic propositionem quam primo posuit, et est quod A et B dicuntur de se adinvicem, et dixit *A autem et B cadunt super idem*, idest et utrumque illorum iam positum est converti cum suo compari cum dicantur de eodem quod est eorum subiectum. Deinde dixit *Consequetur igitur H ad B, quod est impossibile*, idest cum posuerimus quod H consequatur A et B consequatur A, necesse est ut H consequatur B, idest quia si H fuerit, B erit; sed iam posita fuerunt opposita non convenientia eidem; quod est impossibile. Et cum declaravit quod si A et B convertantur, necesse est ut C et H convertantur, incepit declarare conversum, scilicet quod si C et H convertuntur, necesse est ut A et B convertantur, et dixit *Et similiter erit actio de A et B*, idest et similiter comparabuntur in demonstratione quod A et B convertuntur adinvicem, cum posuerimus quod C et H convertantur.

90ra37 **130.** [131.] *Dico igitur quod non generabile scilicet A, differt a generabili scilicet C, et quod non est corruptibile scilicet B differt a corrup-* 283a1

53 quod[1]] et *W* | proposuimus] -suerimus *Y* : posuimus *Z* 55 et[2]] *om. BXY* 56 cum] *om. B* | erit] *om. BJ* : est *WZ* 57 cum—se] *om. homoiotel. WYZ* | posuerimus] -suimus *J* 58 ad] *om. Z* 59 Sit] si *B* 60 sequebatur] -quitur *X* : consequatur *W* | sequebatur—non] *om. homoiotel. B* | cum] de *Z* | de] *om. Z* 60-1 necesse—hoc] *om. XY* 62 denudetur] divide- *WX* 63 ipso] *om. JWZ* 64 propositionem] -tioni *YZ* 66 illorum] alio- *YZ* | iam] etiam W 68 Consequetur] -quitur *WXYZ* 70 fuerunt] fuerint *W* : sunt *B* 71 eidem] eisdem *Z* : ei *X* 73 C ... H] *inv. BX* 75 comparabuntur] operabun- *BX* | in] *om. YZ* 77 convertantur] -tuntur *BXY*

130. [131.] 1 Dico] -cit *Z* : -camus *J* 2 et] *om. Z* | est] *om. BJXY*

90rb *tibili, scilicet H, et A et C non cadunt insimul super idem, | neque B et H cadunt etiam insimul super idem, sicut diximus superius. Impossibile est enim ut idem sit factum nove et semper eternum, et impossibile est etiam ut idem sit corruptibile et non corruptibile scilicet semper infinitum.* *283a3*

Cum declaravit hanc conversionem in oppositis secundum quod sunt opposita, vult demonstrare quod in generabili et non generabili et corruptibili et non corruptibili inveniuntur conditiones que, cum **T108b** inveniantur in oppositis, invenietur in eis conversio ut concludat quod in generabili et non generabili et corruptibili et non corruptibili inveniatur conversio, et dixit *Dicamus modo quod non generabile* etc., idest quod generabile et non generabile sint opposita, et similiter corruptibile et non corruptibile. Deinde dixit *Et non cadunt A et C super idem*, idest et dicamus quod generabile et non generabile sunt opposita, quoniam non cadunt in idem insimul neque est aliquid absque altero, et similiter de corruptibili et non corruptibili. Deinde affirmavit ea esse non convenientia eidem, et dixit *Impossibile est enim ut idem sit factum et eternum* etc., idest et diximus quod generabile et non generabile non conveniunt in eodem, quia notum est per se quod impossibile est ut idem sit factum nove et non factum, et similiter de corruptibili et non corruptibili, idest quia impossibile est ut idem sit corruptibile et incorruptibile in eodem tempore.

131. [132.] *ʽManifestum est igitur quod C et H consequuntur se et cadunt super idem: idem enim potest esse generabile sub corruptione; omne ergo generabile est corruptibile et omne corruptibile est generabile.* *[add. ad 283a3]* *90va* *| Et similiter A et B cadunt super idem et consequuntur se, quoniam*

---

3 et A] *om. BJWZ* | C] *om. JZ* | cadunt] + A et C *JZ* 4 cadunt] *om. W* | insimul] similiter *Z* : *om. BWXY* | idem] + neque et H cadit etiam super idem *Z* 5 et[2]] *om. WZ* 6 est] *om. W* | etiam] enim *Z* | sit] + ut sit *W* 9 et] + in *WY* 10 et[1]] *om. JYZ* | non] *ante* corruptibili[1] *Y* | corruptibili[2]] + non *Z* 11 conversio] –versis *Y* 13 modo] ergo *BX* 16 dicamus] -ximus *XY* 17 insimul] simul *WXY* 20 factum] + nove *JZ* 21 quia] quod *Z* : neque *B* 22 ut] quod *Y*

131. [132.] 1-7 Manifestum—generabile] *paraphrasis textus Aristotelis translationem Arabicam secuta est* 2 sub] sine *B sed corr.* B_2 3 est[2]] *om. WZ* *idem potest esse non generabile et non corruptibile; sicut igitur est non generabile, erit non corruptibile, et sicut est non corruptibile erit etiam non generabile*⌉.

Cum declaravit hanc conversionem in oppositis simpliciter, deinde **T109a** in generabili et non generabili et corruptibili et non corruptibili, dedit conclusionem quam intendebat declarare, et dixit *Manifestum est igitur* etc., idest et intendebat per C et H generabile et corruptibile, non opposita simpliciter sicut fecit ante. Et signum huius est hoc quod dixit *Quoniam idem erit generabile sub corruptione*, idest declaratum est igitur ex hoc quoniam omne generabile est corruptibile et omne corruptibile est generabile, cum declaratum est quod termini qui conveniunt ex talibus oppositis in eodem, si unum convertitur et reliquum par convertetur. Deinde dixit *Et similiter A et* **T109b** *B*, idest et similiter apparet quod non generabile et non corruptibile convertuntur ita quod omne non generabile est non corruptibile et econverso.

132. [133.] *Dicens autem non esse impossibile ut aliquid generatum sit* *283a4* *non corruptibile, et aliquod antiquum corruptum, destruit quasdam propositiones quas posuimus et sillogismos quos fecimus; facit enim non corruptibile generatum, et antiquum, non generabile facit corruptum.* *283a7*

Contradicit per hunc sermonem dicentibus generabile esse eternum aut corruptibile esse eternum, et dixit quod qui hoc fingit, *90vb* necessarium est ei destruere propositiones et sillo|gismos quod declaravit; dicet enim contradictorium illarum duarum propositionum, quoniam propositiones et sillogismi quod induximus dant ut eternum impossibile est esse generabile aut corruptibile.

---

5 idem] *om. YZ* 6 erit[1]] est *W* | etiam] *om. WXYZ* 9 et[2]] *om. JZ* 10 dedit conclusionem] *inv. Z* | intendebat] -bant *B* 11 idest] *om. BX* 13 quod] et *Z* | Quoniam] quod *Z* | erit] *om. W* 14 quoniam] quod *JZ* 15 declaratum] demonstra- *BXY* 16 termini] eternum *B* | unum] + per *JX* 17 convertitur] -tuntur *Z* | par] *om. Y* 19-20 et econverso] *om. Y*

132. [133.] 2 et] *om. B* | destruit quasdam] *inv. BJ* 3 et] in *YZ* 6 aut—eternum] *om. homoiotel. WZ sed suppl.* W_2 | dixit] -cit *BX* 7 necessarium] -cesse *WXY* : -cessario *Z* | declaravit] -ravimus *BX* 8 contradictorium] -dicet *Z* | illarum duarum] *inv. JZ* : illarum *B* : suarum *X* 9 quod] quos *XY* | induximus] –duxit *Y* | ut] et *Y*

133. [134.] *Perscrutandum est igitur de hoc, dicendo quod quilibet* 283a7 *habet potentiam aut patiendi aut agendi, aut ut sit ens aut ut sit non ens, et hoc aut in tempore infinito aut in tempore finito, habente principium et finem. Tempus autem infinitum est etiam terminatum, cum post ipsum non sit tempus neque invenitur tempus maius illo. Infinitum vero in aliquo tempore non est eternum infinitum, neque est etiam terminatum habens principium et finem.* 283a10

Vult perficere rationem istius capituli scilicet quod impossibile est ut aliquid sit generatum et post eternum, et ut aliquid sit eternum et post corruptum; et hoc intendebat cum dixit *Perscrutandum est*, idest perficiende sunt rationes; et dixit *Quilibet habet potentiam aut* T 110a *patiendi* etc., idest potentia et actus attribuuntur rei quatuor modis, potentia vero duobus modis, aut potentia que agit scilicet activa, aut qua patitur scilicet passiva. Deinde dixit *Aut est non ens aut est ens*, idest actus vero aut datus est ad esse aut ad non esse scilicet aut est non ens in actu aut ens in actu, quasi igitur diceret quia res aut erit in actu aut in potentia, et que in actu aut ens aut non ens, et que in potentia aut patiendi aut agendi. Deinde dixit *Et hoc aut in tempore* 91ra | *infinito* etc., idest quod hec divisio est nota per se, scilicet quod inter finitum et infinitum non est medium.

Et quia dicentes generatum remanere eternum et non corrumpi in postremo et eternum in preterito et corrumpi in futuro, faciunt tempus infinitum tribus modis, incepit destruere hanc opinionem de infinito, et dixit *Et tempus infinitum est terminatum* etc., idest et impossibile est ut sit tempus maius tempore infinito, quoniam si esset, esset infinitum maius infinito; et hoc sequitur dicentem esse

---

133. [134.] 2 ut sit[2]] *om. Z* 3 finito] terminato *BX* | habente] –bentem *Z* 3-4 habente—terminatum] *om. homoiotel. B* 4-7 Tempus—finem] *om. homoiotel. Y* 4 autem] *om. J* | cum] etiam *praem. B* 5 neque] + etiam *WZ* 6 vero] *om. X* : non *WZ sed corr.* W_2 : autem *J* | est etiam] *inv. BJ* 8 Vult] + etiam *BX* | istius] huius *W* 9 et ut] aut *BX* | sit[2]] *om. BX* 10 dixit] + et *BX* 12 modis] *om. WXY* 13 activa] accidentia *Y* 14 qua] que *BJ* : quia *Y* | est[1]] erit *BX* | non] *om. W* | est[2]] *om. XY* : erit *B* : non *W* 15 datus] actus *W* | esse[1]] + est ad esse *W* | non] *post* ad[1] *BX* | est[2]] erit *XY* 16 actu[1, 2]] actum *Z* | erit] est *Z* 17 in[2]] *om. JWX* | et[1]] + quia *W* 18 patiendi … agendi] *inv. BX* 19 est nota] *inv. J* : nota *B* 20 finitum … infinitum] *inv. BX* 21 et[2]] *om.* Y | non] eternum *BXY* | corrumpi] -ruptum *B* 22 eternum] *om. Z* | preterito … futuro] *inv. JZ* | corrumpi] *om. XY* | futuro] –turum *BW* 24 est] + etiam *J* | terminatum] + etiam *BX* 25 est] *om. Z* 26 sequitur] super *Y* | dicentem] -cente *Z* infinitum in preterito et finitum in futuro, aut econverso. Et non debet aliquis dicere quod hec intentio non est deleta in tempore, quoniam si invenimus hoc tempus presens, invenimus quod tempus quod est ante ipsum est infinitum et quod est post ipsum est infinitum, secundum sententiam dicentis mundum esse eternum quoniam non invenietur hoc in tempore nisi accidentaliter, cum nihil ex tempore existit in actu.

Quod autem est impossibile est ut infinitum sit in actu maius infinito, et hoc aut potentia aut in esse, quoniam potentia que remanet in infinito in tempore preterito absque futuro, aut in futuro absque preterito, est potentia que potest facere aliquid infinitum; impossibile est igitur ut ponatur in preterito absque futuro aut in futuro absque preterito, nisi potentie infinite adderent supra se, quod est T110b impossibile. Et ideo dixit *Quod tempus infinitum est terminatum*, idest impossibile est ut aliquid addatur supra ipsum scilicet tempus eiusdem potentie infinite; et cum ita sit, omnis ergo potentia aut agit in tempore finito aut infinito, et que agit in tempore infinito eadem potentia tantum; et manifestum est quod hec demonstratio differt a demonstrationibus predictis. Deinde dixit *Infinitum autem in aliquo* 91rb *tempore non est eternum infinitum, neque etiam | est terminatum habens principium et finem*, idest illud autem quod habet potentiam infinitam in tempore terminato, aut in preterito tantum aut in futuro tantum, non est in tempore eterno quod non habet terminum neque in tempore terminato; et inter eternum et terminatum non est medium.

134. [135.] *Et etiam quod in primo est eternum ens, deinde fuit cor-* 283a11 *ruptibile in tempore eius quod est infinitum, scilicet in tempore eius quod non fuit primo non ens, deinde in postremo fuit non ens et in alio*

---

27 infinitum] -nito *Z* : + et hoc *J* | in[1]] + tempore *JZ* | finitum] futurum *Y* 29 si] cum *BX* 31 secundum] *om. Z* 32 invenietur] -nitur *XY* 34 est[2]] *om. X* : + impossibile *Z* | ut] + sit *WZ* 35 que] qua *Y* 38 aut] et *Z* 40 ideo] non *Y* 41 supra] super *BX* 43 aut] + in tempore *W* | tempore[2]] + finito aut *Y* | infinito[2]] + est *J* 45 demonstrationibus predictis] *inv. JZ* 46 terminatum] eternum *Y* 47 autem] *om. JZ* | quod] *om. BXYZ* | habet] *om. XY* : autem *B* 49 quod] quia *Y* | quod—terminum] *om. J* | neque] *om. BZ* : nec *J*

134. [135.] 1 etiam] + illud *JZ* | quod] quia *XY* 2 quod] quia *YZ* 3 fuit[1]] fuerit *BJ* | non[2]] *om. XY* | alio] aliquo *Z* *tempore. Si igitur tempora non diversantur quia sunt infinita, manifestum est ex quo diximus quod aliquid generatum fuit in tempore infinito et quod aliquid fuit corruptibile et est eternum infinitum in tempore; quod est impossibile. Generabile enim impossibile est ut sit eternum, et* *283a15* *similiter corruptibile impossibile est ut sit eternum, sed tamen possunt esse ens et non ens, quoniam corruptibile habet potentiam ut non sit ens in postremo, et generabile habet potentiam ut sit non ens in primo.* *283a17*

Idest quod illud quod est eternum in preterito, deinde in postremo est corruptum, habet potentiam ad corruptionem in tempore infinito, et est tempus in quo fuit eternum; et hoc intendebat cum dixit *Scilicet in tempore eius quod non fuit primo non ens deinde in postremo fuit non ens*; et dixit hoc quia illud quod primo fuit non ens deinde post fuit ens est generatum.

Deinde dixit *Si igitur tempora non diversantur* etc., idest et si ali- T111a quid inveniatur generatum postquam fuit non ens in tempore infinito, aut aliquid non ens postquam fuit ens in tempore infinito, tunc *91va* invenietur in generato potentia generationis in tempore infinito, | et in corruptibili potentia corruptionis in tempore infinito: sed tempus infinitum non diversatur ita quod possumus dicere quod tempus potentie ad aliquid in eo est aliud a tempore actionis, quapropter necessarium est ut illud quod est non ens semper non sit generabile, et quod est ens semper non sit corruptibile. Distinguendum est enim tempus esse a tempore eius quod est non esse; et cum posuerimus tempus esse aut non esse infinitum, non invenietur tempus quod dignius sit ad esse quam ad non esse, aut econverso; necesse est igitur ut illud quod est non ens in tempore infinito sit non ens semper scilicet in toto tempore, et similiter illud quod est ens in tempore infinito.

Deinde dixit *Generatum enim impossibile est ut sit semper* etc., potest intelligi Et propter hoc generatum impossibile est ut semper sit non ens quoniam tunc non esset generatum; et etiam eternum

---

7 Generabile] -ratum *BJ* | enim] igitur *WZ* 8 impossibile est] *inv. Z* 15 quia] quod *BYZ* | quod] + in *WXY* 17 tempora] *om. WYZ* | et] *om. B* 19 aut—infinito] *om. J* 20 invenietur] -venit *Z* 23 ad aliquid] *om. B* 24 non^1] *om. XY* 26 posuerimus] + ipsos *B* : + quod *X* 27 tempus esse] *inv. B* 30 est] + non *Z* 31 infinito] + sit ens semper *JW* 32 impossibile] quod possibile *Y* 33 Et] *om. YZ* 34 non esset] *inv. JX*

impossibile est ut sit corruptum quoniam tunc esset non corruptum; et potest intelligi quod generatum impossibile est ut sit semper ens, quia tunc esset non generatum, quia tempus non esse in eo non distinguitur a tempore esse, cum tempus esse est infinitum.

Et cum dixit quod generatum impossibile est ut semper sit ens, et quod illud quod semper est ens impossibile est ut sit corruptum, demonstravit quod generabili contingit contrarium, et est ut aliquando sit ens et aliquando non ens, et similiter corruptibili, et dixit *Sed tamen possunt esse non ens et ens*, idest aliquando ens aliquando non ens. Deinde dedit differentiam inter generabile et non ens, in hoc scilicet quod in generabili precedit tempus non esse ad tempus eius quod est non esse, et econverso in corruptibili, et dixit *Quoniam corruptibile habet potentiam* etc., idest sed corruptibile differt a generabili
91vb rabili in hoc quod corruptibile habet potentiam ut in po|stremo sit non ens, generatum autem habet potentiam ut in primo sit non ens. Et secundum hoc quod explanavimus erit principium demonstrationis quia cum generabile ponatur ens in tempore infinito, contingit ut non sit generabile, et cum ens in tempore infinito ponatur corruptibile, contingit quod non sit corruptibile: corruptibile enim aut generabile est cum tempus sui esse est distinctum a tempore sui non esse.

Fundamentum autem predicte demonstrationis est quod nulla potentia est media inter finitum et infinitum, cum potentia infinita est una potentia tantum; demonstratio igitur in hoc capitulo super T111b
hoc est alia a demonstratione predicta in predicto capitulo. Themistius vero explanat capitulum predictum alio modo: dicit enim quod primo declarat in eo Aristoteles quod omnis potentia est finita, sive fuerit in tempore finito sive fuerit in tempore infinito, et similiter que est in tempore infinito est etiam finita; deinde posuit quod inge-

36 quod] quia *WXY* : quoniam *B* | sit semper] *inv. Z* 37 quia[1]] quoniam *BX* | generatum] + et *B* 38 esse[1]] + et *B* 41 generabili] -rabile *B* : -raliter *XY* 43 non] *post* et *JZ* | ens[3]] *om. X* : + et *B* 44 non ens[2]] corruptibile *JW* 46 non] *om. JW* 48 quod] quia *BX* | ut] *om. Z* 49 ut] *om. BZ* | in] *om. WXY* 50 erit] est *WZ* | principium demonstrationis] *inv. JZ* 51 quia] quod *BJX* | infinito] finito *B* 53 contingit] -get *BX* 54 tempus] -pore *Z* 55 predicte] prime *Z* 56 potentia est] *inv. BX* | finitum] –nitam *BX* | infinitum] -nitam *BX* 57 potentia tantum] *inv. J* : potentia *Z* 58 predicta] *om. B* | predicto] -cedenti *JZ* 59 vero] autem *JZ* 59-60 quod primo] *inv. Z* : quod in primo *B* 60 declarat] -ravit *JZ* 61 finito—tempore] *om. homoiotel. Y* | fuerit[2]] *om. Z* 62 etiam] *om. YZ* | finita] in- *XZ*

nerabili eterno et incorruptibili eterno contingit ut potentia sit infinita; et cum omnis potentia est finita, ergo positio istius potentie est impossibilis, et cum non sit potentia, non est generatio neque corruptio; potentia enim est generabilis et corruptibilis. Deinde declaravit in secundo capitulo quod potentia, que est ingenerabili et eterno corruptibili necesse est ut sit infinita, et est una propositionum acceptarum in demonstratione unius modi, igitur erunt due demonstrationes.

Et secundum hanc explanationem non intendebat (cum dixit in primo capitulo Et tempus infinitum est terminatum) declarare quod impossibile est quod potentia infinita sit maior potentia infinita, ut
92ra nos explanavimus, sed | intendebat docere quod potentia, que est in tempore infinito in duobus extremis, est aliquo modo finita; quasi igitur dicat quod omnis potentia est finita quoniam aut erit in tempore finito aut in tempore infinito, et quocumque modo sit, erit terminata: tempus enim infinitum est terminatum aliquo modo. Et secundum hoc est intentio sui sermonis etiam *Infinitum autem in aliquo tempore* etc., idest illud autem quod est eternum in aliquo tempore potentia eius non est finita sicut potentia eterni in omnibus temporibus, neque potentia eius est sicut potentia eius quod est in tempore infinito.

Deinde incepit post hoc in eo quod est eternum in aliquo tempore, et dixit *Et etiam illud quod in primo est eternum ens* etc., idest et dicamus etiam quod potentia eius quod semper fuit in preterito, et in postremo est corruptum ad corruptionem, est potentia infinita, et similiter potentia eius quod generatur postquam non fuit in tempore infinito ad generationem est infinita, ut concludat ex hoc quod in hoc modo entis non est potentia ad generationem T112a
neque ad corruptionem; et hec explanatio est propinqua nostre

63 eterno[1]] *om. WYZ* | incorruptibili] cor- *XY* 65 potentia] *om. Z* | est] *om. J* : erit *BXY* 66 potentia] -tialis *WY* : possibilis *Z* | enim] ei *praem. WZ* | generabilis] -rabili *WXY* : in generabile *Z* | corruptibilis] -tibili *WXY* : -tibile *Z* 67 que est] *inv. W* : est *Z* 69 unius] *om. W* : primi *J* | modi] + non *J* 72 tempus] tunc *W* 73 potentia infinita[1]] *inv. Z* 74 docere] dic- *JZ* 75 aliquo] alio *XY* | quasi] a. *(obsc.) YZ* 77 finito—tempore] *om. homoiotel. Z* | aut—infinito] *om. homoiotel. X* | in tempore] *om. W* | erit] est *WZ* 79 est] erit *BXY* | etiam] *om. YZ* 82 est sicut] *inv. W* 84 aliquo] alio *Y* 85 quod] + est *Z* 86 dicamus] -cimus *W* : -catur *Z* 87 potentia infinita] *inv. JZ* 91 hec—est] hec expositione *B* : est explanatio *Y*

explanationis, sed differunt in predicto capitulo; nos vero posuimus ipsum capitulum declarationem per se, Themistius autem posuit ipsum dependens cum secundo; et bene tolerabiles sunt ambe explanationes.

135. [136.] *Dicamus etiam quod contraria non conveniunt insimul in* 283a17
aliqua hora temporis infiniti, quoniam si in aliqua hora convenirent, tunc idem haberet potentiam ut esset et non esset in tempore infinito;
quod est impossibile, sicut declaravimus superius. 283a20

92rb Cum induxit hanc declarationem, confirmavit ipsam per declarationem super quam sustentatur in hoc, et est quod ponenti hoc contingit ut possibile sit impossibile, et dixit *Quod contraria non conveniunt* etc., idest quod cum positum fuerit tempus esse unius dictorum duorum contrariorum, et potentie ad esse alterius contrarii tempus infinitum, continget ut duo tempora eorum sint idem; si igitur ponatur potentior illorum in aliquo tempore esse ens, contingit ut contrarium eius in illo tempore sit ens, quia suum esse est etiam infinitum in tempore; et cum ita sit, positio eius erit ad esse ens in actu falsa impossibilis, et illud cuius exitus in actu fuit impossibilis, nullam potentiam habebit.

Deinde dixit *Quoniam si conveniunt in aliquo tempore, tunc illud idem habebit potentiam ut sit et ut non sit in eodem tempore*, idest et contingit ex hac positione duo contraria in actu esse, quia cum posuerimus aliquod ens habere potentiam ad esse in tempore infinito et ad non esse, erit tempus exitus duarum potentiarum in actu idem tempus, erit igitur idem habens potentiam ut sit et ut non sit in eodem tempore, verbi gratia quia si in Sorte est potentia ad surgen-

---

92 in] a *JZ* 94 dependens] -dere *JZ*

135. [136.] 1 Dicamus] Et *praem. BX* 2 in—convenirent] conveniunt in aliquo tempore *J* 3 esset[1, 2] sit *J* | et—esset] *om. Z* 4 impossibile] + sed *W* | sicut] sed *Z* 5 induxit] dixit *B* | hanc declarationem *Z* | ipsam] *om. WYZ* 6 ponenti] potenti *Z* 7 possibile … impossibile] *inv. XY* | sit] et *Z* 8 quod] *om. B* 9 dictorum] *om. WXY* 11 potentior] -tia *JW* | contingit] -get *BX* 13 erit] *ante* positio *BX* | ad] *om. BXYZ* | esse] *om. YZ* | ens] *om. W* 16 conveniunt] -nirent *Z* | illud] *om. YZ* 17 idem habebit] *inv. Z* | eodem tempore] *inv. JW* 17-8 idest—esse] *om. Z* 17 et[2]] *om. BX* 18 esse] + in eodem tempore *WXY* 21 erit] est *W* 22 quia] quod *B* | Sorte] sor *X* : Socrate *J* : secundo *BYZ* : sectio *(obsc.) W*

dum et sedendum in eodem tempore, necessario invenitur in eodem tempore surgens et sedens insimul, et si posuerimus hoc in eo cadens in tempore alio a tempore sui casus, erit falsum possibile non falsum impossibile; quod est inconveniens.

136. [137.] *Et etiam dicamus quod potentia est ante actum scilicet quod* 283a20 T112b
res habet potentiam ut sit aliquid antequam faciat ipsum. Et cum ita sit, erunt res secundum quod dicunt semper sine fine, scilicet illud quod sem-
92va *per fuit ens | et illud quod in postremo fuit ens, ut semper sint sine fine, semper erunt res habentes potentiam ut sint et non habent potentiam ut sint ens in tempore et ut sint non ens in tempore infinito insimul.* 283a24

Videtur quia cum declaravit quod impossibile est ut aliquid quod semper fuit et semper erit habeat potentiam ad corruptionem, aut quod aliquid quod semper fuit non ens et semper erit non ens habeat potentiam ad esse; continget enim ex hoc quod possibile revertatur ad impossibile, vult mutare hanc declarationem ad illud quod positum est semper esse in preterito sed in postremo corrumpitur, et ad illud quod ponitur generatum quod semper est ens in futuro. Et preponit huic propositionem concessam, et est quod necesse est ut potentia eius quod generatur aut corrumpitur sit precedens secundum tempus generationi aut corruptioni; et hoc intendebat cum dixit *Et dicimus etiam quod potentia*, etc.

Deinde dixit *Et cum ita sit* etc., idest et cum potentia generationis aut corruptionis precedit tempore generationem aut corruptionem, continget ut res que semper fuerint et in postremo corrumpuntur habeant potentiam ad corruptionem in tempore infinito, et similiter necesse est ut res que generantur postquam non erant in tempore

---

23 necessario] -cesse *B* | invenitur] -niretur *Z* : -nietur *XY* 24 surgens … sedens] *inv. JW* 25 alio] *om. Y* : aliquo *Z* | possibile] + et *B*

136. [137.] 1 etiam dicamus] *inv. JZ* 2 ut] *om. Z* | sit] faciat *B* 3 secundum] eedem *Z* : hee eedem *Y* 4 sint] fuit *Z* : sit *BX* 5 habentes] + habuerint *XY* | sint] sit ens in tempore *B* 6 in tempore[1]] *om. Z* 7 quia] quod *JXZ* | cum] *om. B* 8 aut] ut *YZ* 9 quod[1]] *om. BWY* : ad *Z* : ut *X* | erit] est *Z* | habeat] -bet *B* 10 ex] ut *Z* | quod] ut *BX* 11 mutare] -tari *Z* 13 quod[2]] quia *Y* : sed *BX* | est] erit *BXY* | preponit] propositionum *Z* 14 propositionem] -tioni *WZ* 16 generationi … corruptioni] –ratum … -ruptum *JZ* 19 precedit] -cedunt *B* | tempore] -pus *Z* 20 fuerint] -runt *Z* : -rant *B* 21 similiter] *om. Z* 22 erant] erunt *YZ* infinito habeant potentiam ad generationem in tempore infinito. Et quia hoc non contingit eis nisi propter hoc quod ponunt generatum semper remanere in futuro et corruptum quod semper fuit in preterito, dixit *Secundum quod dicunt semper sine fine*. Et quia dixit *Erunt res* demonstravit in quibus rebus contingit hoc, et dixit *Scilicet illud* *92vb* *quod semper fuit ens* etc., idest et intelligo per res in quibus con|tingit ut sit potentia aut res que semper fuerunt et in postremo non sunt, et rem aut res que semper non fuerunt et in postremo sunt, sci- **T113a** licet quod per potentiam eorum in hiis duobus ut sint semper et sine fine, contingit ut sit in eis potentia infinita, in uno vero eorum post generationem, in altero autem ante corruptionem. Et hoc intendebat cum dixit *Ut sint semper sine fine*, et hoc quod dixit *Semper habent potentiam ut sint* est predicatum illius quod dixit *Erunt res secundum quod dicunt*, quasi ergo dicat Et cum ita sit, erunt res secundum quod dicunt semper sine fine habentes semper potentiam ut sint; et intelligo per res illud quod semper fuit ens et in postremo corrumpitur et illud quod semper fuit non ens et in postremo generatur.

Et intelligo per hoc quod dixit *Secundum quod dicunt semper sine fine* ut sint huiusmodi semper infinite: una autem earum est non ens in preterito in infinitum et ens in futuro in infinitum, alterum vero ens in preterito in infinitum et non ens in futuro in infinitum; et vult declarare per hoc quod contingit huic positioni ut tempus potentie et actus sit idem et non idem, non idem vero quia tempus potentie est aliud a tempore actus, idem autem quia tempus actus est infinitum et similiter tempus potentie. Et forte intendebat per hoc quod dixit complere declarationem predictam scilicet quod positio

---

24 non] *om. Z* | contingit] + in *B* | nisi] + postquam *Z* : + quod ponunt XYZ_2 25 et—preterito] *om. B* 26 Secundum] quod *B* : quis *YZ* | quod] qui *B* | dicunt] -cit *J* : -xerunt *B* | dixit Erunt] erunt erit *B* : dicit erunt *J* : dixit erit *X* 27 quibus] quibusdam *B* | Scilicet] sed *Z* 28 et] quod *Z* 29 aut] *om. JZ* 30 et^{1}—sunt] *om. homoiotel. XY* | rem] *om. J* 31 per potentiam] *om. BX* 32 sit] *post* eis *Z* 33 generationem] -tiones *B* | autem] *om. Z* 34 semper^{1}] *om. Z* 35 sint] sicut *Z* | predicatum] -camentum *Z* : -catur *B* 36 dicunt] -cuntur *YZ* 38 res] hoc *Z* | et] *om. BX* 39 semper] *om. Z* | et^{2}] + intelligo *B* | generatur] -ratum *WZ* 40 dixit—quod] *om. homoiotel. B* | dicunt] -cuntur *Z* 41 sint] sit *B* | semper] + habentes semper potentiam ut sint *Z* | ens] + alter in vero ens *Z* 42 in^{2}] *om. B* | in^{4}] *om. X* : non *B* 43 et^{1}—infinitum^{2}] *om. homoiotel. Y* | in^{4}] *om. X* 44 ut tempus] in tempore *Z* 45 sit] sint *JZ* 46 est aliud] *inv. BX* 48 quod dixit] *om. BX* : quod dicit *J* | declarationem] demonstratio- *Z*

possibilis sit impossibilis scilicet positio eius esse actu ens in tempore potentie, et secundum hoc videtur quod non intendebat nisi ad distinguendum tempus potentie a tempore actus ut sit verificatum quod positio eius esse ens est possibilis, quoniam qui ponit potentiam cum actu non ponit possibilitatem ad actum esse ante eius esse, quapropter in hoc non completur actu demonstratio.

93ra 137. [138.] *Et dicemus etiam alium sermonem per quem declarabimus* *283a24* *errorem dicentis quod possibile est aliquid esse non generabile quod possibile* *283a25* *est ut sit non generabile et cum hoc sit corruptibile, et quod incorruptibile possibile est ut sit generatum. Erit ergo, secundum quod dicunt, corruptibile non corruptum in aliquo tempore, quoniam si non corrumpitur, erit corruptibile et non corruptibile in actu. Tunc igitur possibile est ut idem sit semper ens et non sit semper ens; et si hoc impossibile est, ergo corruptibile corrumpetur in aliquo tempore. Et similiter igitur generabile fuit in aliquo tempore, et non fuit omnino, possibile est ergo ut aliquid sit generabile et non fuit semper ens.* *283a29*

Hec est quasi demonstratio generalis quod impossibile est ut aliquid quod semper fuit habeat potentiam ad corruptionem absque **T113b** hoc quod corrumpatur; hoc enim contingit dicenti aliquid esse generatum sed post eternum, et similiter dicenti aliquid eternum sed post corruptum, immo manifestius est in hoc, et dixit *Dicamus etiam alium sermonem* etc. Deinde dixit *Erit ergo secundum quod dicunt illud quod est corruptibile non corruptum in aliquo tempore*, idest et contingit eis propter hoc quod dicunt quod in aliquo eterno est

---

49 actu] *om. BXY* : non *Z* 51 tempus potentie] *om. YZ* 52 esse—est] est ens *Y* 54 actu] *om. X* : actum *Z* : comparatur *B*

137. [138.] 1 sermonem] *om. Z* 3-4 ut—est] *om. homoiotel. X* 3 et[1]—corruptibile] *om. B* | hoc] + quod possibile est ut sit non generabile et cum hoc quod possibile est igitur *Z* | sit[2]] *om. YZ* : quia *W* | corruptibile et] *om. Y* | quod] *om. Z* | incorruptibile] cor- *BJY* 4 possibile est] *om. Z* | generatum—ergo] *om. B* 5 corruptibile] *om. BWXY* | non] *om. XY* | aliquo] alio *XYZ* 6 non] *post* erit *Z* 7 sit semper[1]] *inv. JZ* 8 ergo] + ut *Z* | corruptibile] *om. B* | corrumpetur] -peretur *B* : -pitur *J* | igitur] *om. WZ* 9 aliquo] *om. BJX* 11 generalis] -rabilis *WZ* | quod] quia *B* 14 aliquid] + esse *JZ* 15 est] *om. JZ* 16 Erit] est *WZ* 17 aliquo] alio *XY* 18 eis] + igitur *BX*

potentia ad corruptionem absque eo quod corrumpatur, ut illud in quo est potentia ad corruptionem non corrumpatur in aliqua hora. Deinde dixit *Quoniam si non corrumpitur* etc., idest et impossibile est ponere aliquid habens potentiam ad corruptionem absque hoc quod corrumpitur, quoniam continget ei, si non corrumpatur in aliquo
93rb tempore, ut non sit in eo | potentia ad corruptionem: hec enim est diffinitio eius in quo non est potentia ad aliquid scilicet ut numquam exeat in actum.

Deinde dixit *Tunc igitur est possibile ut idem* etc., idest et contingit huic positioni, si concesserimus in eo esse potentiam ad corruptionem, ut sit in eo potentia ut semper sit ens et ut non sit ens semper, et hoc est cum posuerimus in eo potentiam ad corruptionem essente in eo potentia ad semper esse, quoniam ex potentia corruptionis non est semper, et ex potentia ad esse est semper, est igitur in eo in potentia ad esse semper et ad non esse semper; quod est impossibile. Deinde dixit *Et si hoc impossibile est* etc., idest et quia cum posuerimus aliquid habens potentiam ad corruptionem absque hoc quod corrumpatur, sequuntur hec impossibilia scilicet ut in eo sit potentia et non sit potentia, aut in eo sit potentia ad duo opposita insimul scilicet ut sit semper et ut non sit; necesse est igitur ut habens potentiam ad corruptionem corrumpatur in postremo.

Deinde dixit *Et similiter igitur generabile fuit* etc., idest et cum T114a
verum est quod omne habens potentiam ad corruptionem omnino corrumpetur, non est verus sermo dicentis esse generatum sed non corruptibile, quoniam generatum in alia hora fuit non ens in aliqua hora in actu scilicet antequam generabatur; et cum ita sit, generatum est ens et in eo est potentia ut non sit semper ens. Et hoc intendebat cum dixit *Possibile est igitur ut aliquid sit generabile et non fuit semper*, idest possibile est igitur ut generabile non sit semper ens et cum in eo sit potentia ut non sit; et iam demonstratum est quod illud in quo

---

19-20 absque—corruptionem] *om. homoiotel. Y* 19 illud] aliquid *BJX* 21 corrumpitur] -petur *Z* 23 aliquo] alio *Y* 27 est possibile] *inv. BXY* : est impossibile *W* 29 sit[1]—eo] in eo sit *BX* : in eo devisit *Y* : sit in eodem *W* | ens semper] *inv. J* 30 est] erit *BX* 32 est[1, 2] erit *BXY* | est[3]] sunt *BX* | in potentia] potentia *X* : semper *Y* 34 impossibile est] *inv. WZ* 36 sequuntur] -quitur *BXY* 40 idest—cum] *om. Y* 43 aliqua] alia *JZ* 44 actu] -tum *Z* 45 sit semper] *inv. YZ* | ens[2]] *om. Y* 47 igitur] *om. JW* | semper] *om. Z* 48 quod] ut *Z* est potentia ut non sit, omnino exibit in actum, ergo omne generabile corruptibile est corruptibile necessario.

93va **138.** [139.] *Volumus etiam perscrutari de sermone eorum alio modo, et* 283a29
dicamus quod impossibile est ut generabile non cadat in aliquo tempore 283a30
sub corruptione, et impossibile est ut non generabile cadat sub corruptione, impossibile est enim ut aliquid non generabile et non corruptibile fiat sine factore casu: facta enim sine factore rara sunt, et cum fuerint,
cito destruuntur et consumuntur: permanens autem in eternum et per- 283b1
manens in tempore infinito non fit sine factore omnino. Et cum ita sit, natura est faciens omnia eterna necessario, et ipsa facit ut idem sit ens et ut sit non ens in tempore et tempore, et ut quedam sint semper entia et quedam non entia semper in tempore infinito; et causa illius est forti-
tudo istius nature <et elementum eius>. Non est igitur possibile ut ali- 283b5
quid semper sit non generabile et semper corruptibile; quod si fuerit possibile, tunc contraria erunt in actu necessario; quod est impossibile. Dicens enim quod modo est annus preteritus et quod annus preteritus est modo, mentitur in suo sermone; et apparet falsitas in suo sermone quod aliquid sit eternum et corruptum in aliquo tempore quia habet potentiam ut sit non ens in aliquo tempore sed non est in tempore preterito
non ens, neque in tempore in quo est ens in actu. 283b10

Abbreviatio huius sermonis est quod si aliquid est non generabile, sive ponatur corruptibile sive non, et aliquid fuerit non corruptibile
93vb sive ge|neratum sive non, manifestum est quod impossibile est quod T114b
sit casu, nam sicut quod fit casu raro fit, ita etiam parum existit, sicut dixit, quod cito consumuntur. Et cum ita sit, illud quod semper est

50 est] *om. BWXY* | corruptibile[2]] *om. BJX*

138. [139.] 1 eorum] illo- *BJXY* 2 generabile non] *inv. JXY* | in aliquo tempore] *om. XY* 3 et] *om. XY* | est] *om. W* : + enim *JZ* 3-4 non—ut] *om. JXY* 4 est enim] *inv. W* | non[2]] *om. JXY* 5 fiat] sit *W* | casu—factore] *om. homoiotel. Y* | facta] -tum *B* : -tu *X* | factore rara] factore tota *X* : actore raro *J* 8 omnia] omnino *YZ* 9 et[1]—et[2]] *om. Y* 11 et—eius] *supplevi (cf. infra, c. 138 [139], v. 41, et transl. Arab.)* 12 fuerit] fuerint *Z* 13 contraria erunt] *inv. Z* 15 apparet] + sua *BX* | falsitas] + eius *J* : factas eius *Z* 16 sit] fit *W* | quia] *om. BJ* : quod *X* | habet] + enim *J* 16-7 habet—ens] *om. B* 16 potentiam] + in tempore *J* 17 in—tempore] *om. BJ* | aliquo] alio *XY* | est] *om. WZ* 18 non ens] *om. JY* 20 corruptibile[1]] *om. XY* | fuerit] fuit *BXZ* | non[2]] *om. XY* 21 quod[2]] ut *BX* 23 dixit] -cit *J* : -ximus *Z* | semper est] *inv. Z*

non est semper nisi ex natura innata ad hoc, et est natura eius per quam est quod est. Nature igitur rerum dant eis aut ut sint semper aut ut non sint semper, aut ut aliquando sint aut aliquando non sint; si igitur invenietur aliquid generabile eternum non corruptum aut eternum corruptum, esset possibile ut natura possibilis transmutaretur in necessariam, et necessaria in possibilem in agente et recipiente; quod est impossibile. Et dixit *Impossibile est enim ut aliquid sit non generabile* etc., idest impossibile est ut aliquid sit non generabile aut non corruptibile casu, sed necessarium est ut hoc sit ex natura faciente, quoniam omnia existentia casu non permanent nisi parum, sicut non fiunt nisi raro. Deinde dixit *Permanens autem in eternum* etc., idest non fit nisi quia substantia et essentia eius dat ei permanentiam.

Deinde dixit *Et si ita sit, natura est* etc., idest et cum causa esse aut T115a
non esse rerum naturalium sit ex natura naturalia autem sunt tribus modis, aut quorum natura dat eis ut semper sint, aut ut semper non sint, aut ut sint in duabus dispositionibus, idest et nulla natura dat ut sit semper et non sit nisi necessarium transmutetur in possibile. Deinde dixit *Et causa istius est potentia istius nature et elementum eius*, idest et causa diversitatis entium in istis dispositionibus est causa diversitatis nature in substantia sua. Deinde dixit *Non est ergo possibile*, idest non est possibile ut eadem natura dat eidem esse semper et non esse semper: hoc autem contingit dicenti hoc aliquid esse generatum et post perpetuum, aut aliquid non generabile et post corruptum, scilicet quod necesse est ei ponere aliquid cuius natura dat ei ut
94ra semper | sit et semper non sit.

Deinde confirmavit hoc per demonstrationem per quam sustentatus est in hiis demonstrationibus, et dixit *Si ergo fuerit hoc possibile, erunt contraria in actu necessario, quod est impossibile*, idest et si aliquis posuerit quod possibile est ut in aliquo eterno sit potentia ad cor-

---

24 non] si *Z* 25 est—est] *om. B* | igitur] enim *JZ* : ergo *W* 26 aut[2]] *om. BXY* : + ut *W* 27 generabile] in- *YZ* | eternum[1]] *om. JX* | non] *om. XY* : aut *B* | aut] *om. BX* | eternum[2]] *om. J* : + et *Z* 28 corruptum] *om. B* : in- *Y* 29 possibilem] impossibile *Y* : + est *Z* 30 est enim] *inv. Y* : enim *WZ* 31 idest—generabile] *om. B* 34 fiunt] fuerint *BZ* | in] et *WXY* 36 aut] et causa *Z* 37 autem] *om. BXY* | sunt] *om. Z* 39 ut[1]] *om. WY* 41 istius[1]] illius *BXY* | istius[2]] illius *Z* 42 istis dispositionibus] *inv. Z* 43 substantia sua] *inv. BX* : substantia *YZ* | ergo] igitur *BX* 44 idest—possibile] *om. homoiotel. WYZ* 46 corruptum] -tibile *B* 48 et] + ut *BX* 49 confirmavit] -mat *JZ* | per[2]] super *BX* 50 ergo] + non *BXY* 51 erunt contraria] *inv. J* : erunt *Z* 52 sit] fit *WZ*

ruptionem absque eo quod corrumpatur, contingit ei quod in illo sint due potentie contrarie existentes scilicet ad esse et potentia ad non esse in actu; si ergo actio unius potentie fuerit in actu, necessario est altera in potentia, cum impossibile est ut actiones duarum potentiarum sint insimul, scilicet non esse in actu et esse in actu, nisi idem sit et non sit in actu; quod est impossibile. Et si una illarum fuerit in potentia, erit positio sue actionis quod sit in actu falsa non possibilis, sed est falsa impossibilis quia inducit ad esse actionum duarum potentiarum insimul in actu, ergo illud quod positum fuit, scilicet potentiam esse ad contrarium non est in eternum falsum est et nihil possibile, quoniam impossibile est contrarium possibili, et ex quo sequitur impossibile est impossibile.

Et quia hec destructio non complebitur nisi in eo quod ponit adversarius generatum et post eternum, ita quod concedet quod potentia est in eo apud suum esse respectu futuri, et similiter contingit in corrupto quod ponit ut apud suum non esse fuerit potentia in T115b
eo ad esse respectu futuri. Et quia dicens hos duos sermones potest prohibere quod potentia ad non esse in ente non est in respectu futuri neque potentia generationis in corrupto etiam apud suam corruptionem est in respectu futuri, dixit *Impossibile* est enim quod sequitur ex sermone, et dixit *Dicens enim quod modo* etc., idest et qui
94rb dicunt quod in generabili | non est potentia ad non esse apud suum esse in respectu futuri, dicunt impossibile, idest quasi dicat aliquis quod instans preteritum et presens est idem.

Deinde dixit *Quomodo sequitur hoc impossibile ex suo sermone et apparet falsitas eius* etc., idest et hoc impossibile sequitur dicentem aliquid esse eternum quod in preterito fuit non ens et quod modo, scilicet in sui esse tempore, non est in eo potentia ad non esse in

---

53 quod[2]] ut *Z* | in] *om. BY* | illo] illic *B* : ille *Y* 54 existentes] + semper *JX* : + semper absque eo *B* 56 est[1]] erit *BXY* | est[2]] *om. Z* 56-7 duarum potentiarum] *inv. Z* 57 scilicet] sed *W* | non] *post* et *BX* 58 idem] + ut *Z* 59 positio] primo *W* : potentia *Y* | non] et *JZ* 60 possibilis] impossibile *B* | est] etiam *Z* 62 est[1]] esse *XY* | eternum] -no *BJX* | falsum] factum *Z* 63 nihil] non *Z* | possibile] *om. XY* | est] *om. WYZ* 67 esse] + et *Y* 68 ut] cum *BXY* | potentia] *post* eo *(69) BJX* 69 ad] + non *W* | esse] + in *WY* | duos sermones] propter hoc *Z* 70 esse] + apud scilicet non generatum *W* | in ente] et *Y* 71 generationis] *om. WZ* | apud] *om. WZ* 72 enim] *om. BX* 75 in—futuri] *om. Y* | dicunt impossibile] d. Arist. *Y* 76 preteritum] *om. B* 79 quod modo] quomodo *BX* 80 non[1]] + enim *B*

futuro, quoniam illud quod erat non ens in actu in aliquo tempore, deinde exivit ad actum, fuit apud suum non esse non ens in actu et apud suum esse non ens in potentia. Et cum impossibile est ut hec potentia ad non esse inveniatur in eo in tempore preterito quoniam in tempore preterito fuit non ens in actu, neque etiam in tempore presenti in respectu futuri quoniam contingit predictum impossibile, remanet ergo ut sit in eo in presenti in respectu preteriti; et qui dicit quod aliquid habet potentiam ad tempus preteritum, facit de preterito futurum aut presens; potentia enim non intelligitur nisi in respectu futuri. Et dixit *Ut aliquid sit eternum et corruptibile in tempore*, idest et est cadens sub non esse in aliquo tempore sicut contingit dicenti esse generatum et perpetuum, quoniam ipse ponit eternum quod in aliquo tempore fuit non ens; et cum dixit hanc positionem, dixit quomodo contingit dicenti ipsam dicere quod eternum habet potentiam ad non esse, et dixit *Habet enim potentiam ut sit in tempore non ens*, idest et necesse est dicere in tali eterno quod in eo est potentia ad non esse, quia iam fuit non ens in actu in alio tempore. **T116a**
Deinde dixit *Sed non est in tempore preterito neque in tempore in quo*
94va *est ens* | *in actu*, idest sed ista potentia non est in respectu preteriti temporis neque temporis in quo est ens in actu: quasi igitur intendit quod contingit huic sermoni ut non sit in eo potentia ad preteritum sit futurum.

139. [140.] *Si igitur aliquis dixerit quod illud quod est in potentia pos-* *283b11*
sibile est ut sit in hoc, dicemus quod si hoc possibile: potentia enim non est in tempore preterito, sed in tempore in quo dicitur modo et in tempore futuro; et similiter dicimus in eo quod est primo eternum ens, deinde sit post non ens. Habet enim potentiam in tempore in quo non est *283b15*
actu ut sit non ens; si igitur hoc aliquis dixerit esse possibile, dicemus quod modo est annus preteritus. *283b16*

---

82 ad actum] in actum *X* : in actum suum *J* 84 in[2]] *om. B* 87 ergo] igitur *YZ* | in[3]] *om. JZ* | in[3]—preteriti] *om. Y* | preteriti] futuri *XZ* 90 Ut] quod *J* 94-5 dixit—potentiam[1]] *om. Z* 95 ut sit] *post* tempore *JZ* 99 in[1]] *om. WYZ* | non est] *inv. Z* 99-100 preteriti temporis] *inv. BX* 100 neque temporis] *om. Y* 101 ad] aut *WXY* 102 sit] sed *B* : sed ad *J*

139. [140.] 2 est] + etiam *J* | sit] + etiam *BX* | possibile] + est *J* 3 et] *om. W* 5 post] postremo *JG* : primo *WYZ* | est] + in *JX* 6 non] *om. BXY* | possibile] + est *J*

Hoc quod incepit dicere est quasi obiectio contra illud quod predixit, scilicet quia posuerit potentiam ad non esse que est in generato perpetuo ex potentia si non fuerit posita in respectu futuri, continget ut non exibit ad actum, et cum non exibit ad actum non erit potentia; quasi ergo hoc dicens contradicit huic sermoni, dicendo quod *94vb* bene scimus quod omnis potentia possibilis est ut exeat ad actum, | et eius exitus in generato eterno iam preteriit, et non est venturus. Et nulla differentia est inter hoc quod exitus eius ad actum sit antequam non sit, quia potentia non dicitur nisi in respectu actus, quapropter ipse dixit respondendo *Dicemus modo quod si hoc possibile est* etc., quia potentia non est in tempore preterito, idest dicemus respondendo si omne quod est in potentia possibile est ut exeat ad actum, ergo potestas et potentia non sunt nisi in respectu futuri, quoniam nihil potest in respectu preteriti quoniam actio que fuit in preterito iam preteriit et impossibile est ut revertatur nunc, quapropter nulla potentia est ad ipsum omnino, et potentia non est nisi ad eius simile in futuro secunda vice et indifferenter, sive possibile fuerit ad esse aut ad non esse.

Et quia hoc quod dixit de potentia manifestum est in eo quod **T116b** ponitur eternum et in postremo corruptum, scilicet quod apparet in eo quod potentia ad non esse est in eo in respectu futuri econtra illi quod estimant de generato perpetuo, dixit *Et similiter dicimus de eo quod est primo semper ens, deinde in postremo non ens*, idest et similiter dicemus nos et adversarii nostri in eo quod ponunt non generabile et in postremo corruptum scilicet quod potentia ad corruptionem non intelligitur in eo nisi in respectu futuri. Deinde dixit *Habet enim potentiam in tempore in quo non est in actu ut sit non ens*, idest quoniam potentia que est in eo ad corruptionem non est nisi respectu temporis in quo non est ens in actu, scilicet in futuro. Deinde dixit *Si igitur aliquis dixerit quod possibile est*, idest et si aliquis dixerit quod possibile est ut potentia ad non esse in generato perpetuo sit

---

8 obiectio contra] abstractio circa *Y* 9 quia] + qui *XY* 10 ex potentia] *Z* 12 quasi] *om. Z* 14 venturus] futurus *JZ* 15 differentia est] *inv. Z* 16 non[1]] *om. WZ* | in respectu] *om. B* 17 etc.] *om. BXY* 20 potestas ... potentia] *inv. JW* | in] *om. W* 23 est[2]] *om. XY* 26 de] quod *Y* 28 futuri] + et *BX* | econtra] -trario *Z* : contra *X* 29 estimant] + est *BY* | similiter] *om. Z* 33 nisi] *om. Z* 34 actu] -tum *Z* | idest] + et *Y* 35 quoniam] quia *BJ* | nisi] + in *JZ* 36 ens] *om. WYZ*

95ra potentia ad non esse pre|cedens, dicemus quod possibile est tunc ut
preteritum sit futurum aut presens.

140. [141.] *Et possumus perscrutari de hoc sermone et declarare errorem* *283b17*
dicentis ipsum sermone naturali non universali, sicut fecimus ante. Dicamus igitur quod impossibile est ut aliquid primo sit eternum ens, deinde post corrumpatur, aut ut aliquid primo sit non ens deinde post sit ens
eternum. Omnia enim generabilia et corruptibilia alterantur et trans- *283b20*
mutantur, et non alterantur nisi ex suis contrariis ex quibus componuntur et generantur generatione naturali, et per ea consumuntur et cor-
rumpuntur etiam. *283b22*

Vocavit istam demonstrationem naturalem et predictam universalem, quoniam hec est ex propositionibus nature, illa autem ex propositionibus logicis, ut hoc quod dixit quoniam falsum impossibile non sequitur ad falsum possibile. Hee autem propositiones sunt naturales, quod omne quod generatur a suo simili generatur, et omne
95rb quod corrum|pitur a suo contrario corrumpitur; omne igitur genera- T117a
bile habet contrarium, et omne corruptibile simile: et cum coniungitur huic, quod omne generabile habet contrarium et omne habens contrarium est corruptibile, contingit quod omne generabile est corruptibile. Et similiter si coniungitur huic, quod omne corruptibile habet simile et omne habens simile est generabile, contingit quod omne corruptibile est generabile: habens igitur contrarium impossibile est ut remaneat per tempus infinitum neque in preterito neque in futuro. Si ergo mundus fuit primo non ens, deinde generatus fuit post non esse, necesse est huic contrarium ex quo fuit non ens ante-

---

40 aut presens] *om. Y*

140. [141.] 1 declarare] + sermonem *YZ* | errorem] -ris *Z* 2 dicentis] + sermonem *W* | sermone] -monem *WX* | non] si *Z* | universali] -saliter *J* : nulli *Z* 4 aut] *om. BYZ* | primo] post *YZ* 4-5 ens eternum] *inv. Z* 5 Omnia] omnino *YZ* 7 ea] eas *WXY* 9 istam] illam *Z* 10 hec] hoc *Z* 11 quoniam] quod *B* 12 autem] enim *Z* 13 naturales] + scilicet *JW* | a—generatur] *om. homoiotel. Y* 15 simile] -liter *JYZ* 17 contingit—corruptibile] *om. homoiotel. B* 18 coniungitur] -iungit *BWY* : -tingit *Z* 19 et—generabile] *om. BWXYZ sed suppl.* W_2 | contingit] -get *XY* 20 igitur] + impossibile *B* 22 Si ergo] *inv. Z* : si igitur *BJ* | generatus] -ratum *XY* 23 necesse] -ario *XY* | est] erit *BXY* | non ens] movens *WYZ* quam generabatur, et cum ei est contrarium, necessario corrumpetur post generationem; et quod hec curret secundum ordinem, et sequitur ex hoc ut iste mundus sit ex alio mundo, quoniam si ita non esset, quare ergo dominatur in aliquo suum contrarium in aliquo tempore et corrumpitur, et in alio generatur? Continget ergo necessario ut mundus reducatur ad causas ad quas ascendunt omnia generabilia et corruptibilia, scilicet secundum quod est dictum in libro de Generatione et Corruptione; dicens ergo quod mundus generatus fuit postquam non fuit omnino, et quod est corruptum corruptibile sine regressu, non naturaliter loquitur, et similiter dicens ipsum esse generatum non corruptibile aut non generabile et corruptum postremo.

24 corrumpetur] -pitur *BW* 25 hec] hoc *Z* : *obsc.* *BWY* | et sequitur] consequitur *B* 26 ut] quod *BX* | iste—sit] sit mundus *B* | mundo] + continget ergo necessario mundus ut reducatur ad causas ad quas ascendunt *Z* 27 aliquo[1]] alio *YZ* 28 alio] aliquo *Z* | Continget] -git *B* 32 est] erit *BJXY* 33 regressu] ingr- *BYZ* 34 et] *om.* *WXY* : in *Z* | corruptum] -rumpitur *XY* 35 postremo] + explicit commentum libri primi, incipit secundus tractatus cum auxilio Dei W_2

PRINTED ON PERMANENT PAPER • IMPRIME SUR PAPIER PERMANENT • GEDRUKT OP DUURZAAM PAPIER - ISO 9706

N.V. PEETERS S.A., WAROTSTRAAT 50, B-3020 HERENT